Fundamental Theories of Physics

Volume 184

The international monograph series "Fundamental Theories of Physics" aims to stretch the boundaries of mainstream physics by clarifying and developing the theoretical and conceptual framework of physics and by applying it to a wide range of interdisciplinary scientific fields. Original contributions in well-established fields such as Quantum Physics, Relativity Theory, Cosmology, Quantum Field Theory, Statistical Mechanics and Nonlinear Dynamics are welcome. The series also provides a forum for non-conventional approaches to these fields. Publications should present new and promising ideas, with prospects for their further development, and carefully show how they connect to conventional views of the topic. Although the aim of this series is to go beyond established mainstream physics, a high profile and open-minded Editorial Board will evaluate all contributions carefully to ensure a high scientific standard.

More information about this series at http://www.springer.com/series/6001

Michael J.W. Hall · Marcel Reginatto

Ensembles on Configuration Space

Classical, Quantum, and Beyond

 Springer

Michael J.W. Hall
Centre for Quantum Dynamics
Griffith University
Brisbane, QLD
Australia

Marcel Reginatto
Physikalisch-Technische Bundesanstalt
Braunschweig
Germany

ISSN 0168-1222   ISSN 2365-6425   (electronic)
Fundamental Theories of Physics
ISBN 978-3-319-81692-0   ISBN 978-3-319-34166-8   (eBook)
DOI 10.1007/978-3-319-34166-8

Printed on acid-free paper

This Springer imprint is published by Springer Nature
The registered company is Springer International Publishing AG Switzerland

This book is dedicated to Robyn, and Mila and Aldo

Preface

Much effort in theoretical physics goes towards building mathematical models that can describe as a wide a variety of physical systems as possible. To build such models, it is necessary to introduce just the right amount of formalism: the mathematics must be able to capture the essential properties of the physical system, while keeping the amount of mathematical structure which does not have a direct physical interpretation to a minimum.

This book is concerned with the description of physical systems in terms of *ensembles on configuration space*. As will be seen, this is an approach which introduces very few physical and mathematical assumptions. As a consequence, the formalism has wide applicability: it can be used to describe physical systems that are deterministic as well as systems subject to uncertainty; discrete systems, particles, and field theories; classical and quantum theories. It also allows for theories that are difficult to formulate using other approaches, such as hybrid quantum-classical theories where there is an interaction between quantum and classical sectors, including the coupling of quantum matter to classical gravity. Finally, it provides insights into classical and quantum physics that not only lead to unified approaches to concepts such as thermodynamics, weak values, locality and superselection, but to novel reconstructions of quantum theory from physical and geometric axioms.

We therefore believe that a detailed account of the formalism and the physics of ensembles on configuration space is valuable in providing a useful (and beautiful) reformulation of existing theories, and in suggesting various generalisations and directions for formulating new theories, and hope that ideas from this book will be incorporated into the standard toolkit of theoretical physicists.

The book is structured into four main parts. Part I deals with general concepts and properties of ensembles on configuration space. Part II examines how quantum mechanics emerges naturally from three very different axiomatic scenarios, based respectively on an exact uncertainty principle, information geometry on configuration space, and local representations of rotations on discrete configuration spaces. Part III develops a theory of hybrid quantum-classical interactions, which overcomes various no-go theorems in the literature and provides an explicit model of interaction between quantum systems and classical measuring apparatuses. Finally, Part IV extends these ideas to show how quantum fields can be consistently coupled to classical gravity. While much of the material is based on publications by the authors and colleagues over the past 15 years or so, many results are presented here for the first time.

The authors would like to thank a number of people and organisations. We began our collaboration on the physics of ensembles on configuration space by email correspondence, and worked on two papers together before finally meeting face-to-face in Germany in 2001, courtesy of travel funding provided by the Alexander-von-Humboldt Foundation. Our collaboration continued, and was bolstered by again being able to meet in person at the Perimeter Institute in 2009, courtesy of travel funding for a conference organised by Philip Goyal. In addition to these organisations and individuals, MH would also like to thank Howard Wiseman at the Centre for Quantum Dynamics, for permitting time to be spent on this book project, and wife Robyn and children Conan and Seriden for their support. MR is grateful to Hans-Thomas Elze and Dieter Schuch for invitations to the very stimulating DICE and Symmetries in Science meetings, where some of these results were presented for the first time, and would like to thank Ruth and Ava for putting up with the late nights and long hours. Finally, we thank Angela Lahee at Springer for her support throughout all the stages of preparation of this book.

Brisbane, Australia   Michael J.W. Hall
Braunschweig, Germany   Marcel Reginatto
March 2016

Contents

Part I
General Properties of Ensembles on Configuration Space

Chapter 1
Introduction

Abstract Ensembles on configuration space have wide applicability. They may be used to describe classical, quantum and hybrid quantum-classical systems, physical systems that are deterministic or subject to uncertainty, discrete systems, particles and fields. They also lead to novel reconstructions of quantum theory from physical and geometric axioms. We introduce the basic elements of the theory, discuss a number of classical and quantum examples, and provide an overview of the many generalizations and applications that form the subjects of later chapters. The approach introduces very few physical and mathematical assumptions. The basic building blocks are the configuration space of the physical system, an ensemble of configurations, and dynamics generated from an action principle. An important role is played by the ensemble Hamiltonian which determines the equations of motion. It must satisfy certain requirements which we discuss in detail. We provide examples of classical and quantum systems and show that the primary difference between quantum and classical evolution lies in the choice of the ensemble Hamiltonian.

1.1 The Description of Physical Systems in Terms of Ensembles on Configuration Space

As noted in the Preface, there is great value in building a formalism for physical theories which is both widely applicable and has a minimal amount of uninterpreted mathematical structure. For example, we may treat the motion of a classical particle subject to an external force using Newtonian, Lagrangian, Hamiltonian or Hamilton–Jacobi formulations [1, 2]. These different formalisms are equivalent in the sense that they lead to the same predictions, but the amount of effort that goes into solving a particular problem will depend on how well suited the formalism is to that problem. More importantly, different formalisms emphasize different aspects of a system, thus leading to different physical pictures, and will suggest different types of generalizations. For example, the Hamiltonian formalism for classical particles was well suited for the early development of quantum mechanics, while Feynmann diagrams in quantum field theory arose from a Lagrangian formulation. Thus, there are very good reasons for considering more than one formulation of a physical theory.

© Springer International Publishing Switzerland 2016

M.J.W. Hall and M. Reginatto, *Ensembles on Configuration Space*,
Fundamental Theories of Physics 184, DOI 10.1007/978-3-319-34166-8_1

This book develops and applies a formalism for physical theories that is extremely broad in scope—including both classical and quantum mechanics as particular examples—and which is underpinned by a very simple physical picture: ensembles evolving on a configuration space. The mathematical structure is correspondingly simple (simpler than that of C^*-algebras for example). Yet it is sufficiently nontrivial to be able to guarantee (unlike generalised probabilistic theories for example) the existence of useful objects such as a Lie bracket for observables, canonical transformations, measurement interactions, weak values and thermal ensembles .

The aims of this introductory chapter are to (i) introduce basic elements of the approach, largely via a number of classical and quantum examples, and (ii) preview the many generalisations and applications that form the subjects of later chapters.

1.2   Basic Concepts and Examples

The core of the formalism may be summarised in the following

Central idea: Physical systems are described by ensembles on configuration space, the dynamics of which is governed by an action principle.

This central idea makes reference to three familiar concepts: configuration space, ensembles, and Hamilton's principle of least action, briefly discussed in turn below.

The first basic concept is the *configuration space* of a physical system. For example, in classical mechanics a particle may be in different locations of space, and the configuration space is three-dimensional Euclidean space. By contrast, the configuration space of a two-level system, corresponding to the outcomes of a classical coin toss or measurement on a qubit, is a discrete space with only two elements, such as {heads, tails} or {up, down}. The primary role of a physical theory is to make predictions about system configurations and their evolution. Thus, the mathematical description of a physical system will always involve a choice of configuration space.

The second concept is that of an *ensemble* on configuration space. The underlying physical idea here is that the formalism should be general enough to cover systems that are subject to uncertainty. Thus, in general, the system will be described by some probability distribution P defined over the configuration space.

The third concept is that the dynamics of the ensemble is specified by an *action principle*. Since the principle of least action is a generic feature of all known fundamental physical theories, this is not contentious or surprising in itself. However, in the context of ensembles evolving on configuration space, the requirement that the probability distribution P remains positive and normalised under evolution will be seen to be an important and nontrivial constraint on allowed ensemble Hamitonians.

As a first example, consider a non-relativistic classical particle of mass m moving in a potential $V(x)$. Our starting point in applying the above concepts will be one of the standard formulations of classical dynamics, the Hamilton–Jacobi equation

$$\frac{\partial S}{\partial t} + \frac{|\nabla S|^2}{2m} + V = 0. \tag{1.1}$$

A solution S of this equation provides a complete description of the motion when there is no uncertainty: if the particle is known to be at position x at time t, then it has momentum $m\dot{x} = \nabla S(x, t)$ and energy $-\partial_t S$ [1, 2].

It is not always appreciated that the Hamilton–Jacobi formalism is fundamentally a theory of ensembles [3]. The gradient of S/m, in defining a velocity vector at every point on the configuration space, allows any uncertainty regarding the position of the particle to be naturally described by a probability density over the configuration space, $P(x, t)$, with the conservation of probability ensured by a continuity equation of the usual form

$$\frac{\partial P}{\partial t} + \nabla \cdot \left(P \frac{\nabla S}{m} \right) = 0. \tag{1.2}$$

Equations (1.1) and (1.2) are the equations of motion of an ensemble on configuration space, for the case of a classical particle. But what is the corresponding action principle for the ensemble described by P? Equivalently, what is the corresponding Langrangian or Hamiltonian that generates these equations of motion? Since P (not x) is the fundamental quantity describing the ensemble, any suitable 'ensemble Hamiltonian' specifying its evolution should depend on P and some canonically conjugate quantity Π [1]. In fact, for the above classical equations of motion one may take S as the quantity canonically conjugate to P, and the corresponding classical ensemble Hamiltonian is given by the functional

$$\mathcal{H}_C[P, S] := \int dx \, P \left(\frac{|\nabla S|^2}{2m} + V \right). \tag{1.3}$$

In particular, it may be checked that Eqs. (1.1) and (1.2) are equivalent to the Hamiltonian equations of motion

$$\frac{\partial P}{\partial t} = \frac{\delta \mathcal{H}_C}{\delta S}, \qquad \frac{\partial S}{\partial t} = -\frac{\delta \mathcal{H}_C}{\delta P}, \tag{1.4}$$

where $\delta/\delta f$ denotes the variational derivative with respect to the function f[1]. The classical ensemble Hamiltonian in Eq. (1.3) may immediately be recognised as the average energy of the ensemble.

---

[1]For readers unfamiliar with variational derivatives, more details are given in Appendix A of this book. It is sufficient to recall here that for $F = \int dx \, g(x, f, \nabla f)$, one has $\delta F/\delta f = \partial g/\partial f - \nabla \cdot \partial g/\partial(\nabla f)$. The case of discrete configuration spaces is mathematically simpler, as discussed in Sect. 1.3.

The formulation of the equations of motion for a classical particle in Eqs. (1.3) and (1.4) may appear unfamiliar, and it certainly does not appear in standard textbooks on mechanics. However, the form of the classical ensemble Hamiltonian $\mathscr{H}_C$ is in fact well known in the theory of ideal fluids (dating back to the 19th century), where the quantity S/m is reinterpreted as a velocity potential [4]. Moreover, the fundamental nature of this formulation quickly becomes apparent when it is realised that the equations of motion for a quantum particle have a remarkably similar form.

In particular, let $\psi(x)$ denote the wave function for a quantum particle of mass m subject to potential $V(x)$, and define the real functions P and S on configuration space via the polar decomposition $\psi = \sqrt{P}\,e^{iS/\hbar}$. The equations of motion for P and S then follow via substitution into the standard Schrödinger equation for ψ, yielding the continuity equation

$$\frac{\partial P}{\partial t} + \nabla \cdot \left(P\frac{\nabla S}{m} \right) = 0 \tag{1.5}$$

as before, and the modified Hamilton–Jacobi equation

$$\frac{\partial S}{\partial t} + \frac{|\nabla S|^2}{2m} + V - \frac{\hbar^2}{2m}\frac{\nabla^2 P^{1/2}}{P^{1/2}} = 0. \tag{1.6}$$

These equations are equivalent to the Hamiltonian equations of motion

$$\frac{\partial P}{\partial t} = \frac{\delta \mathscr{H}_Q}{\delta S}, \qquad \frac{\partial S}{\partial t} = -\frac{\delta \mathscr{H}_Q}{\delta P}, \tag{1.7}$$

analogous to Eq. (1.4) above, where $\mathscr{H}_Q$ is the *quantum* ensemble Hamiltonian

$$\mathscr{H}_Q[P, S] := \int dx\, P \left(\frac{|\nabla S|^2}{2m} + V + \frac{\hbar^2}{8m}\frac{|\nabla P|^2}{P^2} \right). \tag{1.8}$$

Equations (1.5) and (1.6), together with the ensemble Hamiltonian $\mathscr{H}_Q$, were first given by Madelung in 1926 [5]. Madelung further showed that

$$\mathscr{H}_Q[P, S] = \int dx\, \psi^*(x) \left(-\frac{\hbar^2}{2m}\nabla^2 + V \right) \psi(x) = \left\langle -\frac{\partial S}{\partial t} \right\rangle, \tag{1.9}$$

which may be recognised as the average quantum energy for the wave function ψ. Thus, just as for the classical case, the ensemble Hamiltonian is equal to the average energy of the ensemble, and also to the average of $-\partial_t S$. Further, in the limit $\hbar \to 0$ one has $\mathscr{H}_Q \to \mathscr{H}_C$, and hence the quantum and classical equations of motion become identical.

The above examples show that the dynamics of classical and quantum non-relativistic particles can be treated using the common framework of ensembles on configuration space, where the primary difference between quantum and classical evolution lies in the choice of the ensemble Hamiltonian. We explore this framework a little further in this chapter, to give examples for discrete configuration spaces and to note some fundamental properties of ensemble Hamiltonians. In later chapters we will considerably develop the framework, to give a unified treatment of concepts such as observables, fields, interaction, measurement, superselection, weak values, and thermodynamics on configuration space, without reference to any particular theory. This general framework is capable of attacking a number of interesting problems that either cannot be formulated in the language used by other theories or are extremely difficult to formulate, such as consistent descriptions of measurements on a quantum system by a classical measuring device and the coupling of quantum fields to a classical gravitational field. It is also shown to be a powerful starting point for the derivation of the quantum formalism from suitable axioms (see Sect. 1.5 for an indication of the material covered in later chapters).

1.3 Further Examples: Discrete Configuration Spaces

The formalism of ensembles on configuration space is very broadly applicable. It is worthwhile in this introductory chapter to see how it is able to incorporate classical and quantum systems with discrete configuration spaces. Indeed, the formalism in this case is somewhat simpler than the one for particles described in the previous section, as no variational derivatives are involved.

1.3.1 Classical Rate Equations

Consider the evolution of a classical probability distribution $P \equiv \{P_j\}$ on a discrete set of points labelled by $1, 2, 3, \ldots$. This is a ubiquitous problem in the theory of stochastic processes [6]. If T_{jk} denotes the rate at which probability flows from site k to site j, then the evolution of P is given by the transition rate equation

$$\frac{\partial P_j}{\partial t} = \sum_k \left(T_{jk} P_k - T_{kj} P_j \right) \tag{1.10}$$

(sometimes called a classical master equation or Kolmogorov equation). The first part of the sum generates the total probability flowing into site j, and the second part generates the total probability flowing out. Note that the transition rates can depend on P, and that T_{kk} may be chosen arbitrarily as it makes no contribution to the rate equation. The equation is often represented pictorially by a transition graph: the points correspond to the vertices of the graph, and for each pair of vertices (j, k) there is a directed edge from vertex k to vertex j labeled by T_{jk} [6].

Equation (1.10) is a discrete analog of the continuity equation (1.2) for probability flow on a continuous configuration space, and similarly ensures the conservation of probability. This may be checked via summation of both sides over j, and swapping the dummy indices in the second term.

To model this case within the formalism of ensembles on configuration space, we require the existence of a canonically conjugate quantity $S \equiv \{S_j\}$ on the configuration space, and a suitable ensemble Hamiltonian $H(P, S)$, such that the equations of motion are given by

$$\frac{\partial P_j}{\partial t} = \frac{\partial H}{\partial S_j}, \qquad \frac{\partial S_j}{\partial t} = -\frac{\partial H}{\partial P_j}. \tag{1.11}$$

It is straightforward to construct such a Hamiltonian system: choose

$$H(P, S) = \sum_{j,k} T_{jk} P_k (S_j - S_k). \tag{1.12}$$

The Hamiltonian equations of motion then generate the transition rate equation (1.10) as desired. It should be noted that this is a purely formal construction: transition rate equations are typically obtained from an underlying microscopic description that incorporates the fundamental physics, with a given transition rate equation being compatible with many possible such descriptions. Nevertheless, the construction shows that such equations, while only providing a phenomenological model, can easily be incorporated into the formalism of ensembles on configuration space.

An important feature of the above ensemble Hamiltonian $H(P, S)$ is that it only depends on S via the differences between its components. It will be seen in Sect. 1.4 that this feature is generic for discrete configuration spaces, and is related to the conservation of probability. It may further be noted that, for the special case of constant transition rates, the above ensemble Hamiltonian further generates the equation

$$\frac{\partial S_j}{\partial t} = -\sum_k T_{jk} (S_j - S_k) \tag{1.13}$$

for the conjugate quantity S. As this is independent of the evolution of P, it has no direct physical import (at least, not without a specific underlying microscopic model). However, by multiplying this equation by P_j and summing over j one immediately obtains the connection

$$H(P, S) = -\sum_j P_j \frac{\partial S_j}{\partial t} = \left\langle -\frac{\partial S}{\partial t} \right\rangle \tag{1.14}$$

between the ensemble Hamiltonian and the average of $-\partial_t S$, which similarly held for the classical and quantum ensembles of the previous section. Hence, $-\partial_t S_j$ may be interpreted as the energy associated with site j in this case. It will be seen in Sect. 1.4 that this connection is guaranteed for ensemble Hamiltonians satisfying a homogeneity property. Other physical roles for S will be seen in later chapters (most notably in regard to generating physical transformations of the ensemble such as translations and reflections).

1.3.2 Finite-Dimensional Quantum Systems

For quantum systems evolving on a finite Hilbert space, the configuration space may be chosen to be the set of discrete outcomes of any complete measurement on the system. For example, for nondegenerate Hermitian operator $\hat{A}$, with orthormal eigenstates $\{|a_j\rangle\}$, the configuration space is $\{a_j\}$. The choice of measurement is in principle arbitrary, and corresponds to the choice of the 'computational basis' in quantum computing theory.

For quantum state $|\psi\rangle = \sum_j \psi_j |a_j\rangle$, define P_j and S_j via the polar decomposition $\psi_j = \sqrt{P_j} e^{iS_j/\hbar}$. Thus, $P \equiv \{P_j\}$ is the probability distribution on configuration space for the ensemble of systems described by $|\psi\rangle$. Guided by the previous examples, an obvious candidate for the corresponding ensemble Hamiltonian is the average energy of the ensemble,

$$H(P, S) := \langle \psi | \hat{H} | \psi \rangle = \sum_{j,k} h_{jk} \sqrt{P_j P_k} e^{-i(S_j - S_k)/\hbar}, \qquad (1.15)$$

where $\hat{H}$ is the quantum Hamiltonian operator and $h_{jk} := \langle a_j | \hat{H} | a_k \rangle$. It is then straighforward to check that the Hamiltonian equations of motion (1.11) are indeed equivalent to the real and imaginary parts of the Schrödinger equation

$$\hat{H} |\psi\rangle = i\hbar \frac{\partial |\psi\rangle}{\partial t}. \qquad (1.16)$$

It may also be checked that Eq. (1.14) holds for the average energy, i.e., $H(P, S) = \langle -\partial_t S \rangle$.

Just as for the classical rate equations, the ensemble Hamiltonian in Eq. (1.15) only depends on differences between the components of S, implying that the equations of motion are invariant under the addition of an arbitrary constant to the components of S. In quantum mechanics this is usually interpreted in terms of the physical irrelevance of the global phase of the wave function. However, in the more general formalism of ensembles on configuration space, this property follows as a fundamental consequence of the conservation of probability (see Sect. 1.4).

It is straightforward to generalise the above results to any complete basis set $\{|a\rangle\}$ for a quantum system, whether finite or infinite, discrete or continuous, orthonormal or otherwise [7] (one only requires that $\sum_a |a\rangle\langle a| = \hat{1}$, with summation replaced by integration over any continuous ranges of a). One may similarly generalise the results to quantum field theory (see for example Chaps. 5 and 11, where fields are discussed).

1.4   Fundamental Properties of Ensemble Hamiltonians

We may abstract from the examples of the previous sections to rewrite the central idea in Sect. 1.2 more formally, as

> **Central idea (formal version):** Physical systems are described by a probability density P on configuration space, a canonically conjugate quantity S, and an ensemble Hamiltonian $\mathcal{H}(P, S)$.

This idea is very simple but surprisingly powerful. It not only incorporates and unifies the description of standard classical and quantum dynamics, but provides a general and useful framework for formulating more general physical theories, as will be seen in later chapters.

The state of the system is described by the configuration ensemble (P, S). The role of the ensemble Hamiltonian $H(P, S)$ is to specify the dynamics of this ensemble.[2] Hence, in particular, it must evolve probabilities to probabilities. This places some important physical constraints on the choice of possible ensemble Hamiltonians, as will now be discussed.

1.4.1   Conservation of Probability

Probabilities must always sum to unity, and hence the evolution of an ensemble must respect this property. It will be shown here that the conservation of probability corresponds to invariance of the dynamics under the addition of an arbitrary constant to S.

First, for an ensemble on a discrete configuration space with probability distribution P_j, note that conservation of probability is equivalent to $\sum_j P_j(t) = \sum_j P_j(t + \varepsilon) = 1$. Hence, to first order in ε, one has via Eq. (1.11) that

$$0 = \sum_j \left[P_j(t + \varepsilon) - P_j(t)\right] = \sum_j \varepsilon \frac{\partial P_j}{\partial t} = \varepsilon \sum_j \frac{\partial \mathcal{H}}{\partial S_j} = \mathcal{H}(P, S + \varepsilon) - \mathcal{H}(P, S). \quad (1.17)$$

Hence, by considering a sequence of infinitesimal evolutions, it follows that

---

[2] For continuous configuration spaces P and S are functions, and hence one should more properly write the ensemble Hamiltonian as a functional, $H[P, S]$, in this case. However, it is convenient to use the notation $H(P, S)$ when referring to the general case.

$$\mathcal{H}(P, S + c) = \mathcal{H}(P, S) \tag{1.18}$$

as claimed. In particular, this implies for discrete configuration spaces that the ensemble Hamiltonian only depends on S via the differences $M_{jk} := S_j - S_k$: only relative values of S are physically relevant. Note for quantum systems that this property corresponds to (and explains) the irrelevance of a global phase factor of the quantum state (see previous section).

It follows, writing $\mathcal{H}(P, S) \equiv f(P, M)$ for some function f, that the equation of motion for P reduces to the form of a transition rate equation,

$$\frac{\partial P_j}{\partial t} = \sum_k \left(\frac{\partial f}{\partial M_{jk}} - \frac{\partial f}{\partial M_{kj}} \right) = \sum_k \left(T_{jk} P_k - T_{kj} P_j \right), \tag{1.19}$$

with corresponding transition rates

$$T_{jk} := (P_k)^{-1} \frac{\partial f}{\partial M_{jk}} \tag{1.20}$$

(and $T_{jk} := 0$ for $P_k = 0$). Note that, unlike the classical rate equation in Eq. (1.10), these transition rates will generally depend on S. In particular, for the discrete quantum ensemble Hamiltonian in Eq. (1.15) one finds

$$T_{jk} = \hbar^{-1} \sqrt{P_j/P_k} \operatorname{Im} \left\{ h_{jk} e^{-i(S_j - S_k)/\hbar} \right\}. \tag{1.21}$$

Such quantum transition rate equations have been used by Bell to formulate a theory of beables for fermionic fields [8], and by others to formulate modal interpretations of quantum dynamics [9].

For a continuous configuration space the ensemble Hamiltonian is more properly written as a functional, $H[P, S]$. Replacing partial derivatives by variational derivatives and summation by integration in the derivation of Eq. (1.18), one then obtains[3]

$$0 = \int dx \, [P(x, t + \varepsilon) - P(x, t)] = \int dx \, \varepsilon \frac{\partial P}{\partial t} = \varepsilon \int dx \, \frac{\delta \mathcal{H}}{\delta S} = \mathcal{H}[P, S + \varepsilon] - \mathcal{H}[P, S], \tag{1.22}$$

to first order in ε. Thus, in complete analogy to the discrete case in Eq. (1.18), we have

$$\mathcal{H}[P, S + c] = \mathcal{H}[P, S]. \tag{1.23}$$

Note that this property is guaranteed for the classical and quantum ensemble Hamiltonians $\mathcal{H}_C[P, S]$ and $\mathcal{H}_Q[P, S]$, in Eqs. (1.3) and (1.8) respectively, because they only depend on S via its derivative ∇S. The general result is worth highlighting:

---

[3]Here the defining property of the variational derivative, $F[f + \delta f] - F[f] = \int dx (\delta F/\delta f)\delta f$ for arbitrary infinitesimal variations δf, has been used (see Appendix A of this book).

Conservation of probability implies that

- The ensemble Hamiltonian is invariant under $S \rightarrow S + c$ for any constant c.
- Only relative values of S have dynamical significance.

1.4.2 Positivity of Probability

A second fundamental property of probability is that it is positive, i.e., $P \geq 0$. This enforces a further nontrivial constraint on the ensemble Hamiltonian. For example, while $H(P, S) = S_2 - S_1$ satisfies Eq. (1.18), and hence conserves probability, the equations of motion (1.11) yield $P_1(t) = P_1(0) - t$, which eventually becomes negative.

A necessary condition for ensuring positivity is obtained by expanding the probability about a given time t:

$$P(t + \varepsilon) = P(t) + \varepsilon \frac{\partial P}{\partial t} + \frac{1}{2}\varepsilon^2 \frac{\partial^2 P}{\partial t^2} + \cdots . \tag{1.24}$$

In particular, one must have $\frac{\partial P}{\partial t} = 0$ and $\frac{\partial^2 P}{\partial t^2} \geq 0$ whenever $P(t) = 0$, as otherwise P will be negative at times either just before or just after time t. The first of these conditions corresponds to the constraints

$$\frac{\partial \mathcal{H}}{\partial S_j} = 0 \quad \text{for } P_j = 0, \qquad \frac{\delta \mathcal{H}}{\delta S} = 0 \quad \text{for } P(x) = 0, \tag{1.25}$$

on the ensemble Hamiltonian, for discrete and continuous configuration spaces respectively. One can similarly write down a corresponding constraint for the second condition, involving derivatives of the ensemble Hamiltonian to second order.

For example, for classical and quantum particles the positivity condition (1.25) reduces, via the continuity equation (1.2), to

$$\frac{\partial P}{\partial t} = -\nabla \cdot \left(\frac{P \nabla S}{m} \right) = -\frac{P \nabla^2 S}{m} - \frac{\nabla P \cdot \nabla S}{m} = 0 \quad \text{for } P(x) = 0. \tag{1.26}$$

To verify this condition holds, note that if $P(x) = 0$ for some point x (at some fixed time t), then this is necessarily a global minimum of $P(x)$, implying that $\nabla P = 0$ at point x. Thus $\frac{\partial P}{\partial t} = 0$ as required. One may similarly check that $\frac{\partial^2 P}{\partial t^2} \geq 0$.

For the discrete examples in Sect. 1.3, positivity condition (1.25) reduces, via either of the transition rate equations (1.10) and (1.19), to $\sum_k T_{jk} P_k = 0$ for $P_j = 0$. For the classical transition rate equation (1.10) the condition must be checked in each case, but will of course hold for equations derived from physical microscopic models. In the quantum case its validity follows immediately from Eq. (1.21) for the quantum transition rates—indeed, for this case one has the stronger result $T_{jk} = 0$ for $P_j = 0$.

1.4.3 Homogeneity

The examples of ensemble Hamiltonians in Eqs. (1.3), (1.8) and (1.15) all satisfy the simple homogeneity property

$$\mathscr{H}(\lambda P, S) = \lambda \mathscr{H}(P, S), \qquad \lambda \geq 0. \tag{1.27}$$

This property, while not mathematically necessary, is of fundamental interest due to its physical implications.

First, note that taking the derivative of Eq. (1.27) with respect to λ and evaluating the result at $\lambda = 1$ yields

$$\mathscr{H}(P, S) = \left\langle -\frac{\partial S}{\partial t} \right\rangle \tag{1.28}$$

for both discrete and continuous configuration spaces (using the Hamiltonian equations of motion $\partial_t S_j = -\partial \mathscr{H}/\partial P_j$ and $\partial_t S = -\delta \mathscr{H}/\delta P$ respectively). Second, consider the decomposition of P into a mixture of two ensembles on the configuration space, i.e.,

$$P = w_1 P^{(1)} + w_2 P^{(2)}, \qquad w_1, w_2 \geq 0, \qquad w_1 + w_2 = 1. \tag{1.29}$$

It then follows from Eq. (1.28), assuming a discrete configuration space for definiteness, that

$$\mathscr{H}(P, S) = -\sum_j P_j \frac{\partial S_j}{\partial t} = -w_1 \sum_j P_j^{(1)} \frac{\partial S_j}{\partial t} - w_2 \sum_j P_j^{(2)} \frac{\partial S_j}{\partial t}$$

$$= w_1 \mathscr{H}(P^{(1)}, S) + w_2 \mathscr{H}(P^{(2)}, S). \tag{1.30}$$

Hence, the numerical value of the ensemble Hamiltonian is just the weighted average of the values for the two subensembles, as required for interpreting it as an average energy. It further follows from Eq. (1.28) that $-\partial_t S$ may be interpreted as a corresponding local energy density on the configuration space (however, note that the first result does *not* require that this local energy density is a physical energy—see also Sect. 2.4.2). Thus:

The homogeneity property implies that

- The ensemble Hamiltonian can be interpreted as the average energy of the ensemble.
- The quantity $-P\partial_t S$ is a local energy density for the ensemble.

We will see in Chap. 2 that the above results leads to a substantial generalization of the concept of weak values in quantum mechanics, to any theory with observables satisfying the homogeneity property.

The usefulness of similar homogeneity properties (for evolving the quantum wave function) has been noted previously in nonlinear extensions of quantum theory [10–12]. The above results show that homogeneity can be motivated at the more general level of ensembles on configuration space, without any reference to wave functions or to quantum mechanics. In Chaps. 3 and 9 this assumption will also be seen to be important for the consistent description of independently evolving ensembles.

1.5 Outline of This Book

In the following chapters, we will further develop the general formalism of ensembles on configuration space, and apply it in a wide variety of contexts, including measurement, thermodynamics, axiomatic approaches to quantum mechanics, and the coupling of classical spacetime to quantum matter. Along the way many concepts in classical and quantum theory will be unified and generalised via the overarching framework provided by the configuration ensemble approach. While some of the material has previously appeared in some form in the literature, as indicated below, many results are presented for the first time.

We continue Part I of the book, i.e., the exposition of general properties of the formalism, in Chaps. 2–4. First, in Chap. 2 we introduce a definition of observables for arbitrary ensembles on configuration space, and show that these encompass both classical and quantum observables [7, 13, 14, 24]. We also show that the formalism allows for the generalisation of certain quantum concepts, such as eigenstates, eigenvalues, weak values and transition probabilities, to arbitrary configuration ensembles.

In Chap. 3, we describe composite systems via joint ensembles [13, 14]. These may consist of subsystems which are, e.g., independent or entangled, interacting or noninteracting. A precise formulation of these properties is provided. We define the extension of single-system observables to joint ensembles, and discuss their algebraic and separability properties. The rest of this chapter is devoted to a description of measurement interactions, starting with basic measurement models, then more elaborate ones describing weak measurements and measurement-induced collapse.

In Chap. 4, we consider mixtures of configuration space ensembles, and generalise the quantum notions of 'proper' and 'improper' mixtures. With the help of such mixtures it becomes possible to unify and generalise traditional classical and quantum approaches to thermodynamics, via the definition of suitable 'thermal mixtures.' Our formulation is very different to standard approaches based on the maximum entropy principle, and is of particular interest in providing a novel Hamilton–Jacobi picture of classical thermodynamics.

In Part II of the book, comprising Chaps. 5–7, we show how the configuration ensemble approach provides a natural starting point for three very different axiomatic approaches to quantum mechanics. First, in Chap. 5 we introduce a quantization procedure for classical ensembles which provides an alternative to standard quantization methods [15–19]. We show that it is possible to formulate an exact form of the uncertainty principle, which provides the single key element that is needed for moving from the equations of motion of a classical ensemble to those of a quantum ensemble. The quantization procedure is used to derive the Schrödinger equation and bosonic field equations.

In Chap. 6, we look at the geometry of ensembles on configuration space for both discrete and continuous systems [20–23]. We show that the theory has a rich geometry, and that the geometrical structures natural to the space can be used to obtain a geometrical reconstruction of quantum mechanics. The basic structures used are the natural metric on the space of probabilities (information geometry) and the description of dynamics using a Hamiltonian formalism (symplectic geometry); requirements of consistency then lead to a Kähler geometry. This geometrical reconstruction of quantum mechanics has some remarkable features. The wave functions of quantum mechanics appear as the natural complex coordinates of the Kähler space, the full group of unitary transformations is derived based on consistency requirements, and a Hilbert space may be associated with the Kähler space of the theory, leading to the standard version of quantum theory.

In Chap. 7, we consider local representations of rotations on discrete configuration spaces, focusing in particular on ensembles of either one or two spin-half systems, which we call rotational bits or 'robits'. In the case of a single robit, the theory is equivalent to that of a single quantum mechanical qubit. The description of a pair of robits is more complicated, in that requirements of locality and subsystem independence must be taken into account. We show that in this case, in addition to a theory which is equivalent to the quantum theory of a pair of qubits, it may also be possible to have non-quantum local models.

Part III of the book, comprising Chaps. 8 and 9, deals with hybrid classical-quantum systems [13, 14, 24–26]. The problem of defining hybrid systems comprising quantum and classical components is highly nontrivial, and the approaches that have been proposed to solve this problem run into various types of fundamental difficulties. The formalism of configuration-space ensembles is able to overcome many of these difficulties, allowing for a general and consistent description of interactions between quantum and classical ensembles. In Chap. 8, we discuss general properties of hybrid ensembles and consider various applications: measurements of a quantum system by a classical apparatus, scattering, harmonic oscillators, and hybrid

Wigner functions. In Chap. 9, we focus on consistency requirements for quantum-classical interactions. We show how the configuration ensemble approach is able to satisfy desirable properties such as a Lie algebra of observables and Ehrenfest relations, while evading no-go theorems based in part on such properties. We then discuss locality aspects of the approach and present a measurement model of wide applicability.

Finally, Part IV, comprising Chaps. 10 and 11, is devoted to ensembles of classical gravitational fields and their interaction with quantum fields [18, 19, 27, 28]. After considering the case of pure gravity in Chap. 10, which we illustrate with the example of ensembles of black holes, Chap. 11 discusses the coupling of classical gravitational fields to quantum matter fields. In the standard approach to this problem (i.e., semiclassical gravity), the energy momentum tensor that serves as the source of the Einstein equations is replaced by the expectation value of the energy momentum operator with respect to a particular quantum state. This approach, however, presents a number of well known difficulties. We show that a viable alternative is provided by the use of ensembles on configuration space, which leads to a theory that is consistent and which does not have any of the problems of semiclassical gravity. We illustrate the power of the approach with two examples: a cosmological model which consists of a closed Robertson–Walker universe with a massive quantum scalar field and a classical CGHS black hole in a collapsing geometry interacting with a quantized scalar field.

References

1. Goldstein, H.: Classical Mechanics. Addison-Wesley, New York (1950)
2. Synge, J.L.: Classical dynamics. In: Flügge, S. (ed.) Encyclopedia of Physics, vol. III/1, pp. 1–225. Springer, Berlin (1960)
3. Landauer, R.: Path concepts in Hamilton–Jacobi theory. Am. J. Phys. **20**, 363–367 (1952)
4. Zakharov, V.E., Kuznetsov, E.A.: Hamiltonian formalism for nonlinear waves. Physics Uspekhi **40**, 1087–1116 (1997)
5. Madelung, E.: Quantentheorie in hydrodynamischer Form. Z. Physik **40**, 322–326 (1926)
6. Beichelt, F.: Stochastic Processes in Science, Engineering and Finance, chapter 5. Taylor & Francis, Boca Raton (2006)
7. Hall, M.J.W.: Superselection from canonical constraints. J. Phys. A **27**, 7799–7811 (2004)
8. Bell, J.S.: Beables for quantum field theory. In: Bell, M., Gottfried, K., Veltman, M., John, S. (eds.) Bell on the Foundations of Quantum Mechanics, pp. 159–166. World Scientific, Singapore (2001)
9. Gambetta J., Wiseman, H.M.: Modal dynamics for positive operator measures. Found. Phys. **34**, 419–448 (2004) (see also references therein)
10. Haag, R., Bannier, U.: Comments on Mielnik's generalized (non linear) quantum mechanics. Commun. Math. Phys. **60**, 1–6 (1978)
11. Kibble, T.W.B.: Relativistic models of nonlinear quantum Mechanics. Commun. Math. Phys. **64**, 73–82 (1978)
12. Weinberg, S.: Testing quantum mechanics. Ann. Phys. (N.Y.) **194**, 336–386 (1989)
13. Hall, M.J.W., Reginatto, M.: Interacting classical and quantum ensembles. Phys. Rev. A **72**, 062109 (2005)

14. Hall, M.J.W.: Consistent classical and quantum mixed dynamics. Phys. Rev. A **78**, 042104 (2008)
15. Hall, M.J.W., Reginatto, M.: Schrödinger equation from an exact uncertainty principle. J. Phys. A **35**, 3289–3303 (2002)
16. Hall, M.J.W., Reginatto, M.: Quantum mechanics from a Heisenberg-type equality. Fortschr. Phys. **50**, 646–651 (2002)
17. Hall, M.J.W., Kumar, K., Reginatto, M.: Bosonic field equations from an exact uncertainty principle. J. Phys. A **36**, 9779–9794 (2003)
18. Reginatto, M.: Exact uncertainty principle and quantization: implications for the gravitational field. Braz. J. Phys. **35**, 476–480 (2005)
19. Hall, M.J.W.: Exact uncertainty approach in quantum mechanics and quantum gravity. Gen. Relativ. Gravit. **37**, 1505–1515 (2005)
20. Reginatto, M., Hall, M.J.W.: Quantum theory from the geometry of evolving probabilities. In: Goyal, P., Giffin, A., Knuth, K.H., Vrscay, E. (eds.) Bayesian Inference and Maximum Entropy Methods in Science and Engineering, 31st International Workshop on Bayesian Inference and Maximum Entropy Methods in Science and Engineering, Waterloo, Canada, 10–15 July 2011. AIP Conference Proceedings, vol. 1443, American Institute of Physics, Melville, New York (2012)
21. Reginatto, M., Hall, M.J.W.: Information geometry, dynamics and discrete quantum mechanics. In: von Toussaint, U. (ed.) Bayesian Inference and Maximum Entropy Methods in Science and Engineering, 32nd International Workshop on Bayesian Inference and Maximum Entropy Methods in Science and Engineering, Garching, Germany, 15–20 July 2012. AIP Conference Proceedings, vol. 1553, American Institute of Physics, Melville, New York (2013)
22. Reginatto, M.: From probabilities to wave functions: a derivation of the geometric formulation of quantum theory from information geometry. J. Phys.: Conf. Ser. **538**, 012018 (2014)
23. Reginatto, M.: The geometrical structure of quantum theory as a natural generalization of information geometry. In: Mohammad-Djafari. A., Barbaresco, F. (eds) Bayesian Inference and Maximum Entropy Methods in Science and Engineering, 34th International Workshop on Bayesian Inference and Maximum Entropy Methods in Science and Engineering, Clos Lucé, Amboise, France, 21–26 Sep 2014. AIP Conference Proceedings, vol. 1641, American Institute of Physics, Melville, New York (2015)
24. Reginatto, M., Hall, M.J.W.: Quantum-classical interactions and measurement: a consistent description using statistical ensembles on configuration space. J. Phys.: Conf. Ser. **174**, 012038 (2009)
25. Chua, A.J.K., Hall, M.J.W., Savage, C.M.: Interacting classical and quantum particles. Phys. Rev. A **85**, 022110 (2011)
26. Hall, M.J.W., Reginatto, M., Savage, C.M.: Nonlocal signaling in the configuration space model of quantum-classical interactions. Phys. Rev. A **86**, 054101 (2012)
27. Albers, M., Kiefer, C., Reginatto, M.: Measurement analysis and quantum gravity. Phys. Rev. D **78**, 064051 (2008)
28. Reginatto, M.: Cosmology with quantum matter and a classical gravitational field: the approach of configuration-space ensembles. J. Phys.: Conf. Ser. **442**, 012009 (2013)

Chapter 2
Observables, Symmetries and Constraints

Abstract The notion of observable is one of the key concepts of a physical theory. We introduce a definition of observables within the framework of ensembles on configuration space, based on the idea of associating observables with generators of canonical transformations acting on the phase space of the fundamental variables P and S. These ensemble observables encompass both classical and quantum observables. Remarkably, for classical observables the Poisson bracket of the ensemble observables is isomorphic to the usual bracket on standard classical phase space, while for quantum observables it is isomorphic to the commutator in Hilbert space. We show that the formalism allows for the generalisation of certain quantum concepts, such as eigenstates, eigenvalues, weak values and transition probabilities, to arbitrary configuration ensembles. We discuss also systems with symmetries, in particular examples which involve representations of the Galilean group for the case of a free particle and rotations defined on discrete configuration spaces. Finally, we generalise and reinterpret quantum superselection rules in terms of constraints on observables.

2.1 Some General Considerations

The description of physical systems in terms of ensembles on configuration space introduces very few physical assumptions. However, there are some issues which concern the *interpretation* of the basic elements that are part of the formalism which are of importance and which we now address.

2.1.1 Fundamental Variables and Ontology

The theory of ensembles on configuration space is a statistical theory which describes states of a system (classical, quantum or hybrid) in terms of the two canonically conjugate variables, P and S, with the time evolution of the conjugate variables being determined by an ensemble Hamiltonian, $\mathscr{H}[P, S]$ (see Chap. 1).

© Springer International Publishing Switzerland 2016

M.J.W. Hall and M. Reginatto, *Ensembles on Configuration Space*,
Fundamental Theories of Physics 184, DOI 10.1007/978-3-319-34166-8_2

The interpretation of P is rather straightforward. We assume that the configuration of a physical system is an inherently statistical concept, in which case the state of the system must be described by an ensemble of configurations, corresponding to some probability density P on the configuration space. The physical interpretation of S requires more care. As we have discussed in the previous chapter, the dynamical significance of S is invariant under addition of an arbitrary constant, and one may define a local energy density, $-P\partial_t S$, for ensemble Hamiltonians satisfying a fundamental homogeneity property. Furthermore, the gradient of S plays an important role for continuous configuration spaces because it is proportional to the velocity vector fields which enters into the continuity equation for P (see Sect. 1.2), and its more general role as a generator of translations will be seen below. It is clear that one could attempt to "complete" the theory, for example by assigning a definite momentum $\mathbf{p} = \nabla S$ and a definite energy $E = -\partial_t S$ to particles that belong to an ensemble. This would lead to the usual deterministic interpretation of the Hamilton–Jacobi equation for the case of a classical system, and to the de Broglie–Bohm formulation for the case of a quantum system. However, we will not take this additional step and we will rely instead on a "minimalist" interpretation in which the theory is treated as a purely *statistical* one (see also the discussion of the classical limit in Sect. 9.3). Thus, for particles and continuous configuration spaces, the fundamental concept is that of a probability density P defined on the configuration space of the system, and the existence of a canonically conjugate quantity S is mandated by the requirement that P evolves according to an action principle.

A configuration ensemble defined by a pair of conjugate variables P and S which satisfy the equations of motion derived from the ensemble Hamiltonian will also be called a *pure ensemble*. This terminology corresponds to the notion of a pure state in quantum mechanics, which is described by a wave function $\psi = \sqrt{P}e^{iS/\hbar}$ (in contrast to a mixed state described by a density matrix). Here we apply this terminology to all configuration ensembles, whether classical, quantum or otherwise. In addition, one may also define *mixtures* of configuration ensembles, of the form $\{P_k, S_k; w_k\}$, where each of the components satisfy the equations of motion determined by the same ensemble Hamiltonian $\mathscr{H}[P, S]$ and $\sum_k w_k = 1$, so that $\sum_k w_k \int dx\, P_k(x, t) = 1$. Mixtures are integral to our discussion of thermodynamics on configuration space in Chap. 4. But the formalism of ensembles on configuration space has a *pure state ontology*. In particular, it treats pure ensembles, rather than more general mixtures, as physically fundamental. The latter are taken to merely reflect ignorance of the 'true' pure state.

2.1.2 The Dual Role of Observables

From quantum mechanics, we are familiar with the dual role of the operators that are associated with observables: they are Hermitian operators which, on the one hand, generate transformations which are unitary and thus preserve the normalization of the probability $|\psi|^2$ and, on the other, have real expectation values and eigenvalues which in principle can be determined from measurement (hence the terminology "observable"). It is often necessary to consider *both* roles when analyzing experiments: typically, an effort is made to prescribe operational procedures which define the observables that are being measured in the experiment via a particular interaction process (e.g., the spin of a particle in a Stern–Gerlach experiment via an interaction with an inhomogeneous magnetic field) while at the same time including other observables in the analysis of the experiment in their role of generators of transformations (e.g., the energy in its role of generator of time translations, if the interaction takes place at one time and the detection at a later time).

As is well known, for most of the Hermitian operators that one can define in quantum mechanics there are no operational procedures that specify how they should be measured. In addition, there are fundamental limitations on the precision with which measurements can be made for observables that do not commute with additive conserved quantities (e.g., linear or angular momentum, or charge) [1], which constitute however the vast majority of the observables that are of interest. This does not create serious difficulties when applying quantum theory to actual experiments, but it does mean that the theory allows for a surplus of possible observables, all of which have well defined properties as far as their role as generators of transformations is concerned, but are problematic in their role of measurable quantities in that operational prescriptions for eigenvalues and expectation values are not always available.

A similar situation, regarding both the dual role of observables and the large number of possible observables allowed by the theory, also holds for standard classical dynamics on phase space: observables are functions of position and momentum on phase space, and are regarded as both measurable quantities and generators of canonical transformations [2].

In the discussion on observables that follows, for the general case of ensembles on configuration space, the same considerations apply. We will first address the more general issue of defining generators of transformations within the theory. We will make the connection to measurements in Chap. 3.

In this chapter we give a precise definition of observables, and discuss examples for both classical and quantum ensembles. We introduce the notion of an eigenensemble, and generalise the quantum mechanical notions of weak values and transition probabilities. We address the representation of symmetries by corresponding groups of observables, independently of whether the ensemble is classical, quantum or otherwise, via examples of Galilean particles and "rotational bits". Finally, we end the Chapter with a discussion of constraints and the formulation of superselection rules.

2.2 Observables

A significant advantage of describing physical systems by ensembles evolving on configuration space is the existence of an action principle (see Chap. 1). In particular, this allows the definition of a Poisson bracket for functions of the fundamental phase space variables P and S, and allows observables to be introduced as generators of canonical transformations with respect to this bracket, just as in standard classical dynamics [2]. We will see that this Poisson bracket is isomorphic to the Poisson bracket on a classical phase space for the case of classical ensembles, and is isomorphic to the quantum commutator for the case of quantum observables. The existence of such a bracket will more generally allow us to define dynamics for hybrid classical-quantum systems, such as the coupling of quantum systems to classical measuring apparatuses and of quantum fields to classical gravity (see Chaps. 8 and 11).

For an ensemble on a discrete configuration space, with ensemble Hamiltonian $\mathscr{H}(P, S)$, the evolution is specified by the Hamiltonian equations of motion (see Chap. 1)

$$\frac{\partial P_j}{\partial t} = \frac{\partial H}{\partial S_j}, \quad \frac{\partial S_j}{\partial t} = -\frac{\partial H}{\partial P_j}. \tag{2.1}$$

Defining the Poisson bracket for two arbitrary functions $A(P, S)$ and $B(P, S)$ by

$$\{A, B\} := \sum_j \left(\frac{\partial A}{\partial P_j} \frac{\partial B}{\partial S_j} - \frac{\partial A}{\partial S_j} \frac{\partial B}{\partial P_j} \right), \tag{2.2}$$

these equations of motion may be rewritten in the form

$$\frac{\partial P_j}{\partial t} = \{P_j, \mathscr{H}\}, \quad \frac{\partial S_j}{\partial t} = \{S_j, \mathscr{H}\} \tag{2.3}$$

in complete analogy to the case of classical phase space dynamics [2]. It immediately follows that any function $A(P, S, t)$ of P, S and t evolves as

$$\frac{dA}{dt} = \sum_j \left(\frac{\partial A}{\partial P_j} \frac{\partial P_j}{\partial t} + \frac{\partial A}{\partial S_j} \frac{\partial S_j}{\partial t} \right) + \frac{\partial A}{\partial t}$$

$$= \{A, \mathscr{H}\} + \frac{\partial A}{\partial t}. \tag{2.4}$$

Similarly, for an ensemble on a continuous configuration space, the Poisson bracket of two arbitrary functionals $A[P, S]$ is defined by

$$\{A, B\} = \int dx \left(\frac{\delta A}{\delta P} \frac{\delta B}{\delta S} - \frac{\delta A}{\delta S} \frac{\delta B}{\delta P} \right). \tag{2.5}$$

Noting that $\delta f(x)/\delta f(x') = \delta(x - x')$ (see Appendix A.1 of this book), it follows that the equations of motion for the ensemble can be rewritten as

$$\frac{\partial P}{\partial t} = \{P, \mathscr{H}\}, \quad \frac{\partial S}{\partial t} = \{S, \mathscr{H}\} \tag{2.6}$$

and that again $dA/dt = \{A, \mathscr{H}\} + \partial A/\partial t$ as per Eq. (2.4).

Transformations of the fundamental phase space variables P and S that preserve the Poisson bracket are called canonical transformations. In particular, every function (or functional) of these variables generates an associated infinitesimal canonical transformation, according to

$$\delta P = \{P, A\}\varepsilon = \frac{\delta A}{\delta S}\varepsilon \qquad (2.7)$$

$$\delta S = \{S, A\}\varepsilon = -\frac{\delta A}{\delta P}\varepsilon, \qquad (2.8)$$

where ε is an infinitesimal parameter [2].

It is natural to associate observables with the generators of such canonical transformations, similarly to the case of standard classical and quantum mechanics (see previous section). In particular, the ensemble Hamiltonian may be interpreted as the generator of time translations. However, it will be recalled from Chap. 1 that ensemble Hamiltonians must satisfy certain fundamental constraints, to ensure the conservation and positivity of probability. Similarly, one cannot associate an arbitrary function $A(P, S)$ with an observable: the infinitesimal canonical transformation generated by A,

$$P \to P + \varepsilon \frac{\delta A}{\delta S}, \quad S \to S - \varepsilon \frac{\delta A}{\delta P}, \qquad (2.9)$$

must preserve the normalization and positivity of P. Hence, just as per Eqs. (1.18) and (1.25) for ensemble Hamiltonians, the conditions

$$A[P, S + c] = A[P, S], \quad \frac{\delta A}{\delta S} = 0 \text{ if } P(x) = 0 \qquad (2.10)$$

must be satisfied for observables on continuous configuration spaces (and corresponding conditions for observables on discrete configuration spaces). The first of these conditions implies that only relative values of S have physical significance.

There is a further fundamental requirement which we will impose on observables, corresponding to the homogeneity property for ensemble Hamiltonians discussed in Sect. 1.4: that they be functionals which are homogeneous of degree one in P, i.e.,

$$A[\lambda P, S] = \lambda A[P, S], \qquad (2.11)$$

where λ is an arbitrary positive constant. In particular, this property implies that A can consistently be interpreted as an ensemble average (see Sect. 1.4.3), i.e., the numerical value of A corresponds to the expectation value of the observable over the ensemble.

We are led therefore to the following definition of observables:

Definition The observables of a configuration ensemble are a set of functions (or functionals) of P and S satisfying the probability conservation, positivity and homogeneity properties in Eqs. (2.10) and (2.11).

Note that each of the conditions in Eqs. (2.10) and (2.11) is preserved by the Poisson bracket. First, defining $I[P, S] := \int dx\, P$, the conservation of probability is simply the requirement that I is invariant under allowed canonical transformations, i.e., that $\delta I = \varepsilon\{I, A\} = 0$. Hence, if it holds for two observables A and B, then it automatically holds for $\{A, B\}$ via the Jacobi identity, since

$$\{I, \{A, B\}\} = -\{A, \{B, I\}\} - \{B, \{I, A\}\} = 0. \tag{2.12}$$

Similarly, the positivity condition may be rewritten as $\delta P = \varepsilon\{P, A\} = 0$ whenever $P(x) = 0$ (otherwise $P(x)$ can be decreased below 0 by choosing the sign of ε appropriately), which again holds for $\{A, B\}$, if it holds for A and B, as a consequence of the Jacobi identity. Finally, it is straightforward to check that the Poisson bracket of two functionals which are homogeneous of degree one in P is also homogeneous of degree one in P. Hence, *it may assumed without loss of generality that the set of observables form a closed Lie algebra under the Poisson bracket.*

2.3 Examples

In the previous section we have given a precise definition of observables. We now consider a number of examples, including classical and quantum observables.

2.3.1 Position and Momentum Observables

Two examples of particular interest for continuous configuration spaces are position and momentum observables. Given the interpretation of observables as expectation values, following from the homogeneity property (2.11), an obvious definition for the ensemble position observable is

$$X[P, S] := \int dx\, P x. \tag{2.13}$$

Note that this observable generates the transformation

$$P \rightarrow P + \varepsilon \frac{\delta X}{\delta S} = P \tag{2.14}$$

via Eq. (2.9), and hence P trivially remains positive and normalised, as required by Eq. (2.10). The homogeneity property (2.11) is also trivially satisfied.

The definition of the ensemble momentum may be motivated by noting that in classical and quantum mechanics the momentum observable generates translations (for example, $\psi(x - \varepsilon) = e^{-i\varepsilon \cdot \hat{p}/\hbar} \psi(x)$ for a quantum wave function $\psi(x)$). Hence, identifying the ensemble momentum as the observable Π which generates translations, one has via Eq. (2.9) that

$$\varepsilon \cdot \frac{\delta \Pi}{\delta S} = \delta P = P(x - \varepsilon) - P(x) = -\varepsilon \cdot \nabla P(x), \tag{2.15}$$

and

$$\varepsilon \cdot \frac{\delta \Pi}{\delta P} = -\delta S = -[S(x - \varepsilon) - S(x)] = \varepsilon \cdot \nabla S(x). \tag{2.16}$$

for arbitrary infinitesimal translations ε on configuration space. The solution of these equations is, up to an arbitrary additive constant,

$$\Pi[P, S] := \int dx\, P \nabla S, \tag{2.17}$$

for the ensemble momentum. In particular, this expression immediately yields $\delta \Pi / \delta P = \nabla S$, while under an infinitesimal variation $S \to S + \delta S$ one has

$$\delta \Pi = \Pi[P, S + \delta S] - \Pi[P, S] = \int dx\, P \nabla (\delta S) = -\int dx\, (\nabla P)\, \delta S, \tag{2.18}$$

which implies, via Eq. (A.1) of the Appendix, that one also has $\delta \Pi / \delta S = -\nabla P$ as required. Note that normalisation and positivity of P is trivially preserved under translations, implying that Eq. (2.10) is satisfied by $\Pi[P, S]$. Further, the homogeneity requirement (2.11) is satisfied by direct inspection.

It follows from Eq. (2.17) that $P \nabla S$ is a local momentum density for continuous configuration spaces, independently of whether the ensemble is classical, quantum, or something more general. This result (together with the identification of $-P \partial_t S$ as a local energy density in Sect. 1.4), helps to establish the physical role played by S in the formalism of ensembles on configuration space. However, to maintain full generality, S should not be regarded as a "momentum potential". This would go beyond what is required of a statistical theory. In particular, for an ensemble of classical particles with uncertainty described by the probability P, it will not be assumed that the momentum of a member of the ensemble is a well-defined quantity proportional to the gradient of S, as it is done in the usual deterministic interpretation of the Hamilton–Jacobi equation. This avoids forcing a similar deterministic interpretation in the quantum case. A deterministic picture can be recovered for classical ensembles precisely in those cases in which trajectories are operationally defined [3].

Finally, the Poisson bracket for the components of the ensemble position and momentum may be calculated from Eqs. (2.5), (2.13) and (2.17) as

$$\{X_m, \Pi_n\} = \delta_{mn} \tag{2.19}$$

which is the same result as for the Poisson bracket of two classical position and momentum observables [2]. A more general correspondence will be seen in the next example. This result is relevant to representations of the Galilean group of observables, as will be discussed in Sect. 2.5.

2.3.2 Classical Observables

In classical mechanics, observables corresponds to functions $f(x, p)$ on the classical phase space. We define the corresponding classical ensemble observable C_f by

$$C_f := \int dx\, P f(x, \nabla S) \tag{2.20}$$

This is similar in form to a classical average, and clearly satisfies the homogeneity property (2.11). Hence, the numerical value of C_f may consistently be identified with the ensemble average of the corresponding function $f(x, p)$. Further, it is easily checked that C_f satisfies the required normalisation condition in Eq. (2.10)—the only dependence on S is via its gradient. The positivity condition in Eq. (2.10) is also satisfied, noting that

$$\frac{\delta C_f}{\delta S} = \frac{\delta}{\delta S} \int dx\, P f(x, \nabla S) = P \frac{\partial f(x, \nabla S)}{\partial S} - \nabla \cdot \left[P \frac{\partial f(x, \nabla S)}{\partial \nabla S} \right]$$

$$= -P \nabla \cdot \left[\frac{\partial f(x, \nabla S)}{\partial \nabla S} \right] - \nabla P \cdot \frac{\partial f(x, \nabla S)}{\partial \nabla S} \tag{2.21}$$

(see the Appendix to this book regarding the calculation of variational derivatives). In particular, since P is non-negative, it must reach a global minimum at any point x for which $P(x) = 0$. Hence $\nabla P(x)$ also vanishes, and thus the last line vanishes at $P(x) = 0$ as required.

The Poisson bracket of any two classical observables C_f and C_g follows, using Eq. (2.5) and integration by parts with respect to x, as

$$\{C_f, C_g\} = \int dx \left[-f \nabla_x \cdot (P \nabla_p g) + g \nabla_x \cdot (P \nabla_p f) \right]$$

$$= \int dx\, P \left(\nabla_x f \cdot \nabla_p g - \nabla_x g \cdot \nabla_p f \right)$$

$$= C_{\{f,g\}}, \tag{2.22}$$

where all quantities in the integrands are evaluated at $p = \nabla_x S$, and $\{f, g\}$ denotes the usual Poisson bracket for phase space functions. Hence, we have the remarkable result that

> The Poisson bracket for classical ensembles on configuration space is isomorphic to the usual Poisson bracket on phase space.

This isomorphism between deterministic observables on phase space and ensemble observables on configuration space makes it possible to formulate thermodynamics on configuration space instead of phase space (see Chap. 4), and is crucial to the construction of hybrid quantum-classical systems (see Chaps. 8 and 9).

2.3.3 Quantum Observables

In quantum mechanics, the fundamental observables are represented by Hermitian operators. For Hermitian operator $\hat{M}$ acting on the Hilbert space spanned by the kets $\{|q\rangle\}$, the configuration space is defined by a choice of computational basis $\{|q\rangle\}$ (see Chap. 1), and we define the corresponding quantum ensemble observable $Q_{\hat{M}}$ by

$$Q_{\hat{M}} := \langle \psi | \hat{M} | \psi \rangle$$
$$= \int dq\, dq'\, (PP')^{1/2} e^{i(S-S')/\hbar} \langle q' | \hat{M} | q \rangle, \tag{2.23}$$

where $\psi(q) := \sqrt{P(q)}\, e^{iS(q)/\hbar}$, $P = P(q)$, $P' = P(q')$, etc. (and where integration with respect to q and q' is replaced by summation over any discrete portions of the quantum configuration space). This is just the quantum expectation value of $\hat{M}$ with respect to the wave function $\psi(q)$, and clearly satisfies the homogeneity property (2.11). Hence, the numerical value of $Q_{\hat{M}}$ may be identified with the ensemble average of the corresponding operator $\hat{M}$.

It follows immediately from Eq. (2.23) that $Q_{\hat{M}}$ also satisfies the normalisation condition in Eq. (2.10) since it only depends on differences of S at different points q and q' of configuration space. Further, the positivity condition is trivially satisfied for a discrete configuration space, while for the continuous case one has, under an infinitesimal variation $S \to S + \delta S$,

$$\delta Q_{\hat{M}} = \int dq\, dq'\, (PP')^{1/2} \frac{i}{\hbar} \left(\delta S - \delta S' \right) e^{i(S-S')/\hbar} \langle q' | \hat{M} | q \rangle$$
$$= \int dq\, dq'\, (PP')^{1/2} \frac{i}{\hbar} \left[e^{i(S-S')/\hbar} \langle q' | \hat{M} | q \rangle - e^{i(S'-S)/\hbar} \langle q | \hat{M} | q' \rangle \right] \delta S,$$

immediately implying that

$$\frac{\delta Q_{\hat{M}}}{\delta S} = -\frac{1}{\hbar} \int dq'\, (PP')^{1/2} \operatorname{Im} \left\{ e^{i(S-S')/\hbar} \langle q' | \hat{M} | q \rangle \right\}, \tag{2.24}$$

which vanishes for $P(q) = 0$ as required.

To evaluate the Poisson bracket of any two quantum observables $Q_{\hat{M}}$ and $Q_{\hat{N}}$, it is convenient to first express the Poisson bracket in terms of the wave function $\psi(q)$ and its complex conjugate $\bar{\psi}(q)$. One has in particular for any real functional $A[P, S]$ that

$$\frac{\delta A}{\delta P} = \frac{\partial \psi}{\partial P}\frac{\delta A}{\delta \psi} + \frac{\partial \bar{\psi}}{\partial P}\frac{\delta A}{\delta \bar{\psi}} = \frac{1}{\bar{\psi}\psi}\,\mathrm{Re}\left\{\psi\frac{\delta A}{\delta \psi}\right\}, \tag{2.25}$$

$$\frac{\delta A}{\delta S} = \frac{\partial \psi}{\partial S}\frac{\delta A}{\delta \psi} + \frac{\partial \bar{\psi}}{\partial S}\frac{\delta A}{\delta \bar{\psi}} = -\frac{2}{\hbar}\,\mathrm{Im}\left\{\psi\frac{\delta A}{\delta \psi}\right\}, \tag{2.26}$$

and hence, noting $-ad + bc = \mathrm{Im}\{(a + ib)(c - id)\}$, that

$$\{A, B\} = \frac{2}{\hbar}\,\mathrm{Im}\left\{\int dq\,\frac{\delta A}{\delta \psi}\frac{\delta B}{\delta \bar{\psi}}\right\}. \tag{2.27}$$

A similar result holds for a discrete configuration space, with integration replaced by summation and variational derivatives by partial derivatives. Recalling that $\hat{M}$ and $\hat{N}$ are Hermitian, so that $\bar{\psi}\hat{M}\psi$ may be replaced by $(\overline{\hat{M}\psi})\,\psi$ in Eq. (2.23), it immediately follows that

$$\{Q_{\hat{M}}, Q_{\hat{N}}\} = \frac{2}{\hbar}\,\mathrm{Im}\left\{\int dq\,(\overline{\hat{M}\psi})\hat{N}\psi\right\} = Q_{[\hat{M},\hat{N}]/(i\hbar)}, \tag{2.28}$$

where $[\hat{M}, \hat{N}]$ denotes the usual quantum commutator $\hat{M}\hat{N} - \hat{N}\hat{M}$. Hence, in analogy to classical observables:

> The Poisson bracket for quantum ensembles on configuration space is isomorphic to the usual commutator on Hilbert space.

Thus, the Poisson bracket for ensemble observables unifies the standard classical and quantum brackets. This result is crucial to the construction of hybrid classical-quantum systems (see Chaps. 8 and 9).

2.4 Eigenensembles, Weak Values and Transition Probabilities

The examples discussed in the previous section show that the notion of ensemble observables encompasses both classical and quantum observables. It also allows for the generalisation of certain concepts which are important in quantum mechanics. In particular, as we show in this section, one may introduce generalisations of quantum mechanical eigenstates, eigenvalues, weak values and transition probabilities.

2.4.1 Eigensembles and Eigenvalues

We now want to introduce the notion of a state that is 'sharp' with respect to a particular observable, which we will call an eigenensemble. We will show that it is possible to give a general definition which fits into the canonical formalism of the theory of ensembles on configuration space. Such states are simply generalizations of stationary ensembles, which we will discuss first.

2.4.1.1 Stationary Ensembles

For ensemble Hamiltonians with no explicit time dependence, 'stationary ensembles' may be defined as those ensembles for which the dynamical properties of the ensemble are also time-independent. Recalling that only relative values of S are dynamically relevant (see Sect. 2.2), such ensembles must satisfy the conditions

$$P(x, t) = P(x, t'), \qquad S(x, t) - S(x', t) = S(x, t') - S(x', t'), \qquad (2.29)$$

for all configurations x, x' and times t, t', which are equivalent to $\partial P / \partial t = 0$ and $S(x, t) = s(x) + f(t)$ for some functions s and f (the latter follows by noting the second condition implies $\partial[S(x, t) - S(x', t)] / \partial t = 0$, yielding $S(x, t) - S(x', t) = k(x, x')$ for some function k). Noting that $f''(t) = \partial^2 S / \partial t^2 = -(\partial / \partial t)(\delta \mathscr{H} / \delta P)$ (where $\delta \mathscr{H} / \delta P$ is replaced by $\partial \mathscr{H} / \partial P_j$ for discrete configuration spaces), and that the last term must vanish if the ensemble is time-independent, it follows that stationary ensembles are characterised by the conditions

$$\frac{\partial P}{\partial t} = 0, \qquad \frac{\partial S}{\partial t} = -E, \qquad (2.30)$$

for some constant E.

The above conditions clearly generalise the concept of a stationary state in quantum mechanics, where E is a corresponding energy eigenvalue. In particular, for this case Eq. (2.30) reduces to, using Eqs. (2.1), (2.6) and (2.23) with $\hat{M} = \hat{H}$, the stationary Schrödinger equation $i\hbar \partial_t |\psi\rangle = \hat{H} |\psi\rangle = E |\psi\rangle$. In classical mechanics, these conditions are equivalent to postulating a stationary state with time-independent P and a solution of the Hamilton–Jacobi theory of the special form $S(x, t) = -Et + W(x)$, where E is the energy of the state and $W(x)$ is sometimes called Hamilton's characteristic function [2]. We will meet stationary ensembles again in Chap. 4 (for thermal mixtures) and Chap. 8 (for hybrid quantum-classical ensembles).

2.4.1.2 General Eigensembles

Just as stationary ensembles generalise quantum stationary states, we may generalise
the notion of quantum eigenstates as follows.

Definition For a given observable A, the configuration ensemble (P, S) is defined
to be an 'eigenensemble' of A if and only if the physical properties of the ensemble
are invariant under the canonical transformation generated by A.

To apply this definition, note first that the probability density P is in principle
measurable, and hence must be invariant, i.e.,

$$\delta P(x) = \varepsilon\{P(x), A\} = 0. \tag{2.31}$$

Second, since physical properties are invariant under addition of a constant to S, only
relative values of S are required to be invariant under transformations generated by
A, i.e.,

$$\delta S(x) - \delta S(x') = \varepsilon\{S(x) - S(x'), A\} = 0 \tag{2.32}$$

for all x and x'. It follows that (P, S) is an eigenensemble of observable A if and only
if

$$\frac{\delta A}{\delta S} = 0, \qquad \frac{\delta A}{\delta P} = \text{constant} = \alpha. \tag{2.33}$$

The constant α will be called the *eigenvalue* of A for such an eigenensemble.

A solution of Eq. (2.33) for a particular eigenvalue α will be denoted by (P_α, S_α).
It will be seen in Sect. 2.4.2 below that the value of A on an eigenensemble is equal
to the corresponding eigenvalue, i.e.,

$$A(P_\alpha, S_\alpha) = \alpha. \tag{2.34}$$

Note that Eq. (2.33) reduces to the definition of a stationary state in Eq. (2.30) when
one identifies A with the ensemble Hamiltonian $\mathscr{H}$, and α with the energy E. For the
quantum observable $Q_{\hat{M}}$ in Eq. (2.23) it reduces to the definition of an eigenstate of
$\hat{M}$. Of course, for more general functions A of P and S there may be no corresponding
eigensembles.

2.4.2 Weak Values and Local Densities

Differentiating the homogeneity property $A[\lambda P, S] = \lambda A(P, S)$ in Eq. (2.11) with
respect to λ, and setting $\lambda = 1$, yields the numerical equivalence

$$A[P, S] = \int dx \, P \frac{\delta A}{\delta P}. \tag{2.35}$$

Thus, each observable A has an associated *local density* $P(\delta A/\delta P)$ on the configuration space. For the case of the ensemble Hamiltonian this is a local energy density, $-P\partial_t S$, as noted previously in Chap. 1.

The existence of such local densities may be used to show that A may be consistently interpreted as an ensemble expectation value (the argument is identical to that in Sect. 1.4.3 for ensemble Hamiltonians, and does *not* require any interpretation for the local density itself). Further, Eqs. (2.33) and (2.35) immediately yield Eq. (2.34) for eigensembles of A.

Equation (2.35) leads to a further remarkable result: a far-reaching generalisation of the notion of the 'weak value' of an observable in quantum mechanics [4, 5]. In particular, we will define the weak value of an observable A, for an *arbitrary* configuration ensemble (P, S), by the function

$$A^w(x) := \frac{\delta A}{\delta P} \tag{2.36}$$

on the configuration space (with the variational derivative replaced by a partial derivative for discrete configuration spaces).

Note first that the average of the weak value over the ensemble follows immediately from Eq. (2.35) as

$$\langle A^w \rangle := \int dx \, P(x) \, A^w(x) = A[P, S]. \tag{2.37}$$

Thus, the expectation values of A and A^w are equal. For eigensembles of A the stronger result $A^w = \alpha$ holds via Eq. (2.34).

Second, for the classical ensemble observable C_f defined in Eq. (2.20), the corresponding weak value follows as

$$C_f^w(x) = f(x, \nabla S). \tag{2.38}$$

Thus, the classical weak value is equal to the classical phase space function $f(x, p)$ evaluated at $p = \nabla S$.

Third, for the quantum ensemble observable $Q_{\hat{M}}$ defined in Eq. (2.23), the corresponding weak value follows via Eqs. (2.25) and (2.36) as

$$
\begin{aligned}
Q_{\hat{M}}^w(q) &= \frac{1}{\overline{\psi}(q)\psi(q)} \, \mathrm{Re} \left\{ \psi(q) \frac{\delta Q_{\hat{M}}}{\delta \psi} \right\} \\
&= \mathrm{Re} \left\{ \frac{\langle q | \hat{M} | \psi \rangle}{\langle q | \psi \rangle} \right\},
\end{aligned} \tag{2.39}
$$

where the property $\delta Q_{\hat{M}}/\delta\psi = \overline{\hat{M}\psi(q)}$ has been used, following from the expression

$$Q_{\hat{M}} = \int dq\,\bar{\psi}(q)\hat{M}\psi(q) = \int dq\,\overline{\hat{M}\psi(q)}\psi(q) \tag{2.40}$$

for Hermitian operators. Equation (2.39) may be recognised as the quantum weak value of $\hat{M}$ [4, 5]—thus justifying the use of the terminology 'weak value' for the general case in Eq. (2.36).

As originally introduced by Aharonov and Vaidman, weak values correspond to the average outcome of an apparatus weakly coupled to $\hat{M}$ and postselected by measurement result $\hat{Q} = q$ in the computational basis $\{|q\rangle\}$[1] [4, 5]. An alternative characterisation of $Q_{\hat{M}}^{w}$ is that it provides the best possible estimate of the value of $\hat{M}$ from a measurement in the computational basis on state $|\psi\rangle$ [6–8]. An excellent review on the interpretation of quantum weak values has been given recently by Dressel [9].

It would be of great interest to assess the degree to which the above interpretations can be applied in the general context of arbitrary observables for ensembles on configuration space. We do not address this issue in detail here, but note that it is natural, in this context, to consider the *weak observable* $A^{w}[P, S]$, defined by

$$A^{w}[P] := \int dx\,P(x)\,A^{w}(x) \tag{2.41}$$

(with integration replaced by summation for discrete configuration spaces), treating $A^{w}(x)$ as a fixed function on configuration space. The weak observable only depends on the configuration parameter x, and from Eq. (2.37) is numerically equal to $A[P, S]$. The weak observable corresponds to the average weak measurement outcomes in the first interpretation above, while the difference between the weak observable and A is relevant to defining the optimal estimate in the second interpretation above. The connection of weak values to weak measurements is explored further in Sect. 3.5 of Chap. 3.

2.4.3 Transition Probabilities

We have seen that ensemble observables allow for general definitions of eigensembles, eigenvalues and weak values, which generalize the corresponding concepts in quantum theory. We now want to look briefly at how generalised transition probabilities might be defined.

Suppose first that a particular configuration ensemble, (P, S), is an eigensemble with respect to some observable G, with corresponding eigenvalue γ. The notation

---

[1] Weak values are defined by some authors as $\frac{\langle q|\hat{M}|\psi\rangle}{\langle q|\psi\rangle}$; however, it is the real part of this quantity that has a direct interpretation in terms of weak measurements.

(P_γ, S_γ) would be better suited here, but we will simply use (P, S) when it can not lead to confusion, to simplify the notation. We thus have (see Sect. 2.4.1)

$$\frac{\delta G}{\delta S} = 0, \quad \frac{\delta G}{\delta P} = \gamma, \quad G[P, S] = \langle \delta G/\delta P \rangle = \gamma. \tag{2.42}$$

Consider further a second observable F, which has various possible measurement values parameterized by a variable ϕ. It will *not* be assumed at this stage that the values of ϕ are also eigenvalues of F. We can now ask the following question: *What is the probability of obtaining measurement result $F = \phi$ for the eigensemble of G having eigenvalue γ?* This probability will be denoted by $w(\phi|\gamma)$.

To answer this question, consider first some function $f(\phi)$ of the possible measurement outcomes. Then, the corresponding expectation value of this function follows as

$$\langle f(\phi) \rangle = \int d\phi \, w(\phi|\gamma) f(\phi). \tag{2.43}$$

It is natural to now make the assumption that this expectation value *is itself* the expectation value of some observable. This amounts to a 'completeness' assumption for the set of observables. We will call this observable $A_{f(F)}[P, S]$. Thus, the equality $A_{f(F)} = \langle f(\phi) \rangle$ is satisfied.

It follows immediately from Eq. (2.43) that one has the general relation

$$\int d\phi \, w(\phi|\gamma) f(\phi) = \langle f(\phi) \rangle = A_{f(F)} = \int dx P \frac{\delta A_{f(F)}}{\delta P}. \tag{2.44}$$

for an *arbitrary* function f. The task then is to choose a particular set of functions f which allows this relationship to be inverted, so as to solve for the value of the transition probability $w(\phi|\gamma)$. For example, one could choose a set of orthogonal polynomials (e.g., Legendre polynomials) in the case of bounded sets of measurement outcomes. Here we consider another choice, the relatively simple 'Fourier' choice $f_z(\phi) = e^{iz\phi}$. Hence the left hand side of the above relationship is a Fourier transform. We can then apply the inverse transform with respect to z, to obtain the explicit solution

$$w(\phi|\gamma) = \frac{1}{2\pi} \int dz \, dx \, P(x) \, e^{-iz\phi} \frac{\delta A_{f_z(F)}}{\delta P}. \tag{2.45}$$

For discrete-valued observables, a discrete Fourier transform would be appropriate.

However, the solution given by Eq. (2.45) remains formal until we specify how the functional $A_{f(F)}$ is to be constructed from a given observable F and function f. We discuss two approaches for doing this.

The first approach is to give $A_{f(F)}$ an *operational* definition. For example, for both quantum and classical observables weak values can be *measured* following the approach proposed by Aharonov and Vaidman, via a coupling to a weak meter followed by a position measurement [5] (see also Sects. 2.4.2 and 3.5). Now, suppose it is possible to measure weak values more generally, by a similar well-defined procedure—e.g., $A^w_{f(F)}(x)$ might be measurable via coupling to a weak F-meter followed by a position measurement. Equation (2.43) can then be rewritten in the operationally well-defined form

$$w(\phi|\gamma) = \frac{1}{2\pi} \int dz\, dx P(x)\, e^{-iz\phi} A^w_{f(F)}(x). \tag{2.46}$$

A different, more formal approach to inverting Eq. (2.43) requires defining the observable F^k for $k = 2, 3, \ldots$, as this would allow one to construct most observables $A_{f_z(F)}$ of interest. This effectively corresponds to defining a product algebra on the set of observables. We carry out this construction for the classical and quantum observables defined in Sect. 2.3. For the classical observable C_g, where g is some phase space function $g(x, p)$, we define

$$f(C_g) := C_{f(g)}. \tag{2.47}$$

Thus, for example, one has

$$(C_g)^2[P, S] = \int dx\, P\, g(x, \nabla S)^2. \tag{2.48}$$

For the quantum observable $Q_{\hat{M}}$, where $\hat{M}$ is some Hermitian operator, we define

$$f(Q_{\hat{M}}) := Q_{f(\hat{M})}. \tag{2.49}$$

Thus, for example, the observable corresponding to square of the momentum is

$$(Q_{\hat{p}})^2[P, S] = \int dx\, P\left[|\nabla S|^2 + (\hbar^2/4)|\nabla \log P|^2\right]. \tag{2.50}$$

Thus transition probabilities may be calculated via powers of observables in these cases.

2.5 Symmetries and Transformations

The Poisson bracket satisfies all the properties required of a Lie algebra, i.e., linearity, asymmetry and the Jacobi identity:

$$\{A + B, C\} = \{A, C\} + \{B, C\}, \qquad \{A, B\} = -\{B, A\}, \tag{2.51}$$

$$\{A, \{B, C\}\} + \{B, \{C, A\}\} + \{C, \{A, B\}\} = 0, \tag{2.52}$$

as may easily be verified directly from Eqs. (2.2) and (2.5). Hence the set of canonical transformations, generated by a set of observables closed under the Poisson bracket, form a Lie group (for quantum observables this group is of course the usual unitary transformations). This allows us to describe systems with symmetries. We will consider two examples below: nonrelativistic particles and rotational bits.

2.5.1 Nonrelativistic Particles

Consider first the possible descriptions of a free nonrelativistic particle—whether classical, quantum or otherwise. We will take the configuration space to be the Euclidean space R^3. To describe such a particle, we look for a realization of the Galilean group in terms of the algebra of Poisson brackets. The Galilean group has 10 generators: A_i which generate spatial displacements, H which generates time displacements, L_i which generate spatial rotations, and G_i which generate Galilean transformations ("boosts"), with $i = 1, 2, 3$. These generators have to satisfy the Poisson bracket relations [10]

$$\{H, A_i\} = 0, \qquad \{H, L_i\} = 0, \tag{2.53}$$

$$\{L_i, A_j\} = \varepsilon_{ijk} A_k, \qquad \{L_i, L_j\} = \varepsilon_{ijk} L_k, \qquad \{L_i, G_j\} = \varepsilon_{ijk} G_k, \tag{2.54}$$

$$\{A_i, A_j\} = 0, \qquad \{G_i, G_j\} = 0, \qquad \{G_i, A_j\} = m\delta_{ij}, \qquad \{G_i, H\} = A_i, \tag{2.55}$$

where m is the mass of the particle, and $\varepsilon_{ijk} = 1 \, (= -1)$ for even (odd) permutations i, j, k of $1, 2, 3$ and vanishes otherwise. Note the first line implies that H transforms as a scalar under translations and rotations, while the second line implies that A_i, L_i, and G_i transform as vectors.

In the framework of ensembles on configuration space, these generators are represented by suitable observables. For spatial displacements and rotations one finds that

$$A_i = \Pi_i[P, S] = \int d^3x \, P \, (\partial_i S), \qquad L_i = \int d^3x \, P \, \left(\varepsilon_{ijk} \, x_j \, \partial_k S \right), \tag{2.56}$$

up to additive constants. These are the ensemble momentum and angular momentum observables. The former observable, $\Pi[P, S]$, was derived in Sect. 2.3.1, and the latter may be similarly obtained by considering infinitesimal rotations of P and S. Further, for the Galilean boost transformations it is natural to choose the observables

$$G_i = \int d^3x \, P \, (mx_i - t\partial_i S) = mX_i[P, S] - t\Pi_i[P, S], \tag{2.57}$$

where t is the time. This follows from the standard definition $G_i = (mX_i - tA_i)$ in classical mechanics [10], together with the natural choice $X_i = \int d^3x \, P \, x_i$ for the position observable of an ensemble on configuration space as per Eq. (2.13).

The above results do not fully determine the form of H, which will of course, since it generates infinitesimal displacements in time, be identified with the ensemble Hamiltonian $\mathcal{H}$. It is straightforward to check from the above equations that the general solution is of the form

$$H = \mathcal{H}[P, S] := \int dx \, P \, \frac{|\nabla S|^2}{2m} + K[P, S], \qquad (2.58)$$

where K is any observable invariant under translations, rotations and boosts, i.e., K is a Galilean scalar. Solutions include both the classical ensemble Hamiltonian for a free particle (see Sect. 1.2),

$$H = \mathcal{H}_C[P, S] = \int d^3x \, P \, \frac{|\nabla S|^2}{2m} \qquad (2.59)$$

corresponding to $K \equiv 0$, and the quantum ensemble Hamiltonian for a free particle (see Sect. 1.2),

$$H = \mathcal{H}_Q[P, S] = \int d^3x \, P \left[\frac{|\nabla S|^2}{2m} + \frac{\hbar^2 |\nabla \log P|^2}{8m} \right] \qquad (2.60)$$

corresponding to $K = (\hbar^2/8m)F[P]$, where $F[P]$ is the Fisher information of P [11] (see also Chap. 5). A more general solution corresponds to the choice

$$K[P, S] = \int dx \, P \, k(|\nabla \log P|, \nabla^2 \log P, \dots), \qquad (2.61)$$

where k is an arbitrary function of scalars formed by the derivatives of $\log P$. Note that all the above generators satisfy the homogeneity condition, Eq. (2.11), and hence have clear interpretations as expectation values.

2.5.2   Rotational Bits

A quantum mechanical spin-half system may be characterised as having a set of two-valued observables which generate infinitesimal rotations in three dimensions. We want to consider such a two-level system, but now within the formalism of ensembles on configuration space. The generator of rotation about a given direction will be identified with the measurement of spin in that direction. Such a system may be called a *rotational bit* or *robit*, to distinguish it from the standard quantum qubit.

The observable corresponding to a measurement in unit direction **n** thus has the form $L \cdot \mathbf{n}$, where $L = (L_1, L_2, L_3)$ satisfies the $so(3)$ Lie algebra,

$$\{L_j, L_k\} = \varepsilon_{jkl} L_l, \tag{2.62}$$

for $j, k = 1, 2, 3$.

It convenient to define the probability distribution P in terms of the possible measurement outcomes of spin in the z-direction, which may be labelled by $\pm 1/2$. Thus, $P \equiv \{P_+, P_-\}$, where P_α denotes the probability of measuring spin value $\alpha/2$ in the z-direction. The canonically conjugate quantities are therefore labelled as $S \equiv \{S_+, S_-\}$.

Note that the identification of generators with expectation values immediately fixes the form of L_3. In particular, the average value of spin measurements in the z-direction may be calculated directly from the probability distribution,

$$L_3(P, S) = s(P_+ - P_-) = (P_+ - P_-)/2. \tag{2.63}$$

where $s = 1/2$ for spin-half particles (note for the quantum case we are effectively choosing units in which $\hbar = 1$).

We explore robits in detail in Chap. 7, where we develop theories for a single robit and pairs of robits. Here however we restrict to a single robit and focus on the problem of representing rotations on a *discrete* configuration space.

2.5.2.1   Reduced Phase Space for a Two-Level System

The fundamental variables for a two-level system are $\{P_+, P_-, S_+, S_-\}$, thus the phase space is four-dimensional. However, since $\sum_k P_k = 1$ is a quantity that is conserved, it is possible to describe the system in a reduced phase space. To do this, introduce coordinates

$$\begin{aligned}
q_0 &= (P_+ + P_-)/2, \\
q_1 &= (P_+ - P_-)/2, \\
p_0 &= S_+ + S_-, \\
p_1 &= S_+ - S_-.
\end{aligned} \tag{2.64}$$

It is easy to check that this transformation is a canonical transformation. Since the P_k are probabilities we must set $q_0 = 1/2$. In these coordinates, the conditions of probability conservation and homogeneity in Eqs. (2.10) and (2.11) require that observables G be of the form

$$G(q_1, q_2, p_1, p_2) = 2q_0 F(q_0^{-1} q_1/2, p_1), \tag{2.65}$$

where F is an arbitrary function and factors of 2 have been included for convenience. Since p_0 does not appear in G, the equations of motion lead to q_0 being a constant of the motion, as required.

It is now straightforward to describe the system in a phase space of dimension $4 - 2 = 2$. In particular, setting $q_0 = 1/2$ in the expression for G leads to

$$G(q_1, p_1) = F(q_1, p_1). \tag{2.66}$$

Thus we have identified the true degrees of freedom of the system, q_1 and p_1.

2.5.2.2   Two-Level System with $SO(3)$ Symmetry

We now look for the most general representation of $so(3)$ on this two-dimensional phase space. The generators must satisfy the Poisson brackets of Eq. (2.62), with the condition of Eq. (2.63), where the latter corresponds to

$$L_3 = q_1. \tag{2.67}$$

Equations (2.62) and (2.67) lead to

$$L_1 = -\frac{\partial L_2}{\partial p_1}, \tag{2.68}$$

$$L_2 = \frac{\partial L_1}{\partial p_1}, \tag{2.69}$$

$$L_3 = q_1 = \frac{\partial L_1}{\partial q_1}\frac{\partial L_2}{\partial p_1} - \frac{\partial L_1}{\partial p_1}\frac{\partial L_2}{\partial q_1}. \tag{2.70}$$

It is convenient to define $Z := L_1 - i\,L_2$, and write Eqs. (2.68) and (2.69) as the single complex equation

$$\frac{\partial Z}{\partial p_1} = i\,z. \tag{2.71}$$

Equation (2.71) has the general solution $Z = \exp\{ip_1 + a(q_1) + ib(q_1)\}$, where a and b are real functions of q_1. Thus

$$L_1 = e^{a(q_1)}\cos(p_1 + b(q_1)), \qquad L_2 = -e^{a(q_1)}\sin(p_1 + b(q_1)). \tag{2.72}$$

Substitution into Eq. (2.70) then leads to

$$L_1 = \sqrt{c^2 - q_1^2} \, \cos(p_1 + b(q_1)), \tag{2.73}$$

$$L_2 = -\sqrt{c^2 - q_1^2} \, \sin(p_1 + b(q_1)), \tag{2.74}$$

$$L_3 = q_1 \tag{2.75}$$

where c is a constant and b an arbitrary function of q_1.

To fix the value of c^2, we impose the condition that the probability remain positive. We consider an arbitrary ensemble Hamiltonian $H(L)$ which is a function of the generators of the $so(3)$ Lie algebra and calculate the change induced on q_1 by the action of H. Evaluation of the Poisson bracket leads to

$$\frac{\partial q_1}{\partial t} = \sqrt{c^2 - q_1^2} \left[-\frac{\partial H}{\partial L_1} \sin(p_1 + b) - \frac{\partial H}{\partial L_2} \cos(p_1 + b) \right], \tag{2.76}$$

The condition that the probability remain positive requires

$$\begin{aligned}
\frac{\partial P_+}{\partial t} &= +\frac{\partial q_1}{\partial t} = 0 \quad \text{when} \quad q_1 = -\frac{1}{2}, \\
\frac{\partial P_-}{\partial t} &= -\frac{\partial q_1}{\partial t} = 0 \quad \text{when} \quad q_1 = +\frac{1}{2}.
\end{aligned} \tag{2.77}$$

The positivity conditions have to be valid for all possible choices of H, which leads immediately to $c^2 = 1/4 = q_0^2$.

Thus, the general solution for the L_k is given by Eqs. (2.73)–(2.75) with $c^2 = 1/4$. One can see that the L_k still depend on the arbitrary function $b(q_1)$. However, we can always set $b(q_1) = 0$ via the simple canonical transformation

$$q_1 \rightarrow q_1, \qquad p_1 \rightarrow p_1 - b(q_1), \tag{2.78}$$

which obviously preserves the condition of Eq. (2.67). This allows us to write the generators of $so(3)$ in their simplest form,

$$L_1 = \sqrt{1/4 - q_1^2} \, \cos(p_1) = \sqrt{P_+ P_-} \, \cos(S_+ - S_-), \tag{2.79}$$

$$L_2 = -\sqrt{1/4 - q_1^2} \, \sin(p_1) = -\sqrt{P_+ P_-} \, \sin(S_+ - S_-), \tag{2.80}$$

$$L_3 = q_1 = (P_+ - P_-)/2. \tag{2.81}$$

We will show in Chap. 7 that a single robit is equivalent to a single quantum mechanical qubit. Notice however that we have derived the theory of a single robit without making any assumptions which are particular to quantum mechanics. We will develop this theme further in Chap. 7. In the case of a pair of robits, which we also discuss in Chap. 7, such an equivalence is no longer automatically fulfilled, but one may introduce further assumptions involving locality and a restriction of the functional form of the generators to obtain a similar equivalence to a pair of qubits.

2.6   Constraints and Superselection Rules

In the usual quantization of a classical system subject to constraints, each classical constraint is mapped to a linear operator constraint on the wavefunction, of the form $\hat{C}\psi = 0$ [12–14]. Thus, in standard quantum mechanics, one usually restricts to constraints that are *linear* in the wave function. However, in the more general context of ensembles on configuration space, it is natural to consider general constraints formulated in terms of P and S [15]—which, for quantum ensembles, will typically be *nonlinear* in the wave function. In particular, we define a constraint in a very general way, as any equation of the form

$$C(P, S) = 0 \tag{2.82}$$

that is required to hold at all times. This is completely analogous to the treatment of constraints in classical phase space physics [2, 13], and restricts the evolution to a submanifold of the fundamental variables (P, S). It will be seen that constraints of the above form have a fundamental role to play in quantum theory, even when they cannot be rewritten in the linear form $\hat{C}\psi = 0$. More generally, they may be interpreted as a generalisation of quantum superselection rules [15].

The Schrödinger equation for a quantum system is linear, implying that the superposition of any two solutions is also a solution. However, some combinations of states have never been observed, including coherent superpositions of integer and half-integer spins, electric charges, and Schrödinger's cat. Possible explanations for why such superpositions are not observed fall into two logical categories:

1. *measurement superselection rules*: such superpositions may be allowed, but physical limitations on measurement prevent their observation;
2. *state superselection rules*: such superpositions are not physically allowed.

State superselection rules are stronger than measurement superselection rules (one cannot observe what does not exist), and are clearly constraints on possible states of quantum ensembles. However, they are not *linear* constraints. For example, a superselection rule restricting possible wave functions of a quantum system to a set of orthogonal subspaces of Hilbert space (corresponding, e.g., to different spin values), with corresponding projection operators $\{\hat{E}_j\}$, is equivalent to the nonlinear constraint

$$\sum_j \langle \psi | \hat{E}_j | \psi \rangle^2 = 1 \tag{2.83}$$

on the wave function.

In this section, we present an example of a simple constraint on P and S that may be applied to both classical and quantum ensembles, and which in the latter case acts to rule out superpositions of energy eigenstates. Thus, this example shows how constraints of the form of Eq. (2.82) may be interpreted as generalised state superselection rules.

In particular, consider the rather simple canonical constraint

$$J := P \nabla S = 0. \tag{2.84}$$

This constraint is local, invariant under the transformation $S \to S + c$, and may be physically interpreted as the requirement that the ensemble momentum density vanishes everywhere. Note that for quantum ensembles it can be re-expressed as $\text{Im } \bar{\psi} \nabla \psi = 0$, which clearly cannot be put in the linear form $\hat{C} \psi = 0$ of the standard approach to quantum constraints.

To investigate constraint (2.84) for a *classical* ensemble of particles, note that consistency with the equations of motion requires $\partial J / \partial t = 0$. The equations of motion for the ensemble (see Sect. 1.2) yield the secondary constraint

$$
\begin{aligned}
0 = \partial(P\nabla S)/\partial t &= (\partial P/\partial t)\nabla S + P(\partial(\nabla S)/\partial t) \\
&= -[\nabla \cdot (P\nabla S)]\nabla S - P \nabla[|\nabla S|^2/(2m) + V] \\
&= -P\nabla V. \tag{2.85}
\end{aligned}
$$

Hence the classical force, $-\nabla V$, vanishes over the support of the ensemble, i.e., the ensemble is constrained to be stationary. Note in particular that if the potential energy has a single minimum, then the constraint requires the ensemble to be concentrated solely at this minimum, i.e., the ensemble must occupy the classical ground state.

In contrast, for a *quantum* ensemble of particles, Eq. (2.84) requires that ∇S vanishes on the support of the wavefunction, and hence that S has no spatial dependence for $P \neq 0$. Secondary constraints arising from consistency with the equations of motion can be determined similarly to the classical case above. However, it is simpler to directly substitute the ansatz $S(x, t) = -f(t)$ into the Schrödinger equation and use the continuity equation to obtain the secondary constraints

$$\dot{f}P^{1/2} = \left[\frac{-\hbar^2}{2m}\nabla^2 + V\right]P^{1/2}, \qquad \partial P/\partial t = 0 \tag{2.86}$$

respectively. Differentiating the first of these with respect to time and applying the second implies that $\dot{f} = E =$ constant, and hence these constraints are equivalent to the time-independent Schrödinger equation

$$\frac{-\hbar^2}{2m}\nabla^2\psi + V\psi = E\psi. \tag{2.87}$$

Thus the quantum ensemble is required to be in an energy eigenstate.

It is seen that in both the classical and quantum cases, the primary constraint in Eq. (2.29) leads to the requirement that the ensemble is stationary as per the definition in Eq. (2.30). In the quantum case this immediately yields a state superselection rule: *superpositions of states of different energy are forbidden*. Thus this constraint provides a very simple example of how canonical constraints can lead to superselection-type rules for quantum ensembles. Further examples and discussion of constraints are given in [15].

References

1. Wigner, E.P.: The problem of measurement. Am. J. Phys. **31**, 6–15 (1963)
2. Goldstein, H.: Classical Mechanics. Addison-Wesley, New York (1950)
3. Hall, M.J.W., Reginatto, M.: Interacting classical and quantum ensembles. Phys. Rev. A **72**, 062109 (2005)
4. Aharonov, Y., Albert, D.Z., Vaidman, L.: How the result of a measurement of a component of the spin of a spin-1/2 particle can turn out to be 100. Phys. Rev. Lett. **60**, 1351–1354 (1988)
5. Aharonov, Y., Vaidman, L.: Properties of a quantum system during the time interval between two measurements. Phys. Rev. A **41**, 11–20 (1990)
6. Hall, M.J.W.: Exact uncertainty relations. Phys. Rev. A **64**, 052103 (2001)
7. Johansen, L.M.: What is the value of an observable between pre- and postselection? Phys. Lett. A **322**, 298–300 (2004)
8. Hall, M.J.W.: Prior information: how to circumvent the standard joint-measurement uncertainty relation. Phys. Rev. A **69**, 052113 (2004)
9. Dressel, J.: Weak values as interference phenomena. Phys. Rev. A **91**, 032116 (2015)
10. Finkelstein, R.J.: Nonrelativistic Mechanics. W. A. Benjamin, Reading, Massachusetts (1973)
11. Hall, M.J.W.: Quantum properties of classical Fisher information. Phys. Rev. A **62**, 012107 (2000)
12. Dirac, P.A.M.: Lectures on Quantum Field Theory, Chaps. 14–15. Academic, New York (1966)
13. Henneaux, M., Teitelboim, C.: Quantization of Gauge Systems, Chap. 1. Princeton University Press, New Jersey (1992)
14. Weinberg, S.: The Quantum Theory of Fields, vol. I. Cambridge University Press, Cambridge (1995)
15. Hall, M.J.W.: Superselection from canonical constraints. J. Phys. A **27**, 7799–7811 (2004)

Chapter 3
Interaction, Locality and Measurement

Abstract Given two systems with configuration spaces X and Y, we consider their joint description on the configuration space given by the set product $X \times Y$. In the formalism of ensembles on configuration space, this description requires a probability distribution $P(x, y)$ defined over the joint configuration space, the corresponding conjugate quantity $S(x, y)$, and an ensemble Hamiltonian $\mathcal{H}_{XY}[P, S]$. Once a composite system is defined, it becomes necessary to introduce a number of new concepts which must be defined carefully. For example, such systems may consist of subsystems which are independent or entangled, non-interacting or interacting, and one must give a precise mathematical formulation of these properties. Issues of locality must be taken into consideration. Observables which are ascribed to one of the subsystems (and are therefore initially defined on only one of the initial configuration spaces, X or Y) must be extended to the joint ensemble, but this can not be done in an arbitrary way. These concepts play an important role in the description of composite systems, and we address them in the first sections of this chapter. The remaining sections are devoted to a description of interactions between subsystems that model measurements, starting with basic measurement procedures followed by more elaborate procedures that describe weak measurements and measurement-induced collapse.

3.1 Introduction

To describe interactions and correlations between physical systems, such as two particles, one must consider their joint description. In particular, if X and Y denote the configuration spaces of two such systems, then their joint configuration space is given by the set product $X \times Y$. It follows that a corresponding joint ensemble of systems is described by a joint probability density $P(x, y)$ on $X \times Y$, a conjugate quantity $S(x, y)$, and a joint ensemble Hamiltonian $\mathcal{H}_{XY}[P, S]$, as described in Chap. 1.

Several questions of interest immediately arise: How are observables for ensembles on X to be represented on $X \times Y$? What form does the joint ensemble Hamiltonian take for noninteracting ensembles? How are measurement interactions

© Springer International Publishing Switzerland 2016

M.J.W. Hall and M. Reginatto, *Ensembles on Configuration Space*,
Fundamental Theories of Physics 184, DOI 10.1007/978-3-319-34166-8_3

described? Can weak measurements be generalised, from quantum ensembles to arbitrary ensembles? What constraints does locality place on interactions between ensembles? Is there a notion of entanglement beyond quantum entanglement?

The above questions are addressed in this chapter. Various developments and applications, including to the representation of pairs of rotational bits, hybrid quantum-classical interactions, and the interaction of classical spacetimes with quantum fields, are given in Chaps. 7–11.

3.2 Joint Ensembles

3.2.1 Independent Ensembles

Our physical formulation of 'interaction' is based on the idea that interactions lead to correlations. This first requires a definition of 'uncorrelated' or 'independent' ensembles [1].

Definition Two ensembles, with respective configuration spaces X and Y, are defined to be *independent* at some given time if the joint ensemble on $X \times Y$ is fully described by an ensemble $[P_X(x), S_X(x)]$ on X and an ensemble $[P_Y(y), S_Y(y)]$ on Y.

It follows for independent ensembles that no physical distinction is possible between the joint probability density $P(x, y)$ assigned to the joint ensemble, and the pair of individual probability densities $P_X(x)$ and $P_Y(y)$: all physical properties are equally described by either. Hence, from basic probability theory, one must have

$$P(x, y) = P_X(x)\, P_Y(y). \tag{3.1}$$

Similarly, there can be no physical distinction between the joint function $S(x, y)$ and the pair of individual functions $S_X(x)$ and $S_Y(y)$. Now, as discussed in Chap. 2, the conservation of probability requires that all physical quantities are insensitive under addition of a constant to S. More precisely, observables are independent under any transformation of the form $S \to S + c$, where c is independent of the configuration space of the ensemble. It follows for independent ensembles that $S(x, y)$ must be equivalent to $S_X(x)$, up to some additive function of y, and to $S_Y(y)$ up to some additive function of x, implying that

$$S(x, y) = S_X(x) + S_Y(y), \tag{3.2}$$

up to an arbitrary additive constant of no physical significance.

Equations (3.1) and (3.2) fully characterise independent ensembles. It is worth noting that in the case of a quantum ensemble, with wavefunction $\psi = \sqrt{P}e^{iS/\hbar}$ (see Chap. 1), these independence conditions reduce to the factorisability condition

$$\psi_{XY}(x, y) = \psi_X(x)\,\psi_Y(y) \tag{3.3}$$

for the joint wavefunction $\psi_{XY}(x, y)$, up to an arbitrary constant phase factor of no physical significance.

3.2.2 Interaction Versus Noninteraction

Intuitively, interaction between physical systems leads to correlated behaviour. Thus, for example, heavenly bodies orbit a common centre in Newtonian gravity, while the position of an atom may be determined via the direction of a photon scattered from it. We therefore take the creation of statistical correlations as the hallmark that distinguishes interacting from noninteracting ensembles [1].

Definition Two ensembles are non-interacting, under a given joint ensemble Hamiltonian, if and only if all initially independent ensembles remain independent under evolution.

It follows that, since two independent ensembles remain independent if they do not interact, their evolution will be described by two corresponding ensemble Hamiltonians $\mathscr{H}_X$ and $\mathscr{H}_Y$. Recalling the general property $\mathscr{H}[P, S] = -\langle \partial S/\partial t \rangle$ following from homogeneity (see Sect. 1.4), this yields the necessary condition

$$\mathscr{H}_{XY}[P_X P_Y, S_X + S_Y] = -\int dx dy\, P_X P_Y \frac{\partial(S_X + S_Y)}{\partial t}$$

$$= \int dy\, P_Y(y)\,\mathscr{H}_X[P_X, S_X] + \int dx\, P_X(x)\,\mathscr{H}_Y[P_Y, S_Y] \tag{3.4}$$

for their joint Hamiltonian $\mathscr{H}_{XY}$. Here integration is replaced by summation over any discrete parts of the configuration spaces. It is easy to check that this condition, imposed as a constraint on the functional form of $\mathscr{H}_{XY}$ for all $P_X(x)$, $P_Y(y)$, $S_X(x)$ and $S_Y(y)$, is also sufficient for two ensembles to be non-interacting.

It is not difficult to find joint ensemble Hamiltonians that describe noninteracting ensembles as per Eq. (3.4). For example, if

$$\mathscr{H}_X(P_X, S_X) = \int dx\, P_X\, F(\nabla_x \log P_X, \nabla_x S_X, \dots), \tag{3.5}$$

$$\mathscr{H}_Y(P_Y, S_Y) = \int dy\, P_Y\, G(\nabla_y \log P_Y, \nabla_y S_Y, \dots), \tag{3.6}$$

for two continuous configuration spaces X and Y and arbitary functions F and G, where '...' denotes possible higher derivatives of $\log P$ and S, then

$$\mathcal{H}_{XY}[P, S] := \int dx\, dy\, P\, \left[F(\nabla_x \log P, \nabla_x S, \dots) + G(\nabla_y \log P, \nabla_y S, \dots)\right]$$

$$(3.7)$$

satisfies Eq. (3.4).

Note that the ensemble Hamiltonians $\mathcal{H}_C$ and $\mathcal{H}_Q$ in Eqs. (1.3) and (1.8) are of the forms of Eqs. (3.5) and (3.6). Hence, they can be combined to give joint ensemble Hamiltonians corresponding to any one of non-interacting classical-classical, quantum-quantum, and quantum-classical ensembles of particles. Examples of *interacting* ensemble Hamiltonians are considered further below (see also Chaps. 8–11).

3.2.3 Independence Versus Entanglement

Quantum ensembles on configuration space have a remarkable property, first identified by Schrödinger: they are either independent or entangled [2]. For example, if the independence condition Eq. (3.3) does not hold, then the wave function $\psi_{XY}(x, y)$ necessarily violates some Bell inequality, i.e., the joint ensemble exhibits statistical correlations that have no underlying hidden variable model [3–5].

Surprisingly, the concept of entanglement remains meaningful in the general case—even for classical ensembles. However, it is important to note that the notion of 'entanglement' referred to here is not in the strong sense of Bell inequality violation, but in Schrödinger's original weaker sense that the properties of a joint ensemble cannot be decomposed into properties of the individual ensembles [2]. Spekkens has similarly used this weaker sense to define entanglement for a class of 'epistemic' models of statistical correlation [6].

For example, consider a classical joint ensemble, corresponding to two classical particles described by respective configuration spaces X and Y, with probability density $P(x, y)$ and conjugate quantity $S(x, y)$. Noting that the product of two classical phase space functions $f(x, p_x)$ and $g(y, p_y)$ is itself a classical phase space function for the two particles, we recall from Sect. 2.3.2 that the expectation value of this product corresponds to the classical observable

$$C_{fg} = \langle fg \rangle = \int dx dy\, P(x, y)\, f(x, \partial_x S)\, g(y, \partial_y S). \qquad (3.8)$$

Now, there is clearly a trivial hidden variable model for any such observable, whether or not the independence conditions (3.1) and (3.2) are satisfied. In particular, defining $\lambda := [x, y, S(x, y)]$, $P(\lambda) := P(x, y)$, $F(\lambda) := f(x, \partial_x S)$, and $G(\lambda) := g(y, \partial_y S)$, one has

$$\langle fg \rangle = \int d\lambda\, P(\lambda)\, F(\lambda)\, G(\lambda). \qquad (3.9)$$

Hence, no Bell inequality can be violated via such observables [5].

Nevertheless, if the independence condition (3.2) is not satisfied, then the 'hidden value' of the observable $f(x, p_x)$ for the *first* particle, i.e., $F(\lambda)$, will in general depend on the position of the *second* particle, via $p_x = \partial_x S(x, y)$. That is, while knowledge of the position and momentum of the first particle at a given time is sufficient to determine all observables for the particle at that time, it will not be sufficient to determine them at any later time: one needs to know the evolution of the joint quantity $\partial_x S(x, y)$. Moreover, if one locally perturbs the position of the second particle, from y to y', the corresponding perturbation of $S(x, y)$ to $S(x, y')$ will typically perturb the value of p_x in this model.

Hence, a kind of nonlocality, or inseparability, can be associated with joint ensembles even in the classical case. We will, by analogy with Schrödinger's original discussion [2], refer to this property as 'entanglement':

Definition A joint ensemble on the configuration space $X \times Y$ is entangled if and only if $S(x, y) \neq S_X(x) + S_Y(y)$, up to some additive constant.

Note that entanglement, as defined here, is relative to particular configuration spaces X and Y. For quantum ensembles, this corresponds to a particular choice of computational basis for the component ensembles (see Sect. 1.3.2 of Chap. 1). For this reason, our definition of entanglement is stronger than the standard definition for quantum ensembles (where the latter requires only that $S(x, y) \neq S_X(x) + S_Y(y)$ for *some* choice of computational basis, i.e., that the ensembles are not independent). We require a stronger definition because for general configuration ensembles one does not have a similar freedom to arbitrarily choose between configuration spaces.

It follows that while entanglement implies non-independence, the converse does not hold in general. In particular, violation of the independence condition (3.1) by a classical joint ensemble is not sufficient to give rise to any nonlocality of the sort discussed above, as will now be shown.

For simplicity, we consider the condition $S(x, y) = S_X(x) + S_Y(y)$ for the case of a classical ensemble which consists of two one-dimensional, non-relativistic, non-interacting particles. It turns out that this condition is sufficient to ensure no nonlocality can arise, whether or not the joint probability density factorises. In particular, all classical observables for the first ensemble can be determined from the knowledge of $P_X(x) = \int dy\, P(x, y)$ and $S_X(x)$ alone, via Eq. (3.8), and similarly for observables of the second ensemble. Further, the condition is preserved at all later times by the equations of motion, as we now prove.

The ensemble Hamiltonian for two such noninteracting particles is a slight generalization of the one discussed in Sect. 1.2 and is given by

$$\mathscr{H}_C[P, S] = \int dx dy\, P \left[\frac{1}{2m_X} \left(\frac{\partial S}{\partial x} \right)^2 + \frac{1}{2m_Y} \left(\frac{\partial S}{\partial y} \right)^2 + V_X + V_Y \right], \quad (3.10)$$

where m_X, m_Y are the masses of the particles and we have assumed that there is no interaction via the potential term and therefore $V(x, y) = V_X(x) + V_Y(y)$. This ensemble Hamiltonian may also be obtained via a direct application of Eq. (3.7) to classical ensembles, and hence automatically satisfies the non-interaction condition in Eq. (3.4).

Variation of $\mathcal{H}_C[P, S]$ with respect to P leads to a Hamilton–Jacobi equation which decouples in a trivial manner into two independent Hamilton–Jacobi equations, one for $S_X(x)$ and one for $S_Y(y)$. Variation with respect to S leads to a continuity equation of the form $C[P, S] = 0$ which is linear in P. One can check that integrating $C[P, S] = 0$ on both sides with respect to y leads to the continuity equation $\partial_t P_X + \partial_x(P_X \partial_x S_X / m_X) = 0$ for the probability density $P_X(x)$ associated with the first particle, and a similar result is obtained for $P_Y(y)$ by integrating with respect to x. Thus, in the case of a classical ensemble of two non-interacting particles, the condition $S(x, y) = S_X(x) + S_Y(y)$ leads to a decoupling of the equations of motion, with separate equations for each of the two particles. Hence, non-entanglement is preserved under evolution.

It is interesting that an *additional* condition is required for the corresponding quantum case, with ensemble Hamiltonian

$$\mathcal{H}_Q[P, S] = \mathcal{H}_C[P, S] + \frac{\hbar^2}{4} \int dx dy \, \frac{1}{P} \left[\frac{1}{2m_X} \left(\frac{\partial P}{\partial x} \right)^2 + \frac{1}{2m_Y} \left(\frac{\partial P}{\partial y} \right)^2 \right].$$
$$(3.11)$$

In this case, variation with respect to P leads to a modified Hamilton–Jacobi equation which now depends on P, and both $S(x, y) = S_X(x) + S_Y(y)$ and $P(x, y) = P_X(x) P_Y(y)$ are required to ensure decoupling of the equations of motion and nonentanglement. Thus in the *quantum* case, non-entanglement is preserved under evolution only for *independent* ensembles.

It is seen that the conditions under which noninteracting ensembles remain unentangled are different for classical and quantum ensembles. In the former case it is guaranteed, while in the latter case it is not. More generally:

> Independence is a sufficient but not a necessary condition for two noninteracting ensembles to remain unentangled under evolution.

Development of the notion of entanglement for general configuration ensembles remains a topic of interest for future work, but is not investigated further in this book.

3.3 Extending Observables to Joint Ensembles

3.3.1 General Definition

If an observable quantity can be measured on some configuration ensemble, then it is natural to expect it can be measured independently of whether the ensemble is actually part of a larger joint ensemble. In the same vein, regarding observables as generators of canonical transformations (see Chap. 2), it is natural to expect that such transformations on a given ensemble can be carried out independently of whether or not the ensemble is part of a joint ensemble.

These considerations suggest that, for a given observable $A_X[P, S]$ defined on an ensemble with configuration space X, it should be possible to define a corresponding *extended* observable, $A_{XY}[P, S]$, on an ensemble with configuration space $X \times Y$. This is indeed the case.

First, for a probability distribution P_{XY} and conjugate quantity S_{XY} on $X \times Y$, define the conditional functions

$$P_{X|y}(x) := P_{XY}(x, y)/P_Y(y), \qquad S_{X|y}(x) := S_{XY}(x, y) \qquad (3.12)$$

on X, with $P_Y(y) := \int dx\, P_{XY}(x, y)$. Note that these may be interpreted as a type of Bayesian updating of the ensemble on X, given knowledge of y. The conditional probability $P_{X|y}(x)$ depends on the value of y whenever $P_{XY}(x, y) \neq P_X(x)P_Y(y)$.

Second, for an observable $A_X[P, S]$ on X, define its extension (or promotion) to $X \times Y$ by

$$A_{XY}[P, S] := \int dy\, P_Y\, A_X[P_{X|y}, S_{X|y}]. \qquad (3.13)$$

Thus, the extended observable corresponds to the average expectation value of A_X when conditioned on knowledge of y.

To illustrate the application Eq. (3.13), consider the case $A_X[P, S] = \mathcal{H}_Q[P, S]$, the quantum ensemble Hamiltonian of Eq. (1.8). Then

$$A_{XY}[P, S] = \int dy\, P_Y \int dx\, \frac{P_{XY}}{P_Y} \left\{ \frac{1}{2m} \left(\frac{\partial S_{XY}}{\partial x} \right)^2 + V + \frac{\hbar^2}{8m} \frac{P_Y^2}{P_{XY}^2} \left[\frac{\partial}{\partial x} \left(\frac{P_{XY}}{P_Y} \right) \right]^2 \right\}$$

$$= \int dy\, dx\, P \left[\frac{1}{2m} \left(\frac{\partial S}{\partial x} \right)^2 + V + \frac{\hbar^2}{8m} \frac{1}{P^2} \left(\frac{\partial P}{\partial x} \right)^2 \right], \qquad (3.14)$$

where in the last equality we introduced the simpler notation $P = P_{XY}$, $S = S_{XY}$. We will use this simplified notation when it is clear that P and S refer to the configuration space of the joint ensemble.

We now show that the definition given by Eq. (3.13) satisfies several desirable properties. For convenience we will assume the configuration spaces are continuous. Analogous properties are easily obtained for discrete configuration spaces, with variational derivatives replaced by partial derivatives and integration by summation where appropriate.

3.3.2 Two Invariance Properties

First, for two independent ensembles, on configuration spaces X and Y respectively, the extension of A_X may be calculated via Eqs. (3.1), (3.2), (3.12) and (3.13) as

$$
\begin{aligned}
A_{XY}[P_X P_Y, S_X + S_Y] &= \int dy\, P_Y(y)\, A_X[P_X, S_X + S_Y(y)] \\
&= \int dy\, P_Y(y)\, A_X[P_X, S_X] \\
&= A_X[P_X, S_X].
\end{aligned}
\tag{3.15}
$$

Here the second line follows from the property $A[P, S + c] = A[P, S]$ of ensemble observables (corresponding to conservation of probability: see Chap. 2). Thus, for independent ensembles, the extended observable is equivalent to the original observable.

Second, if $C_X := \{A_X, B_X\}$ denotes the Poisson bracket for two observables on configuration space X (see Chap. 2), the corresponding extended observables satisfy the relation

$$
\{A_{XY}, B_{XY}\} = C_{XY},
\tag{3.16}
$$

where the Poisson bracket of Eq. (3.16) is the one that is defined on the configuration space XY. This important property, proved further below, is worth highlighting:

> The Lie algebra of observables on a given configuration space is preserved under the extension to a joint configuration space.

This property implies, for example, that for hybrid quantum-classical systems the classical observables remain classical and the quantum observables remain quantum, in the well-defined sense that their respective Lie algebras are preserved (see Chaps. 8 and 9).

To demonstrate Eq. (3.16), we will consider the effect on Eq. (3.13) of carrying out variations $P \to P + \delta P$ and $S \to S + \delta S$. We first consider the effect of the variation $P \to P + \delta P$, keeping in mind that in Eq. (3.13) A_X is a function of $P_{X|y}, S_{X|y}$ only, and that $P_{X|y} = P_{XY}/P_Y$ as per Eq. (3.12). Then we find that

$$
\begin{aligned}
\delta A_{XY} &= \int dy\, (\delta P_Y\, A_X + P_Y\, \delta A_X) \\
&= \int dy\, \Bigg[\delta P_Y\, A_X + P_Y \int dx dy'\, \frac{\delta A_X}{\delta P_{X|y}} \frac{\delta P_{X|y}}{\delta P_Y(y')} \delta P_Y(y') \\
&\quad + P_Y \int dx dx' dy'\, \frac{\delta A_X}{\delta P_{X|y}} \frac{\delta P_{X|y}}{\delta P_{XY}(x', y')} \delta P_{XY}(x', y') \Bigg]
\end{aligned}
$$

$$= \int dy \left[\delta P_Y A_X + P_Y \int dx \frac{\delta A_X}{\delta P_{X|y}} \left(-\frac{P_{XY}}{P_Y^2} \delta P_Y + \frac{1}{P_Y} \delta P_{XY} \right) \right]$$

$$= \int dy \left[A_X - \int dx \frac{\delta A_X}{\delta P_{X|y}} P_{X|y} \right] \delta P_Y + \int dy dx \frac{\delta A_X}{\delta P_{X|y}} \delta P_{XY}$$

$$= \int dy dx \frac{\delta A_X}{\delta P_{X|y}} \delta P_{XY},$$

where the first and second equalities follow from the definition of the variational derivative in Eq. (A.1) in Appendix A of this book and the chain rule; the third and fourth equalities from the definition $P_{X|y} = P_{XY}/P_Y$; and the last equality from the homogeneity property (2.35) of observables. It follows immediately that

$$\frac{\delta A_{XY}}{\delta P} = \frac{\delta A_X}{\delta P_{X|y}(x)}. \tag{3.17}$$

Similarly, under a variation $S \rightarrow S + \delta S$ in Eq. (3.13), one finds

$$\delta A_{XY} = \int dy \, P_Y \, \delta A_X = \int dx dy \, P_Y \frac{\delta A_X}{\delta S_{X|y}} \delta S_{X|y}.$$

Hence, recalling $S_{X|y}(x) = S(x, y)$ from Eq. (3.12),

$$\frac{\delta A_{XY}}{\delta S} = P_Y \frac{\delta A_X}{\delta S_{X|y}}. \tag{3.18}$$

Finally, substituting Eqs. (3.17) and (3.18) into the definition of the Poisson bracket in Eq. (2.5) gives Eq. (3.16), as desired.

3.3.3 Configuration Separability and Strong Separability

For an arbitrary joint ensemble on the configuration space $X \times Y$, consider the infinitesimal canonical transformation generated by an observable of the first ensemble. We first show that such a local transformation does not affect the statistics of the configuration of the second ensemble.

If we compute the effect on P of a canonical transformation generated by an observable A_{XY} that is derived from A_X according to Eq. (3.13), we obtain, via Eq. (2.7) of Sect. 2.2,

$$\delta_{A_{XY}} P = \varepsilon \frac{\delta A_{XY}}{\delta S} = \varepsilon P_Y \frac{\delta A_X}{\delta S_{X|y}} = P_Y \delta_{A_{XY}} P_{X|y} \tag{3.19}$$

where $\delta_{A_{XY}}$ denotes the variation generated by A_{XY} and the second equality follows from Eq. (3.18). However, Eq. (3.12) leads to the general expression

$$\delta P = P_{X|y}\delta P_Y + P_Y\delta P_{X|y}, \tag{3.20}$$

where the variation denoted by δ is arbitrary. This implies that we must have

$$\delta_{A_{XY}} P_Y = 0, \tag{3.21}$$

as required. We formalise this property as follows:

Configuration separability: Any canonical transformation carried out on the first member of a joint ensemble does not affect the configuration statistics of the second member (and vice versa).

Configuration separability plays an important role in modeling interactions between quantum and classical systems (see Chaps. 8–11).

A stronger separability property holds for independent ensembles. For such ensembles one has from Eqs. (3.1) and (3.2) that $P_{X|y}(x) = P_X(x) = P(x)$, $P_{Y|x}(y) = P_Y(x) = P(y)$ and $S_{X|y}(x) = S_{Y|x}(y) = S_X(x) + S_Y(y)$. Hence, denoting the extensions of observables A_X and B_Y on these ensembles by A_{XY} and B_{YX} respectively, it follows via the definition of the Poisson bracket in Eq. (2.5), and using Eqs. (3.17) and (3.18), that

$$\begin{aligned}
\{A_{XY}, B_{YX}\} &:= \int dxdy \left(\frac{\delta A_{XY}}{\delta P} \frac{\delta B_{YX}}{\delta S} - \frac{\delta A_{XY}}{\delta S} \frac{\delta B_{YX}}{\delta P} \right) \\
&= \int dxdy \left(P_X(x) \frac{\delta A_X}{\delta P_{X|y}(x)} \frac{\delta B_Y}{\delta S_{Y|x}(y)} - P_Y(y) \frac{\delta A_X}{\delta S_{X|y}(x)} \frac{\delta B_Y}{\delta P_{Y|x}(y)} \right) \\
&= \int dx\, P_X \frac{\delta A_X}{\delta P_X(x)} \int dy \frac{\delta B_Y}{\delta S_{Y|x}(y)} - \int dy\, P_Y \frac{\delta B_Y}{\delta P_Y(y)} \int dx \frac{\delta A_X}{\delta S_{X|y}(x)} \\
&= 0.
\end{aligned} \tag{3.22}$$

Here the last line follows since the second integral in each term of the preceding line vanishes via the conservation of probability under canonical transformations (e.g., $\int dy\, (\delta B_Y/\delta S_{Y|x}) = \int dy\, \{P_Y, B_Y\} = \int dy\, \delta P_Y = 0$).

Thus, a local canonical transformation generated via some observable on one ensemble has no influence on any observable of a second independent ensemble. This may be regarded as a strong separability property of independent ensembles. It generalises to arbitrary joint ensembles in the case that the observables are both classical or both quantum, as reviewed in Chap. 8. The consequences of taking strong separability as an axiom for arbitrary observables are considered in Chap. 7 for rotational bits.

3.4 Measurement Interactions: A Simple Example

This section considers an important example of interaction, in the context of a very simple measurement model in which the configuration of one ensemble becomes correlated with the configuration of a second 'pointer' ensemble. The model is a straightforward generalisation of a quantum model of position measurement, first given by von Neumann in 1932 [7], to arbitrary configuration ensembles.

In particular, let X denote the configuration space of an ensemble that is to be measured, and Y denote the configuration space corresponding to the position of a pointer of a measurement apparatus. We will allow the configuration space X to be discrete or continuous, or some combination thereof, but we will assume that Y is continuous. Indeed, since we are only interested in the pointer degree of freedom, we will for convenience restrict Y to be one-dimensional (it is straightforward to lift this restriction if desired).

Guided by von Neumann's quantum measurement model, we now construct a measurement interaction that may be interpreted as coupling the momentum of the pointer to an observable of the ensemble undergoing measurement. First, recall from Sect. 2.3 that the momentum observable associated with configuration space Y is given by

$$\Pi_Y = \int dy \, P_Y(y) \frac{\partial S_Y(y)}{\partial y}, \tag{3.23}$$

and that the observable corresponding to an arbitrary function $f(x)$ of the configuration space X is given by

$$C_f = \int dx \, P_X(x) \, f(x). \tag{3.24}$$

Noting the definition of extended observables in Eq. (3.13), this then suggests that a suitable generalisation of von Neumann's measurement model is given by an interaction ensemble Hamiltonian of the form

$$\mathcal{H}_I[P, S, t] := \kappa(t) \int dx dy \, P(x, y) \, f(x) \frac{\partial S(x, y)}{\partial y}. \tag{3.25}$$

Here $\kappa(t)$ determines the strength and duration of the interaction, and the integral is seen to correspond to the average of the product of the local densities of C_f and Π_Y. The integration over x is to be replaced by summation over any discrete ranges of x.

To investigate this model, assume for simplicity that $\kappa(t)$ is non-negligible only over a timescale sufficiently short that any other contributions to the ensemble Hamiltonian can be ignored. It follows that the evolution of the joint ensemble during the measurement interaction is given by the Hamiltonian equations of motion (see Chap. 1)

$$\frac{\partial P}{\partial t} = \frac{\delta \mathcal{H}_I}{\delta S} = -f(x) \frac{\partial P}{\partial y}, \qquad \frac{\partial S}{\partial t} = -\frac{\delta \mathcal{H}_I}{\delta P} = -f(x) \frac{\partial S}{\partial y}. \tag{3.26}$$

These equations can be immediately solved, to give

$$P(x, y, T) = P(x, y - Kf(x), 0), \qquad S(x, y, T) = S(x, y - Kf(x), 0), \quad (3.27)$$

where T is the total interaction time and $K := \int_0^T dt\, \kappa(t)$.

It is seen that the measurement interaction acts to correlate the position y of the pointer with the function $f(x)$: if the configuration of the measured ensemble is taken to be x, then corresponding probability density $P_{Y|x}$ of the pointer position is given by

$$P_{Y|x}(y, T) = P_{Y|x}(y - Kf(x), 0), \qquad (3.28)$$

i.e., it is translated by an amount proportional to $f(x)$ and to the interaction strength K. Further, the average position of the pointer following the measurement is given by

$$\langle y \rangle_T = \int dx dy\, P(x, y, T)\, y = \langle y \rangle_0 + K \langle f \rangle_0, \qquad (3.29)$$

and hence is translated by an amount proportional to C_f. These properties are similar to the quantum case [7], and consistent with the interpretation of the interaction ensemble Hamiltonian $\mathcal{H}_I$ as generating a measurement of the observable C_f.

It is of interest to check that the above ensemble Hamiltonian $\mathcal{H}_I$ satisfies the formal definition of interacting ensembles in Sect. 3.2.2. In particular, consider the case that the joint ensemble is initially independent, i.e.,

$$P(x, y, 0) = P_X(x)\, P_Y(y), \qquad S(x, y, 0) = S_X(x) + S_Y(y).$$

Hence, at time T one has in this case that

$$P(x, y, T) = P_X(x)\, P_Y(y - Kf(x)), \qquad S(x, y, T) = S_X(x) + S_Y(y - Kf(x)), \tag{3.30}$$

which by inspection cannot satisfy the independence condition (3.1) (whereas the independence condition (3.2) is satisfied in the special case that $S_Y(y) =$ constant). Hence the joint ensemble does not remain independent, implying that the components of the ensemble are indeed interacting under $\mathcal{H}_I$.

Finally, it is of interest to consider how the measured ensemble may be updated given a measurement result $y = y_0$, and how it is 'disturbed' by the measurement process. Such updatings generalise the notion of collapse for quantum ensembles, and will be considered further at the end of this chapter.

In particular, assuming that the ensembles are initially independent, it is consistent to take the updated configuration ensemble on X as

$$P_{X|y_0}(x, T) = P_X(x), \qquad S_{X|y_0} = S_X(x) + S_Y(y_0 - Kf(x)), \qquad (3.31)$$

via Eqs. (3.12) and (3.30). Thus, the configuration probability density P on X is unchanged by the measurement, whereas the conjugate quantity S undergoes a shift in general. In the case that X is a continuous configuration space, this shift corresponds a change in the ensemble momentum Π_X defined in Eq. (2.17), with

$$\Pi_X(T) = \Pi_X(0) - K \int dx \, P_X(x) \, S'_Y(y_0 - Kf(x)) \, \nabla f(x). \qquad (3.32)$$

Thus, a measurement of the configuration observable C_f will typically disturb the momentum of the ensemble, with the disturbance proportional to the interaction strength K.

More general interactions are considered in later chapters—e.g., between rotational bits in Chap. 7; between classical and quantum particles in Chaps. 8 and 9; and between classical spacetimes and quantum fields in Chap. 11. In the next section, however, we will consider a different direction of generalisation: weak measurements.

3.5 Weak Measurements

It may be recalled from Sect. 2.4.2 that the weak value of observable $A[P, S]$ on a general configuration space X is defined by

$$A^W(x) := \frac{\delta A}{\delta P}. \qquad (3.33)$$

The weak value satisfies $A = \int dx \, P(x) \, A^W(x)$, and reduces to the usual quantum weak value for quantum ensembles (see Chap. 2 for further discussion).

Quantum weak values originally arose in connection with weak measurements, i.e., measurements in which an apparatus or probe is only very weakly coupled to the quantum system of interest [8–10]. In particular, the quantum weak value corresponds to the average displacement of a pointer weakly coupled to a quantum observable, where the average is postselected on a measurement result $X = x$ on the quantum system. This section shows how weak position and momentum values may be similarly measured via a weak interaction with a pointer, whether or not the configuration ensemble is quantum.

3.5.1 Weak Position Measurement

Consider a measurement of the configuration observable C_f as discussed in Sect. 3.4, where the interaction strength K is sufficiently small to ensure that the average displacement of the pointer, as given in Eq. (3.29), is much less than the width of the initial probability density $P_Y(t, 0)$ of the pointer. Such a measurement will be said to be *weak*, similarly to the quantum case. For the particular case of a continuous configuration space X, with $f(x) = x$, this corresponds to a weak measurement of position. However, we will not restrict our attention to this case, and so will let $f(x)$ be an arbitrary function of the configuration.

It follows immediately via Eq. (3.28) that if the system is postselected to have configuration x following the interaction, then the postselected average of the pointer position is given by

$$\langle y \rangle_x = \int dx\, P_{Y|x}(y)\, y = \langle y \rangle_{t=0} + K f(x) = \langle y \rangle_{t=0} + K\, C_f^W(x). \tag{3.34}$$

Thus, the postselected average displacement of the pointer is directly proportional to the weak value of C_f, in complete analogy to the quantum case.

It may have been noted by the astute reader that the above result does not in fact rely on K being small, i.e., on the measurement interaction being weak. It holds for any nonzero value of K. In this sense, weak measurements of the configuration (when postselected on the configuration) are trivial—although it may be noted that if the interaction is weak, then the post-measurement ensemble on X is only weakly disturbed from the original ensemble, via Eq. (3.31). However, for more general weak measurements the strength of the measurement interaction plays a nontrivial role, as will now be shown for the case of weak momentum measurements.

3.5.2 Weak Momentum Measurement

Consider the case of a pointer that is weakly coupled to the momentum of a one-dimensional particle. By analogy with Eq. (3.25) above, we will model such a weak measurement by an interaction Hamiltonian of the form

$$\mathscr{H}_I[P, S] = \kappa(t) \int dx dy\, P(x, y) \frac{\partial S(x, y)}{\partial x} \frac{\partial S(x, y)}{\partial y}. \tag{3.35}$$

The weakness of the measurement is assured by a short interaction time, δt. The change in the joint probability density over the interaction then follows as

$$\delta P = \delta t \{ P, \mathscr{H}_I \} = \delta t \frac{\delta \mathscr{H}_I}{\delta S} = -\kappa_0 \delta t \left[\frac{\partial}{\partial x} \left(P \frac{\partial S}{\partial y} \right) + \frac{\partial}{\partial y} \left(P \frac{\partial S}{\partial x} \right) \right]$$

to first order in δt, where κ_0 denotes $\kappa(0)$, and all quantities evaluated at $t = 0$. Assuming that the particle and pointer ensembles are initially independent, as per Eqs. (3.1) and (3.2), this simplifies to

$$\delta P(x, y) = -\kappa_0 \delta t \left[P_X'(x)\, P_Y(y)\, S_Y'(y) + P_X(x)\, P_Y'(y)\, S_X'(x) \right], \tag{3.36}$$

where a prime indicates differentiation with respect the argument. Integrating over y further gives

$$\delta P_X(x) = \int \mathrm{d}y \, \delta P(x, y) = -\kappa_0 \delta t \, P_X'(x) \int \mathrm{d}y \, P_Y(y) \, S_Y'(y). \qquad (3.37)$$

We now consider the corresponding change in the probability distribution of the pointer, when postselected on particle position x. Using the above results, we have

$$\delta P_{Y|x}(y) = \delta \left(\frac{P}{P_X(x)} \right) = \frac{\delta P}{P_X(x)} - \frac{P \delta P_X(x)}{P_X(x)^2}$$

$$= -\kappa_0 \delta t \left\{ P_Y'(y) \, S_X'(x) + \frac{P_X'(x)}{P_X(x)} P_Y(y) \left[S_Y'(y) - \int \mathrm{d}y \, P_Y(y) \, S_Y'(y) \right] \right\}.$$

It follows that the average change in the pointer position, postselected on particle position x, is given by

$$\delta \langle y \rangle_x = \int \mathrm{d}y \, y \, \delta P_{Y|x}(y)$$

$$= -\kappa_0 \delta t \left\{ S_X'(x) \int \mathrm{d}y \, y \, P_Y'(y) + \frac{P_X'(x)}{P_X(x)} \left[\langle y S_Y'(y) \rangle_{t=0} - \langle y \rangle_{t=0} \langle S_Y'(y) \rangle_{t=0} \right] \right\}$$

$$= \kappa_0 \delta t \left[S_X'(x) - \frac{P_X'(x)}{P_X(x)} \mathrm{Cov}(y, S_Y')_{t=0} \right], \qquad (3.38)$$

where integration by parts has been used to obtain the first term, and $\mathrm{Cov}(a, b)$ denotes the classical covariance of random variables a and b.

Hence, choosing the initial pointer ensemble such that the covariance between y and $S_Y'(y)$ vanishes, and recalling that the weak value of the particle momentum Π_X is given by $\Pi^W(x) = S_X'(x)$ at time $t = 0$, we obtain our main result:

$$\delta \langle y \rangle_x = \kappa_0 \delta t \, \Pi_X^W(x). \qquad (3.39)$$

Thus, the weak measurement acts to displace the pointer position by an amount proportional to the weak momentum of the ensemble, generalising the known result for quantum ensembles [9, 10].

3.6 Measurement-Induced Collapse

We now want to use the results developed in the previous sections to outline a description of measurement-induced collapse that goes beyond the simple case of updating an ensemble following measurement of a configuration observable as discussed in Sect. 3.4. The approach that we describe here is closely related to the description of measurements in quantum mechanics, and in the case of quantum systems it reduces to the usual description. However, to apply to other systems, it is clear that it must be formulated *without* making the assumption of linearity that seems essential in the quantum mechanical case. We want to show how this can be done.

The experiments that we want to describe are of the following type. We suppose that the system is prepared in a given state. To be more specific, let us assume that this state is an eigenstate of some observable $g[P, S]$, thus it is described by $P(x|\gamma_j)$, $S(x|\gamma_j)$ where γ_j labels the corresponding eigenvalue (see Sect. 2.4). We now consider a measurement of a different observable $f[P, S]$ and assume that the initial state is not necessarily an eigenstate of $f[P, S]$. This is a very general situation which involves the preparation of a state followed by the measurement of a particular observable. Furthermore, we assume that the measurement leaves the system in a new state $P(x|\phi_k)$, $S(x|\phi_k)$ which *is* an eigenstate of $f[P, S]$, and that this happens with probability $w(\phi_k|\gamma_j)$. In quantum mechanics, this last assumption corresponds to measurement-induced collapse.

Measurement-induced collapse (also known as von Neumann's "first intervention" [7] or projection postulate) has been described by Dirac [11] in the following terms:

> When we measure a real dynamical variable ξ, the disturbance involved in the act of measurement causes a jump in the state of the dynamical system. From physical continuity, if we make a second measurement of the same dynamical variable ξ immediately after the first, the result of the second measurement must be the same as that of the first. Thus after the first measurement has been made, there is no indeterminacy in the result of the second. Hence, after the first measurement has been made, the system is in an eigenstate of the dynamical variable ξ, the eigenvalue it belongs to being equal to the result of the first measurement. This conclusion must still hold if the second measurement is not actually made.

Notice that this postulate goes beyond a simple updating of the probability P of finding the quantum system in a given configuration (such an update could be formulated for example via an application of Bayes theorem) because it requires the update of the wavefunction $\psi = \sqrt{P}\, e^{iS/\hbar}$. Thus the generalization of measurement-induced collapse to different types of ensembles on configuration space will require introducing prescriptions for the simultaneous updating of *both* P and S.

In quantum mechanics, it is straightforward to describe such measurements. If the system is prepared in an eigenstate of the operator $\hat{g}$ with eigenvalue γ_j, we write this initial state as $|\gamma_j>$, where $\hat{g}|\gamma_j> = \gamma_j|\gamma_j>$. Taking advantage of the superposition principle, we write $|\gamma_j> = \sum_k c_k|\phi_k>$, where the $|\phi_k>$ are eigenstates of the operator $\hat{f}$ that is being measured and $\sum_k |c_k|^2 = 1$. Measurement-induced collapse amounts to the statement that the measurement leaves the system in the state $|\phi_k>$ with probability $w(\phi_k|\gamma_j) = |c_k|^2$.

Is there a procedure that works *without* the assumption of linearity which seems so fundamental here, so that it can be applied to other types of configuration space ensembles? It turns out that there *is* such a procedure, which amounts to describing the measurement process via the following three steps:

1. The preparation in an eigenstate of the observable $g[P, S]$ with eigenvalue γ_j, corresponding to a state which satisfies

$$\frac{\delta g}{\delta S} = 0, \quad \frac{\delta g}{\delta P} = \gamma_j, \quad g[P, S] = \gamma_j, \tag{3.40}$$

2. The probability $w(\phi_k|\gamma_j)$ of transition to the state $P(x|\phi_k)$, $S(x|\phi_k)$, which is given by the expression in Eq. (2.45) of Sect. 2.4.3,
3. The final state in an eigenstate of the observable $f[P, S]$ with eigenvalue ϕ_k, corresponding to a state which satisfies

$$\frac{\delta f}{\delta S} = 0, \quad \frac{\delta f}{\delta P} = \phi_k, \quad f[P, S] = \phi_k. \tag{3.41}$$

This shows that measurement-induced collapse can be formulated for ensembles on configuration space. These three steps can always be carried out for a quantum system and they lead to a uniquely determined final state. For classical and hybrid systems there may be observables f and g which either do not allow for non-trivial solutions of Eqs. (3.40) and (3.41) or result in preparation and/or final states which are not uniquely determined.

References

1. Hall, M.J.W., Reginatto, M.: Interacting classical and quantum ensembles. Phys. Rev. A **72**, 062109 (2005)
2. Schrödinger, E.: Discussion of probability relations between separated systems. Proc. Camb. Philos. Soc. **31**, 555–563 (1935)
3. Gisin, N.: Bell's inequality holds for all non-product states. Phys. Lett. A **154**, 201–202 (1991)
4. Home, D., Selleri, F.: Bell's theorem and the EPR paradox. Riv. Nuovo Cim. **14**, 1–95 (1991)
5. Horodecki, R., Horodecki, P., Horodecki, M., Horodecki, K.: Quantum entanglement. Rev. Mod. Phys. **81**, 865–942 (2009)
6. Spekkens, R.W.: Evidence for the epistemic view of quantum states: a toy theory. Phys. Rev. A **75**, 032110 (2007)
7. von Neumann, J.: Mathematical Foundations of Quantum Mechanics. Princeton University Press, Princeton (1955)
8. Aharonov, Y., Albert, D.Z., Vaidman, L.: How the result of a measurement of a component of the spin of a spin-1/2 particle can turn out to be 100. Phys. Rev. Lett. **60**, 1351–1354 (1988)
9. Aharonov, Y., Vaidman, L.: Properties of a quantum system during the time interval between two measurements. Phys. Rev. A **41**, 11–20 (1990)
10. Dressel, J.: Weak values as interference phenomena. Phys. Rev. A **91**, 032116 (2015)
11. Dirac, P.A.M.: The Principles of Quantum Mechanics. Oxford University Press, Oxford (1958)

Chapter 4
Thermodynamics and Mixtures on Configuration Space

Abstract We introduce the concept of a mixture of configuration space ensembles. In very general terms, if a physical system is described by the configuration space ensemble (P_j, S_j) with probability w_j, then it may be said to correspond to the mixture $\{(P_j, S_j); w_j\}$. Mixtures for classical and quantum systems are shown to be equivalent to phase space densities and density operators, respectively, and obey corresponding classical and quantum Liouville equations. We also generalise d'Espagnat's distinction between 'proper' and 'improper' quantum mixtures to mixtures of arbitrary configuration ensembles. With the help of mixtures it becomes possible to unify and generalise traditional classical and quantum approaches to thermodynamics, via the definition of suitable 'thermal mixtures' based on two universal primary notions: stationarity and distinguishability. Our formulation is very different to standard approaches based on the maximum entropy principle, and is equivalent to the standard statistical mechanics formulation in each of the quantum and classical cases. The latter case is of particular interest, as it provides a novel Hamilton–Jacobi picture of classical thermodynamics.

4.1 Introduction

The configuration ensemble approach must, if it is to provide a complete formalism for describing physical systems, be able to incorporate statistical mechanics and thermodynamics. However, there is an immediate difficulty in this regard: classical configuration ensembles correspond to probability densities on configuration space, whereas the statistical formulation of classical thermodynamics is in terms of probability densities on phase space. Similarly, quantum configuration ensembles correspond to wave functions (see Chap. 1), whereas the statistical formulation of quantum thermodynamics is in terms of density operators on Hilbert space.

The question therefore arises as to whether or not this difficulty can be overcome—is it possible to give a general formulation of thermodynamics in the configuration ensemble approach? We show here that this question can be answered in the affirmative.

© Springer International Publishing Switzerland 2016
M.J.W. Hall and M. Reginatto, *Ensembles on Configuration Space*,
Fundamental Theories of Physics 184, DOI 10.1007/978-3-319-34166-8_4

The central concept required is that of a *mixture* of configuration ensembles. In very general terms, if a physical system is described by the configuration ensemble (P_j, S_j) with probability w_j, then it may be said to correspond to the mixture $\{(P_j, S_j); w_j\}$. Mixtures are introduced in Sect. 4.2, and for classical and quantum systems are shown to be equivalent to phase space densities and density operators, respectively, with appropriate (Liouville) equations of motion. We also generalise d'Espagnat's distinction between 'proper' and 'improper' quantum mixtures [1] to mixtures of arbitrary configuration ensembles.

With the help of such mixtures it becomes possible to define suitable 'thermal mixtures', thus unifying and generalising the thermodynamics of classical and quantum configuration ensembles. This is a nontrivial undertaking, as discussed in Sect. 4.3.1. For example, the usual statistical mechanics approach to classical and quantum thermodynamics is equivalent to maximising a suitable entropy, subject to an average energy constraint, with the Shannon phase space entropy used for the classical case and the von Neumann entropy for the quantum case [2]. However, while there is in fact a natural phase space entropy for mixtures of general configuration ensembles, arising from the inherent canonical structure of the formalism, a naive maximisation of this entropy leads to results inconsistent with standard thermodynamics. Hence a different approach must be taken. It is shown in the remainder of Sect. 4.3 that the key to a successful general formulation is via mixtures of distinguishable stationary ensembles.

In Sects. 4.4 and 4.5 our proposed formulation is shown to reduce to the standard statistical mechanics formulation in each of the quantum and classical cases. The latter case requires some care and is of particular interest, as it provides a novel Hamilton–Jacobi picture of classical thermodynamics. It is remarkable that this new picture relies heavily on an object more commonly seen in discussions of semiclassical approximations to quantum mechanics, known as the van Vleck determinant [3]. The example of one-dimensional classical systems is considered in detail in Sect. 4.6.

Finally, some remaining open issues are briefly discussed in Sect. 4.7.

4.2   Mixtures

4.2.1   General Definition

Up to now, we have considered the description of physical systems at a given time in terms of a pair of functions P and S, where $P(x)$ is a probability density on the configuration space of the system, and $S(x)$ is a function canonically conjugate to P (see Chap. 1). While this intrinsically provides a statistical description of the system, it is straightforward to generalise this to a description in terms of *mixtures* of configuration ensembles.

Thus, suppose that the state of a system is described by the configuration ensemble (P_α, S_α), with statistical weight w_α. Here α labels some set of ensembles on configuration space, and may be continuous or discrete. This statistical mixture may be denoted by $\mathscr{W}$, with

$$\mathscr{W} \equiv \{(P_\alpha, S_\alpha); w_\alpha\}. \tag{4.1}$$

Recalling that the average of any observable A for the ensemble (P_α, S_α) is $A[P_\alpha, S_\alpha]$ (see Chap. 2), it trivially follows that the average value of A over the mixture $\mathscr{W}$ is given by

$$\langle A \rangle_{\mathscr{W}} = \int d\alpha \, w_\alpha \, A[P_\alpha, S_\alpha] \tag{4.2}$$

(with integration replaced by summation over any discrete ranges of α).

To avoid confusion in what follows, we cannot use the terms 'mixture' and 'ensemble' interchangeably: the latter term, when unqualified, will always refer to a single ensemble on configuration space, (P, S). For notational convenience we will primarily work with continuous configuration spaces below, but the substance of the discussion also applies to the discrete case.

4.2.2 Classical and Quantum Mixtures

The above considerations apply to arbitrary configuration ensembles, whether classical, quantum or beyond. However, mixtures have particularly simple representations in the classical and quantum cases:

Mixtures of configuration ensembles are equivalent to phase space densities for the case of classical systems, and to density operators for the case of quantum systems.

To demonstrate this in the classical case, recall first that observables for classical configuration ensembles have the form

$$C_f[P, S] = \int dx \, P f(x, \nabla S), \tag{4.3}$$

where $f(x, p)$ is any function on classical phase space (see Sect. 2.3.2). Thus, using Eq. (4.2) one has

$$\langle C_f \rangle_{\mathscr{W}} = \int d\alpha dx \, w_\alpha \, P_\alpha f(x, \nabla S_\alpha) = \int dx dp \, \rho_{\mathscr{W}}(x, p) f(x, p), \tag{4.4}$$

for any classical mixture $\mathcal{W}$, where

$$\rho_{\mathcal{W}}(x, p) := \int d\alpha \, w_\alpha \, P_\alpha(x) \, \delta(p - \nabla S_\alpha(x)) \tag{4.5}$$

is a well-defined probability density on classical phase space. Conversely, any given phase space density, $\rho(x, p)$, may be represented by a corresponding mixture $\mathcal{W} \equiv \{(P_\alpha, S_\alpha); w_\alpha\}$ of configuration space ensembles. For example, choosing

$$\alpha \equiv (x_0, p_0), \quad w_\alpha := \rho(x_0, p_0), \quad P_\alpha(x) := \delta(x - x_0), \quad S_\alpha(x) := p_0 \cdot x$$

one has

$$\rho(x, p) = \int dx_0 dp_0 \, \rho(x_0, p_0) \, \delta(x - x_0) \, \delta(p - p_0)$$
$$= \int d\alpha \, w_\alpha \, P_\alpha(x) \, \delta(p - \nabla S_\alpha(x)) = \rho_{\mathcal{W}}(x, p). \tag{4.6}$$

Hence, the equivalence holds as claimed. Another approach for mapping phase space densities to mixtures of configuration space ensembles is presented in Appendix 1 of this chapter.

We remark that for the special case of a single configuration ensemble, (P, S), the corresponding phase space density in Eq. (4.5) has the simple 'pure' form

$$\rho_{\text{pure}}(x, p) = P(x) \, \delta(p - \nabla S(x)). \tag{4.7}$$

It is of interest to note that the value of this phase space density at any given point, (x', p') say, can be written in the form of a classical observable as per Eq. (4.3):

$$\rho_{\text{pure}}(x', p') = C_r[P, S], \quad r(x, p) := \delta(x - x') \, \delta(p - p'). \tag{4.8}$$

Finally, for the quantum case, recall that every configuration ensemble (P, S) corresponds to a pure quantum state $|\psi\rangle$ with (see Sect. 2.3.3)

$$\psi(x) = \langle x | \psi \rangle = \sqrt{P(x)} \, e^{iS(x)/\hbar}. \tag{4.9}$$

Hence the average of any quantum observable $Q_{\hat{M}} = \langle \psi | \hat{M} | \psi \rangle$ over a mixture $\mathcal{W}$ of such ensembles follows via Eq. (4.2) as

$$\langle Q_{\hat{M}} \rangle_{\mathcal{W}} = \int d\alpha \, w_\alpha \, \langle \psi_\alpha | \hat{M} | \psi_\alpha \rangle = \text{tr}[\hat{\rho}_{\mathcal{W}} \hat{M}], \tag{4.10}$$

where $\hat{\rho}_{\mathcal{W}}$ denotes the quantum density operator

$$\hat{\rho}_{\mathscr{W}} := \int d\alpha \, w_\alpha |\psi_\alpha\rangle\langle\psi_\alpha|. \tag{4.11}$$

Conversely, any density operator ρ can be decomposed as a mixture of pure states, and hence as a mixture of corresponding quantum configuration ensembles. Thus, mixtures of quantum configuration ensembles are equivalent to density operators as claimed.

4.2.3 Dynamics and Liouville Equations

The dynamics of a mixture of configuration ensembles can always be determined by evolving the individual components of the mixture in Eq. (4.1). However, for some classes of configuration ensembles one can evolve the mixture directly. For example, the dynamics of quantum configuration ensembles are equivalent to the Schrödinger equation for pure quantum states, $i\hbar\partial_t|\psi\rangle = \hat{H}|\psi\rangle$ (see Chap. 1), which immediately generalises to the quantum Liouville equation

$$i\hbar\partial_t\hat{\rho} = [\hat{H}, \hat{\rho}] \tag{4.12}$$

for mixed quantum states. However, it is less trivial to show that the *classical* Liouville equation

$$\partial_t\rho = \{H, \rho\} \tag{4.13}$$

holds for mixtures of classical configuration ensembles, where $\rho(x, p)$ is a mixture as per Eq. (4.5), $H(x, p)$ is the classical phase space Hamiltonian, and $\{,\}$ denotes the Poisson bracket on phase space.

It is convenient to prove Eq. (4.13) for a configuration ensemble (P, S), with ρ as in Eq. (4.7), as the general result then follows immediately from linearity with respect to ρ. Recall first from Eq. (2.22) that classical observables satisfy the identity $\{C_f, C_g\} = C_{\{f,g\}}$, where $\{f, g\}$ denotes the usual Poisson bracket for phase space functions $f(x, p)$ and $g(x, p)$ (see Sect. 2.3.2). Hence, using Eqs. (4.3) and (4.7),

$$\begin{aligned}
\{C_f, C_g\} &= \int dx \, P(x) \, \{f(x, p), g(x, p)\}|_{p=\nabla S} \\
&= \int dx dp \, P(x)\delta(p - \nabla S) \, \{f(x, p), g(x, p)\} \\
&= \int dx dp \, \rho \, \{f, g\} \\
&= \int dx dp \, \{g, \rho\}f,
\end{aligned}$$

where the last line follows via $\{f, g\} = \nabla_x f \cdot \nabla_p g - \nabla_p f \cdot \nabla_x g$ and integration by parts. Noting that the classical ensemble Hamiltonian has the form $\mathscr{H}[P, S] = C_H$, it then follows from the equation of motion in Eq. (2.4) for general observables, $dA/dt = \{A, \mathscr{H}\} + \partial A/\partial t$, that

$$\frac{dC_f}{dt} = \{C_f, C_H\} = \int dxdp \, \{H, \rho\} f.$$

for any phase space function $f(x, p)$. But from Eq. (4.4) one also has

$$\frac{dC_f}{dt} = \frac{\partial}{\partial t} \int dxdp \, \rho f = \int dxdp \, \frac{\partial \rho}{\partial t} f.$$

Comparing these expressions, the classical Liouville equation (4.13) follows as desired.

Thus, the concept of mixtures allows recovery of the standard quantum and classical dynamics in terms of density operators and phase space densities, respectively.

4.2.4 Proper and Improper Mixtures

A mixture as defined in Eq. (4.1) may clearly be interpreted as representing ignorance as to which one of the configuration ensembles (P_α, S_α) has been prepared, generalising the notion of a 'proper mixture' in quantum mechanics, first defined by d'Espagnat [1]. We will also use the more descriptive term *preparation mixture* when this interpretation applies.

One may similarly generalise d'Espaganat's notion of an 'improper mixture' in quantum mechanics, corresponding to tracing out one component of a joint quantum ensemble [1]. In particular, consider a joint configuration ensemble on a configuration space $X \times Y$, described by probability density $P(x, y)$ and conjugate quantity $S(x, y)$ (see Sect. 3.1). We define the corresponding 'improper mixture' on the configuration space X by

$$\mathscr{W}_X := \{(P_y, S_y); w_y\}, \tag{4.14}$$

with

$$P_y(x) := \frac{P(x, y)}{\int dx \, P(x, y)}, \quad S_y(x) := S(x, y), \quad w_y := \int dx \, P(x, y). \tag{4.15}$$

However, one should be aware that these "improper mixtures" are *not* true mixtures as defined previously in Sect. 4.2.1 because the $(P_y(x, t), S_y(x, t))$ of Eq. (4.15) do not necessarily evolve as ensembles on configuration space: for example, if $(P(x, y), S(x, y))$ satisfy the continuity equation in the configuration space with coordinates (x, y), then in general $(P_y(x), S_y(x))$ will *not* satisfy the continuity equation in the configuration space with coordinate x. Nevertheless, this straightforward generalization of d'Espaganat's notion is both interesting and useful. Such mixtures correspond to reduction of the relevant configuration space from $X \times Y$ to X, and hence we will also use the more descriptive term *reduced mixture*.

For quantum systems, note from Eq. (4.9) that $\psi_y(x) = \psi(x, y)/\sqrt{\int dx P(x, y)}$. Hence, via Eq. (4.11), a quantum reduced mixture corresponds to the density operator

$$
\begin{aligned}
\hat{\rho}_{\mathscr{W}_x} &= \int dy \, w_y |\psi_y\rangle \langle\psi_y| \\
&= \int dy dx dx' \, |x\rangle \langle x'| \, w_y \, \psi_y(x) \, \psi_y^*(x') \\
&= \int dy dx dx' \, \psi(x, y) \, |x\rangle \, \langle x'| \, \psi^*(x', y) \\
&= \int dy \, \langle y|\psi\rangle \, \langle\psi|y\rangle \equiv \mathrm{tr}_Y[|\psi\rangle\langle\psi|],
\end{aligned}
\tag{4.16}
$$

i.e., to the reduced density operator of the pure state associated with (P, S), as expected. This will typically evolve under a master equation, rather than under the quantum Liouville equation.

Reduced mixtures of configuration ensembles have already been implicitly considered in Chap. 3, in the context of separability and measurement properties for joint ensembles. For example the average value of an observable for one component of a joint ensemble, evaluated via Eq. (3.13), is equal to the average value of the observable evaluated for the mixture $\mathscr{W}_X$ via Eqs. (4.2), (4.14) and (4.15). Moreover, a suitable measurement interaction with a pointer can prepare a mixture of the form of $\mathscr{W}_X$, as discussed in Sect. 3.4, providing a link between preparation and reduced mixtures in this case.

Properties of reduced mixtures are also relevant to the discussion of hybrid quantum-classical ensembles and Wigner functions, as will be seen in Chaps. 8 and 9.

4.3 Thermodynamics on Configuration Space

4.3.1 Failure of the Canonical Approach

As remarked in the introduction to this chapter, one cannot take a simple maximum entropy route to describe the thermodynamics of configuration ensembles, despite the existence of a natural phase space entropy for mixtures of such ensembles. It will be seen that a satisfactory solution to this problem is obtained by instead considering mixtures of stationary ensembles. But before we develop that approach, it is instructive to examine the origin of the problem.

Recall first that the quantities P and S, describing a given configuration ensemble, are canonically conjugate by definition (see Sect. 1.4). Thus the configuration ensemble (P, S) represents a point in the "natural" phase space of the formalism, with a corresponding Poisson bracket $\{P_j, S_k\} = \delta_{jk}$ for discrete configuration spaces and $\{P(x), S(x')\} = \delta(x - x')$ for continuous configuration spaces (see Sect. 2.2). Now, in classical phase space mechanics the 'canonical ensemble' is defined by a phase space density $w(x, p) \sim e^{-\beta H(x,p)}$, where H denotes the Hamiltonian on the phase space [2]. Hence, it is tempting to correspondingly define a 'canonical mixture' of configuration ensembles via the prescription

$$w[P, S] \sim e^{-\beta \mathscr{H}[P,S]}, \qquad \text{(wrong!)} \qquad (4.17)$$

where $\mathscr{H}[P, S]$ denotes the ensemble Hamiltonian. Note that this density w maximises the corresponding phase space entropy for a given average energy.

Unfortunately, however, the above 'canonical' prescription does not give physically correct answers. For example, in the quantum case such a mixture is equivalently represented by a corresponding density operator as per Eq. (4.11). Noting that each element (P, S) of the mixture can be equivalently labeled by a pure quantum state $|\psi\rangle$, with $\mathscr{H}[P, S] = \langle \psi|\hat{H}|\psi\rangle$ as per Eq. (1.15), this density operator has the form

$$\rho \sim \int D\psi \, e^{-\beta\langle\psi|\hat{H}|\psi\rangle} |\psi\rangle\langle\psi| \quad \text{(wrong!)} \qquad (4.18)$$

up to a normalisation factor. Here integration is with respect to the (unique) invariant Haar measure on the Hilbert space (this is equivalent to the uniform phase space measure since $(P, S) \rightarrow (\psi, \psi^*)$ is a canonical transformation [4]). In general, however, ρ is *not* equivalent to the density operator

$$\rho' \sim e^{-\beta\hat{H}}, \qquad (4.19)$$

which is the known correct result for the quantum canonical ensemble, corresponding to maximising the von Neumann entropy under an average energy constraint.

We demonstrate this inequivalence by computing a simple example. Consider a 2-dimensional quantum system, i.e., a qubit, with Hamiltonian operator $\hat{H} = \sigma_z$. As is well known, the set of wave functions is represented by the Bloch sphere, parameterized by a unit 3-vector n with $|\psi\rangle\langle\psi| = (1 + \sigma \cdot n)/2$ and $\langle\psi|\hat{H}|\psi\rangle = n_z$, and hence Eq. (4.18) simplifies to

$$\begin{aligned}
\rho &\sim \int dn \, e^{-\beta n_z} (1 + \sigma \cdot n) \\
&= \int_0^\pi d\theta \, \sin\theta e^{-\beta \cos\theta} \int_0^{2\pi} d\phi \left[1 + \sin\theta(\sigma_x \cos\phi + \sigma_y \sin\phi) + \sigma_z \cos\theta \right] \\
&= 2\pi \int_{-1}^1 du \, e^{-\beta u} (1 + u\sigma_z) \\
&= \frac{4\pi}{\beta} \left[\sinh\beta - \sigma_z \left(\cosh\beta - \frac{\sinh\beta}{\beta} \right) \right], \qquad \text{(wrong!)} \qquad (4.20)
\end{aligned}$$

using standard spherical coordinates (θ, ϕ) and defining $u := \cos \theta$. In contrast, Eq. (4.19) simplifies to

$$
\begin{aligned}
\rho' &\sim e^{-\beta \sigma_z} = e^{-\beta} \frac{1 + \sigma_z}{2} + e^{\beta} \frac{1 - \sigma_z}{2} \\
&= \cosh \beta - \sigma_z \sinh \beta.
\end{aligned}
\tag{4.21}
$$

Clearly the operators ρ and ρ' are *not* proportional, and thus the canonical prescription fails. A similar failure obtains for classical configuration ensembles. We must therefore take a different approach to describing thermodynamics on configuration space.

4.3.2 Thermal Mixtures

We now give a prescription for defining 'thermal mixtures' of configuration ensembles, which yields the correct results in the classical and quantum cases. In this way the formalism of configuration ensembles yields a new and conceptually unified approach to thermodynamics, not only incorporating the standard quantum and classical approaches but allowing one to go beyond these.

Our new approach can be regarded as a modification of the 'canonical prescription' in Eq. (4.17), based on two primary notions: *stationarity* and *distinguishability*. These notions provide further natural restrictions on the elements of the mixture appearing in Eq. (4.17), sufficient to ensure compatibility with both classical and quantum thermodynamics.

The notion of stationary ensembles has already been introduced in Sect. 2.4, corresponding to those configuration ensembles for which all physical properties are time independent. To motivate their appearance in a thermodynamical context, we note that thermal equilibrium is an inherently time-invariant concept. Hence, for equilibrium to be described by some mixture of ensembles, the physical properties of this mixture must also be time invariant. Clearly, this is guaranteed if we assume that the mixture comprises only stationary ensembles.

We also require the notion of distinguishable ensembles. In particular, a set of configuration ensembles is defined to be distinguishable if and only if there is some observable which can distinguish unambiguously between its members; i.e., the ranges of the observable for each member are nonoverlapping. For example, two quantum configuration ensembles are distinguishable if and only if the corresponding wave functions are orthogonal, thus allowing them to be distinguished by a quantum observable $Q_{\hat{M}}$ for which these wave functions are eigenfunctions of $\hat{M}$ corresponding to distinct eigenvalues. Similarly, two classical configuration ensembles are distinguishable if and only if their phase space supports $\{(x, \nabla S) : P(x) > 0\}$ are nonoverlapping, thus allowing them to be distinguished by a classical observable C_f for which $f(x, p)$ takes distinct ranges of values over these supports.

A distinguishable set is *maximal* if it cannot be enlarged without losing the distinguishability property.

The motivation for considering such ensembles here is essentially to avoid overcounting. In particular, the failure of the 'canonical prescription' in Eq. (4.17) is due, in large part, to counting redundant statistical contributions from many overlapping configuration ensembles. This is particularly clear when comparing the quantum density operators in Eqs. (4.18) and (4.19): the first comprises integration over all possible wave functions whereas the second is equivalent to a sum over a set of orthogonal energy eigenstates.

We can now give our definition of thermal mixtures.

Thermal mixtures: For a given ensemble Hamiltonian $\mathcal{H}[P, S]$, a thermal mixture is a mixture, $\mathcal{W} \equiv \{(P_\alpha, S_\alpha); w_\alpha\}$, of a maximal set of distinguishable stationary ensembles, with

$$w_\alpha \sim e^{-\beta \mathcal{H}[P_\alpha, S_\alpha]} \tag{4.22}$$

for some constant β.

Note that this definition is indeed a modification of the canonical prescription in Eq. (4.17), corresponding to restricting attention to distinguishable stationary ensembles. The definition is, for our purposes, justified by its consequences for quantum and classical systems, as examined in Sects. 4.4 and 4.5. However, it is of interest to first briefly consider a physical justification for the general form of the thermal weighting distribution w_α in Eq. (4.22).

4.3.3 Thermal Weighting Distribution from the Zeroth Law

The entropy of the thermal weighting distribution in Eq. (4.22) is not simply related to the thermodynamic entropy in general, as will be seen in Sect. 4.5. This motivates finding a non-entropic argument for the form of w_α in Eq. (4.22). A simple such argument is given below, based on the zeroth law of thermodynamics, i.e., on the property that independent systems at the same temperature remain in equilibrium at the same temperature when placed in thermal contact [2].

It is natural to assume, in the absence of any information about a physical system in addition to its ensemble Hamiltonian $\mathcal{H}[P, S]$, that the weight of a given stationary ensemble (P_α, S_α) in a thermal mixture at temperature T is of the form

$$w_\alpha = f_T(\mathcal{H}[P_\alpha, S_\alpha]), \tag{4.23}$$

for some function f_T that may depend on the configuration space of the system.

In our context, the zeroth law implies that a composition of two independent and noninteracting thermal mixtures, each in equilibrium at temperature T, is itself a thermal mixture in equilibrium at temperature T. Now, if α' labels the components of the first mixture, with weighting distribution $w'_{\alpha'}$, and α'' labels the component of the second mixture, with weighting distribution $w''_{\alpha''}$, then independence of the mixtures implies that the composite mixture is labelled by $\alpha \equiv (\alpha', \alpha'')$, with the weighting distribution

$$w_\alpha = w'_{\alpha'} \, w''_{\alpha''}. \tag{4.24}$$

Independence further implies that the joint configuration ensemble labelled by α has the form (see Sect. 3.2.1)

$$(P_\alpha, S_\alpha) = (P'_{\alpha'} \, P''_{\alpha''}, S'_{\alpha'} + S''_{\alpha''}), \tag{4.25}$$

and Eq. (3.4) for noninteracting ensembles then implies the additivity property

$$\mathcal{H}[P_\alpha, S_\alpha] = \mathcal{H}'[P'_{\alpha'}, S'_{\alpha'}] + \mathcal{H}''[P''_{\alpha''}, S''_{\alpha''}] \tag{4.26}$$

for the composite ensemble Hamiltonian $\mathcal{H}$, where $\mathcal{H}'$ and $\mathcal{H}''$ denote the ensemble Hamiltonians of the individual mixtures. The stationarity and maximal distinguishability of $\{(P_\alpha, S_\alpha)\}$ is easily shown to follow from that of $\{(P'_{\alpha'}, S'_{\alpha'})\}$ and $\{(P''_{\alpha''}, S''_{\alpha''})\}$.

The combination of Eqs. (4.23), (4.24) and (4.26) yields

$$f_T(E_{\alpha'} + E_{\alpha''}) = g_T(E_{\alpha'}) \, h_T(E_{\alpha''}), \tag{4.27}$$

where $E_{\alpha'}$ and $E_{\alpha'}$ denote the values of $\mathcal{H}'[P'_{\alpha'}, S'_{\alpha'}]$ and $\mathcal{H}''[P''_{\alpha''}, S''_{\alpha''}]$ respectively, where the functions f_T, g_T and h_T may depend on the corresponding configuration spaces. To find the functional form of w_α, one needs to solve Eq. (4.27), i.e., the functional equation

$$f_T(u + v) = g_T(u) \, h_T(v).$$

Since u and v can each take a continuous range of values (corresponding to the values of suitable ensemble Hamiltonians), then the physically reasonable assumption that f_T, g_T and h_T are continuous implies, via differentiation with respect to u and v respectively, that

$$f'_T(u + v) = g'_T(u) \, h_T(v) = g_T(u) \, h'_T(v).$$

Hence $f'_T/f_T = g'_T/g_T = h'_T/h_T = -\beta_T$ for some constant β_T. Finally, integration and substitution into Eq. (4.23) yields the universal form in Eq. (4.22), i.e.,

$$w_\alpha \sim e^{-\beta_T \mathcal{H}[P_\alpha, S_\alpha]}, \tag{4.28}$$

as desired.

The form of the thermal weighting distribution w_α thus follows as a consequence of the zeroth law and the additivity of ensemble Hamiltonians under the composition of independent ensembles, with no properties of phase space or entropy being required. A generalisation of this argument may be used to find the form of the weighting distribution when the latter is further allowed to depend on additional conserved quantities (such as angular momentum or particle number). Similar arguments may also be used within the standard frameworks of classical and quantum statistical mechanics, but will not be pursued here.

4.4 Quantum Thermal Mixtures

Recall that stationary configuration ensembles are those for which all physical properties are time-independent. Equation (2.30) of Chap. 2 implies that a stationary configuration ensemble (P, S) has the general form

$$P(x, t) = P(x), \qquad S(x, t) = W(x) - Et \tag{4.29}$$

as a function of time, for suitable functions $P(x)$ and $W(x)$ and constant E. The homogeneity property $\mathscr{H}[P, S] = -\langle \partial_t S \rangle$ of ensemble Hamiltonians in Eq. (1.28) immediately yields

$$\mathscr{H}[P, S] = E, \tag{4.30}$$

i.e., the constant E is the corresponding value of the ensemble Hamiltonian.

For quantum configuration ensembles, stationary ensembles correspond to eigenstates of the Hamiltonian operator $\hat{H}$, as discussed in Sect. 2.4.1. Since distinguishability is equivalent to orthogonality for the quantum case (see Sect. 4.3.2), it follows that a maximally distinguishable set of stationary states corresponds to a complete set of mutually orthogonal energy eigenstates. Let $\{|\psi_\alpha\rangle\}$ denote any such set. Thus,

$$\hat{H}|\psi_\alpha\rangle = E_\alpha |\psi_\alpha\rangle, \qquad \langle \psi_\alpha | \psi_{\alpha'} \rangle = 0 \text{ for } \alpha \neq \alpha',$$

where E_α ranges over the eigenvalues of $\hat{H}$, yielding the spectral decomposition

$$\hat{H} = \int d\alpha \, E_\alpha \, |\psi_\alpha\rangle\langle\psi_\alpha|. \tag{4.31}$$

Here integration is replaced by summation over any discrete components of α.

It follows immediately via the definitions in Eqs. (4.5) and (4.22) that quantum thermal mixtures are represented by density operators of the form

$$\rho_\beta \sim \int d\alpha \, e^{-\beta \mathcal{H}[P_\alpha, S_\alpha]} |\psi_\alpha\rangle\langle\psi_\alpha|$$

$$= \int d\alpha \, e^{-\beta E_\alpha} |\psi_\alpha\rangle\langle\psi_\alpha|$$

$$= e^{-\beta \hat{H}}, \tag{4.32}$$

up to a normalisation factor. Thus, thermal mixtures of quantum configuration ensembles recover the standard quantum thermodynamics [2], as desired.

It is of interest to note that the quantum von Neumann entropy of ρ_β is, via the first line of Eq. (4.32), equal to the entropy of the thermal weighting distribution w_α, i.e.,

$$\mathcal{E}_\beta^Q := -\text{tr}[\rho_\beta \log \rho_\beta] = -\int d\alpha \, w_\alpha \log w_\alpha. \tag{4.33}$$

However, an analogous relation does not hold in the classical case, as will be seen in the following section.

4.5 Classical Thermal Mixtures

To show that the definition of thermal mixtures leads to standard thermodynamics in the classical case is less straightforward. The result relies on a beautiful connection between the classical Hamilton–Jacobi equation, canonical transformations, and the so-called van Vleck determinant from semiclassical quantum mechanics [3]. We give our general result here, and consider a specific example in Sect. 4.6.

Recall first that the ensemble Hamiltonian for a classical ensemble on a continuous configuration space has the general form

$$\mathcal{H}[P, S] = C_H[P, S] = \int dx \, P(x) \, H(x, \nabla S) \tag{4.34}$$

as per Eq. (4.3), where $H(x, p)$ is a classical phase space Hamiltonian. The Hamiltonian equations of motion for the ensemble, Eq. (1.4), thus reduce to

$$\frac{\partial P}{\partial t} = \frac{\delta \mathcal{H}}{\delta S} = -\nabla \cdot \left(P \frac{\partial H(x, \nabla S)}{\partial \nabla S}\right), \qquad \frac{\partial S}{\partial t} = -\frac{\delta \mathcal{H}}{\delta P} = -H(x, \nabla S), \tag{4.35}$$

where the functional derivatives have been evaluated via Eq. (A.6) of Appendix A. The first equation is a continuity equation for P, and the second is the classical Hamilton–Jacobi equation for S [9].

A complete solution of the Hamilton–Jacobi equation, on an n-dimensional configuration space, is parameterised (up to an arbitrary additive constant of no physical significance) by n independent constants $\alpha_1, \ldots, \alpha_n$, which we will denote collectively by α [9]. The form of the complete solution $S_\alpha(x, t)$ for stationary ensembles is constrained by Eq. (4.29), which together with the Hamilton–Jacobi equation implies that $H(x, \nabla S_\alpha) = E_\alpha$, and hence via Eq. (4.30) that

$$\mathcal{H}[P_\alpha, S_\alpha] = E_\alpha = H(x, \nabla S_\alpha). \tag{4.36}$$

The complete solution S_α is also associated with a corresponding canonical transformation on phase space, $(x, p) \rightarrow (x', p')$, given by [9]

$$x'_j = \frac{\partial S_\alpha}{\partial \alpha_j}, \qquad p'_j = \alpha_j. \tag{4.37}$$

Thus, the momentum $p'(x, p)$ in the primed coordinates is a constant of the motion, and provides a direct physical interpretation for the label α. Further, the invertibility of the transformation implies that the components of x' are independent functions of x, and hence that [9]

$$D_\alpha := \det \varphi \neq 0, \qquad \varphi_{jk} := \frac{\partial^2 S_\alpha}{\partial x_j \partial \alpha_k}. \tag{4.38}$$

The quantity D_α is well known in semiclassical quantum mechanics as the van Vleck determinant [3, 6–8]. However, here we are interested in exploiting its purely classical properties. In particular, given a complete solution S_α of the Hamilton–Jacobi equation, a solution of the continuity equation in Eq. (4.35) is given by [3, 6, 7]

$$P_\alpha \sim |D_\alpha|, \tag{4.39}$$

over the region of configuration space for which S_α is well defined, with P_α vanishing outside this region. We give an explicit derivation of this result in Appendix 2 of this chapter.

Now, the members of the set of stationary ensembles $\{(P_\alpha, S_\alpha)\}$ are clearly distinguishable, since α labels the possible measurement outcomes of the momentum observable $C_{p'}$ in Eq. (4.37). Moreover, the set is *maximally* distinguishable since α labels the complete set of stationary solutions $\{S_\alpha\}$ to the Hamilton–Jacobi equation by definition, and hence labels the complete set of stationary solutions $\{(P_\alpha, S_\alpha)\}$ via Eq. (4.39). Hence, using the definitions in Eqs. (4.5) and (4.22), classical thermal mixtures are represented by phase space distributions of the form

$$\rho_\beta(x, p) \sim \int d\alpha \, e^{-\beta \mathcal{H}[P_\alpha, S_\alpha]} \, P_\alpha(x) \, \delta(p - \nabla S_\alpha(x))$$

$$= \int d\alpha \, e^{-\beta H(x, \nabla S_\alpha(x))} \, P_\alpha(x) \, \delta(p - \nabla S_\alpha(x)),$$

with the second line following via Eq. (4.36). Making a change of variables from α to $k := \nabla S_\alpha$ (corresponding to mapping the primed momentum $p' = \alpha$ to the unprimed momentum $p = \nabla S$ [3, 9]), and substituting Eq. (4.39) for P_α, then gives

$$\rho_\beta(x, p) \sim \int dk \, J \, e^{-\beta H(x,k)} \, |D_\alpha| \, \delta(p - k),$$

where $J = |\det(\partial \alpha_i / \partial k_j)|$ is the Jacobian of the transformation. Finally, noting that this Jacobian is related to the van Vleck determinant in Eq. (4.38) via $J^{-1} = |\det(\partial k_i / \partial \alpha_j|)| = |D_\alpha|$, one has

$$\rho_\beta(x, p) \sim \int dk \, e^{-\beta H(x,k)} \, \delta(p - k) = e^{-\beta H(x,p)}, \qquad (4.40)$$

up to a normalisation factor. Hence, thermal mixtures of classical configuration ensembles recover the standard classical thermal statistics [2], as desired.

Note that the weighting distribution w_α is parameterised in the classical case by a momentum coordinate, as per Eq. (4.37), and not by a phase space coordinate. Hence the entropy of the classical thermal weighting distribution cannot be equal to the entropy of the phase space distribution $\rho_\beta(x, p)$, in contrast to the analogous quantum relation in Eq. (4.33). We explore this in more detail in the following example.

4.6 Example: One-Dimensional Classical Systems

It is of interest to give an explicit example of a classical thermal mixture, corresponding to a one-dimensional particle of mass m moving in a potential $V(x)$. This includes the case of a harmonic oscillator in particular, with $V(x) = \frac{1}{2} m \omega^2 x^2$.

We begin with the Hamilton–Jacobi equation, which is given by

$$\frac{\partial S}{\partial t} + \frac{1}{2} \left(\frac{\partial S}{\partial x} \right)^2 + V(x) = 0 \qquad (4.41)$$

from Eq. (4.35). A complete set of stationary solutions follows via substitution of Eq. (4.29) for stationary ensembles:

$$S_\alpha(x, t) = \text{sign } \alpha \int dx \, \sqrt{2m(E_\alpha - V)} - E_\alpha \, t, \qquad E_\alpha = |\alpha| \qquad (4.42)$$

up to an arbitrary additive constant, where α is any real number. The Van Vleck determinant in Eq. (4.38) then takes the simple form

$$D_\alpha = \frac{\partial^2 S}{\partial x \partial \alpha} = \frac{\partial}{\partial \alpha} \text{sign } \alpha \, \sqrt{2m(|\alpha| - V)} = \frac{m(\text{sign } \alpha)^2}{\sqrt{2m(E_\alpha - V)}} = \frac{m}{|\partial S_\alpha / \partial x|}. \qquad (4.43)$$

The corresponding stationary probability density on configuration space, in the region in which $S_\alpha(x, t)$ is well defined, i.e., for which $V(x) \le E_\alpha$, follows via Eq. (4.39) and (4.43) as

$$P_\alpha(x) \sim \frac{m}{\sqrt{2m(E_\alpha - V)}} = \frac{m}{|\partial S_\alpha/\partial x|} \tag{4.44}$$

up to a normalisation constant, and vanishes outside this region. Thus, recalling that the classical momentum of the particle is $p = \nabla S$, the probability density is inversely proportional to the speed of the particle.

Assuming a convex potential for simplicity, i.e., $V''(x) > 0$, there are just two classical turning points for any α, $x = a_\alpha$ and $x = b_\alpha$. Hence, using $\partial S/\partial x = m dx/dt$, the normalisation constant can be calculated via

$$\int_{a_\alpha}^{b_\alpha} dx \, \frac{m}{|\partial S_\alpha/\partial x|} = \int_{a_\alpha}^{b_\alpha} \frac{dx}{|dx/dt|} = \frac{1}{2} T_\alpha,$$

where T_α is the period associated with the motion. The stationary probability density thus has the explicit form

$$P_\alpha(x) = \frac{2}{T_\alpha} \frac{m}{\sqrt{2m(E_\alpha - V)}}. \tag{4.45}$$

For the case of a harmonic oscillator of frequency ω, the period reduces to a constant, $T_\alpha = 2\pi/\omega$.

Note that the stationary configuration ensemble (P_α, S_α) defined by Eqs. (4.42) and (4.45) is well known in semiclassical quantum mechanics. In particular, extending the domains of these functions to the whole real line by taking the modulus of any square roots, define the corresponding wave function $\psi_\alpha := \sqrt{P_\alpha} \, e^{iS_\alpha/\hbar}$. The WKB approximation to a quantum stationary state with energy E_α then corresponds to a superposition of ψ_α and $\psi_{-\alpha}$ [10].

The phase space probability density corresponding to a one-dimensional classical thermal mixture follows immediately from Eq. (4.40) as

$$\rho_\beta(x, p) \sim e^{-\beta H(x,p)} = e^{-\beta[p^2/(2m)+V(x)]}, \tag{4.46}$$

up to a normalisation constant, in agreement with classical statistical mechanics [2]. It is well known that the statistical entropy of this phase space density depends explicitly on properties of the potential $V(x)$.

In contrast, the thermal weighting distribution follows from Eqs. (4.22) and (4.42) as $w_\alpha = \frac{1}{2}\beta e^{-\beta|\alpha|}$, which has a statistical entropy

$$\mathscr{E}_w = -\int d\alpha \, w_\alpha \, \log w_\alpha = \log \frac{2e}{\beta}. \tag{4.47}$$

This quantity is completely independent of $V(x)$, and hence can have no simple relationship with the thermodynamic entropy. Indeed it is not even invariant under $1 : 1$ relabellings of the parameter α.

4.7 Discussion

We have shown that the concept of thermal mixtures provides a unified approach to thermodynamics that gives the correct results for both classical and quantum systems. However, a number of interesting questions remain to be studied in future work.

For example, is it possible to make a general connection between thermal mixtures, as defined in Eq. (4.22), and the thermodynamic entropy? One possibility might be to consider the submanifold $\mathscr{S} := \{(P_\alpha, S_\alpha)\}$ of the natural phase space of configuration ensembles, corresponding to a maximal set of stationary configuration ensembles, and determine whether the associated entropy

$$\mathscr{E}(\mathscr{S}) := - \int_{\mathscr{S}} d\mathscr{S}\, w[P_\alpha, S_\alpha] \log w[P_\alpha, S_\alpha] \tag{4.48}$$

is proportional to the thermodynamic entropy for the classical and quantum cases. Here $d\mathscr{S}$ is the induced phase space measure on the submanifold (which may also be used to normalise w_α for this calculation). Another possibility, in the classical case, is to relate thermal mixtures to ergodic properties (see Ref. [5] for preliminary work in this regard).

Furthermore, can the properties of thermal mixtures be explicitly evaluated for a nonclassical and nonquantum example? One obvious candidate in this regard is the hybrid quantum-classical oscillator considered in Chap. 8.

Appendix 1: Representation of Phase Space Densities by Mixtures

In Sect. 4.2.2, we showed that any phase space density can be represented by a classical mixture of configuration space ensembles. Here we present another derivation based on the coordinate transformation generated by the Hamilton–Jacobi relation $p = \nabla S(x, \alpha)$, where the vector α has the same dimension as the vector p. For simplicity, we only consider a two-dimensional phase space. The generalization to more dimensions is straightforward.

We write the phase space density in the form

$$\rho(x, p) = \int dx' dp'\, \rho(x', p')\, \delta(x - x')\, \delta(p - p') \tag{4.49}$$

and carry out the change of coordinates

$$p' = \frac{\partial S(x', \alpha)}{\partial x'}.$$
(4.50)

The only restriction on $S(x', \alpha)$ comes from the requirement that the coordinate transformation of Eq. (4.50) be invertible. We have

$$dx'dp' = dx'd\alpha \left| \frac{\partial^2 S}{\partial x' \partial \alpha} \right|,$$
(4.51)

$$\rho(x', p') = \rho(x', \partial S / \partial x'),$$
(4.52)

$$\delta(p - p') = \delta(p - \partial S / \partial x'),$$
(4.53)

which leads to

$$\rho(x, p) = \int dx' d\alpha \left| \frac{\partial^2 S}{\partial x' \partial \alpha} \right| \rho(x', \partial S / \partial x') \, \delta(x - x') \, \delta(p - \partial S / \partial x')$$

$$= \int d\alpha \left| \frac{\partial^2 S}{\partial x \partial \alpha} \right| \rho(x, \partial S / \partial x) \, \delta(p - \partial S / \partial x).$$
(4.54)

We now evaluate

$$\rho(x) = \int dp \, \rho(x, p)$$

$$= \int dp d\alpha \left| \frac{\partial^2 S}{\partial x \partial \alpha} \right| \rho(x, \partial S / \partial x) \, \delta(p - \partial S / \partial x)$$

$$= \int d\alpha \left| \frac{\partial^2 S}{\partial x \partial \alpha} \right| \rho(x, \partial S / \partial x)$$

$$=: \int d\alpha \, w(\alpha) P(x | \alpha),$$
(4.55)

where the last line defines a pair of new probabilities, $w(\alpha)$ and $P(x|\alpha)$. Thus we can set

$$\left| \frac{\partial^2 S}{\partial x \partial \alpha} \right| \rho(x, \partial S / \partial x) = w(\alpha) P(x | \alpha).$$
(4.56)

It is possible to give explicit expression for both $w(\alpha)$ and $P(x|\alpha)$. Integrating Eq. (4.56) with respect to x on both sides leads to

$$w(\alpha) = \int dx \left| \frac{\partial^2 S}{\partial x \partial \alpha} \right| \rho(x, \partial S / \partial x),$$
(4.57)

where we used $\int dx\, P(x|\alpha) = 1$ (see below). Using Eq. (4.56) again we get

$$P(x|\alpha) = \frac{1}{w(\alpha)} \left| \frac{\partial^2 S}{\partial x \partial \alpha} \right| \rho(x, \partial S/\partial x) = \frac{\left| \frac{\partial^2 S}{\partial x \partial \alpha} \right| \rho(x, \partial S/\partial x)}{\int dx \left| \frac{\partial^2 S}{\partial x \partial \alpha} \right| \rho(x, \partial S/\partial x)}. \tag{4.58}$$

Thus $w(\alpha)$ and $P(x|\alpha)$ are *uniquely* determined by $\rho(x, p)$ and a choice of $S(x, \alpha)$.

Both $w(\alpha)$ and $P(x|\alpha)$ are non-negative and properly normalized, as we now show. Since the integrand of Eq. (4.57) is non-negative, it follows that $w(\alpha) \geq 0$, as required. One can also show that $\int d\alpha\, w(\alpha) = 1$, since

$$1 = \int dx dp\, \rho(x, p) = \int dx d\alpha \left| \frac{\partial^2 S}{\partial x \partial \alpha} \right| \rho(x, \partial S/\partial x) = \int d\alpha\, w(\alpha), \tag{4.59}$$

where in the second equality we carried out the transformation of Eq. (4.50) replacing primed coordinates by unprimed coordinates. Finally, inspection of Eq. (4.58) shows that $P(x|\alpha) \geq 0$ and $\int dx\, P(x|\alpha) = 1$.

Using Eq. (4.56), the expression for $\rho(x, p)$, Eq. (4.54), becomes

$$\rho(x, p) = \int d\alpha\, w(\alpha) P(x|\alpha)\, \delta(p - \partial S/\partial x). \tag{4.60}$$

This shows that $\rho(x, p)$ is indeed mapped to a mixture of configuration space ensembles.

Notice that the functions $P(x|\alpha)$ and $S(x, \alpha)$ are only defined at a given instant of time. Given $P(x|\alpha)$ and $S(x, \alpha)$, one can derive the corresponding time dependent expressions $P(x|\alpha, t)$ and $S(x, \alpha, t)$ by solving the equations of motion with $P(x|\alpha)$ and $S(x, \alpha)$ as initial conditions.

While different choices of $S(x, \alpha)$ lead to different functional forms for $w(\alpha)$, $P(x|\alpha)$ and $\delta(p - \partial S/\partial x)$, these mixtures are all physically equivalent because they map to the same $\rho(x, p)$.

As a simple application, consider any $\rho(x, p)$ together with $S(x, \alpha) = \alpha x$, so that $\partial^2 S/\partial x \partial \alpha = 1$. This leads to

$$w(\alpha) = \int dx\, \rho(x, \alpha),$$

$$P(x|\alpha) = \frac{\rho(x, \alpha)}{\int dx\, \rho(x, \alpha)} = \rho(x|\alpha), \tag{4.61}$$

$$\delta(p - \partial S/\partial x) = \delta(p - \alpha).$$

Appendix 2: Solution of the Classical Continuity Equation

As noted in Sect. 4.5, a complete solution S_α of the classical Hamilton–Jacobi equation in Eq. (4.35) also generates a solution P_α for the corresponding continuity equation via the van Vleck determinant, as per Eq. (4.39) [3, 6, 7]. However, this fact is not discussed in most text books on classical mechanics, and so for completeness we provide the corresponding derivation here.

We will essentially follow Schiller [7], but with a little more detail. Thus, consider a general complete solution $S_\alpha(x, t)$, i.e., one which is not necessarily stationary. First, differentiating the Hamilton–Jacobi equation $\partial S_\alpha/\partial t + H(x, \nabla S_\alpha) = 0$ in Eq. (4.35), with respect to x_j and α_k, yields

$$
\begin{aligned}
0 &= \frac{\partial \varphi_{jk}}{\partial t} + \frac{\partial}{\partial x_j}\left[\frac{\partial H}{\partial(\nabla S_\alpha)} \cdot \frac{\partial(\nabla S_\alpha)}{\partial \alpha_k}\right] \\
&= \frac{\partial \varphi_{jk}}{\partial t} + \left[\frac{\partial}{\partial x_j}\frac{\partial H}{\partial(\nabla S_\alpha)}\right] \cdot \frac{\partial(\nabla S_\alpha)}{\partial \alpha_k} + \frac{\partial H}{\partial(\nabla S_\alpha)} \cdot \frac{\partial^2(\nabla S_\alpha)}{\partial x_j \partial \alpha_k} \\
&= \frac{\partial \varphi_{jk}}{\partial t} + \left[\frac{\partial}{\partial x_j}\frac{\partial H}{\partial(\partial S_\alpha/\partial x_l)}\right]\varphi_{lk} + \frac{\partial H}{\partial(\partial S_\alpha/\partial x_l)}\frac{\partial \varphi_{jk}}{\partial x_l},
\end{aligned}
$$

where φ is defined in Eq. (4.38) and repeated indices are summed over. The property $D_\alpha = \det \varphi \neq 0$ in Eq. (4.38) implies that the inverse matrix φ^{-1} exists, and multiplying the above result by φ_{kj}^{-1} with summation over repeated indices then gives

$$
0 = \mathrm{tr}\left[\frac{\varphi^{-1}\,\partial\varphi}{\partial t}\right] + \nabla \cdot \left(\frac{\partial H}{\partial(\nabla S_\alpha)}\right) + \frac{\partial H}{\partial(\partial S_\alpha/\partial x_l)}\,\mathrm{tr}\left[\varphi^{-1}\frac{\partial\varphi}{\partial x_l}\right]. \tag{4.62}
$$

Second, the determinant $D_\alpha = \det \varphi$ satisfies

$$
\det(\varphi + \delta\varphi) = D_\alpha \det(I + \varphi^{-1}\delta\varphi) = D_\alpha \prod_j [1 + (\varphi^{-1}\delta\varphi)_{jj}] = D_\alpha(1 + \mathrm{tr}[\varphi^{-1}\delta\varphi]),
$$

to first order in $\delta\varphi$, and hence one has the variational property

$$
\delta D_\alpha = D_\alpha\,\mathrm{tr}[\varphi^{-1}\delta\varphi]. \tag{4.63}
$$

Multiplication of Eq. (4.62) by D_α then gives

$$
\begin{aligned}
0 &= \frac{\partial D_\alpha}{\partial t} + D\,\nabla \cdot \left(\frac{\partial H}{\partial(\nabla S_\alpha)}\right) + \frac{\partial H}{\partial(\partial S_\alpha/\partial x_l)}\frac{\partial D_\alpha}{\partial x_l} \\
&= \frac{\partial D_\alpha}{\partial t} + \nabla \cdot \left(D_\alpha \frac{\partial H}{\partial(\nabla S_\alpha)}\right).
\end{aligned} \tag{4.64}
$$

Thus, the van Vleck determinant D_α satisfies the classical continuity equation in Eq. (4.35), over the physical region of configuration space for which S_α is well defined.

Finally, recalling that $D_\alpha \neq 0$ as per Eq. (4.38), there can be no flow from a positive to a negative value of D_α at any point, and vice versa. Hence the absolute value $|D_\alpha|$ must also satisfy the continuity equation. The probability density corresponding to S_α is therefore $P_\alpha \sim |D_\alpha|$, as claimed in Eq. (4.39). We emphasise that this result holds for any complete solution of the Hamilton–Jacobi equation, not just the stationary solutions considered in Sect. 4.5.

References

1. d'Espagnat, B.: Conceptual Foundations of Quantum Mechanics, 2nd edn. Benjamin, Reading (1976)
2. Buchdahl, H.A.: Twenty Lectures on Thermodynamics. Pergamon, Oxford (1975)
3. Van Vleck, J.H.: The correspondence principle in the statistical interpretation of quantum mechanics. Proc. Natl. Acad. Sci. U.S. **14**, 178–188 (1928)
4. Hall, M.J.W., Reginatto, M.: Schrödinger equation from an exact uncertainty principle. J. Phys. A **35**, 3289–3303 (2002)
5. Hall, M.J.W.: Consistent classical and quantum mixed dynamics. Phys. Rev. A **78**, 042104 (2008)
6. Pauli, W.: Selected Topics in Field Quantization. Pauli Lecture on Physics, vol. 6. Dover, Minneola (2000)
7. Schiller, R.: Quasi-classical theory of the nonspinning electron. Phys. Rev. **125**, 1100–1108 (1962)
8. Gutzwiller, M.C.: Chaos in Quantum and Classical Mechanics. Springer, New York (1990) (Chap. 12)
9. Goldstein, H.: Classical Mechanics. Addison-Wesley, New York (1950) (Chap. 9)
10. Merzbacher, E.: Quantum Mechanics, 3rd edn. Wiley, New Jersey (1998)

Part II
Axiomatic Approaches to Quantum Mechanics

Chapter 5
Quantization of Classical Ensembles via an Exact Uncertainty Principle

Abstract The usual Heisenberg uncertainty relation may be replaced by an exact equality, valid for all states. This can be shown by carrying out a decomposition of the momentum of a quantum state into classical and nonclassical components and choosing suitable measures of position and momentum uncertainty. The exact uncertainty relation obtained in this way is sufficiently strong to provide the basis for moving from classical mechanics to quantum mechanics. In particular, the assumption of a nonclassical momentum fluctuation, having a strength which scales inversely with uncertainty in position, leads from the classical equations of motion to the Schrödinger equation. The approach based on the exact uncertainty principle is conceptually very simple, being based on the core notion of statistical uncertainty, intrinsic to any interpretation of quantum theory. This quantization procedure is not restricted to particles but can also be used to derive bosonic field equations. It is remarkable that the basic underlying concept, the addition of nonclassical momentum fluctuations to a classical ensemble, carries through from quantum particles to quantum fields, without creating conceptual difficulties, although significant technical generalizations are needed. This logical consistency and range of applicability is a further strength of the exact uncertainty approach.

5.1 Introduction

The dynamics of classical and quantum non-relativistic particles can be described using the common framework of ensembles on configuration space (see Sect. 1.2), and from this point of view the main difference between quantum and classical evolution lies in the choice of ensemble Hamiltonian. For example, given the classical ensemble Hamiltonian $\mathscr{H}_C$ of a particle of mass m, we can write the corresponding quantum ensemble Hamiltonian $\mathscr{H}_Q$ as

$$s\mathscr{H}_Q[P, S] = \mathscr{H}_C[P, S] + \frac{\hbar^2}{4} \int d^3x\, P\, \frac{|\nabla \log P|^2}{2m}. \tag{5.1}$$

© Springer International Publishing Switzerland 2016
M.J.W. Hall and M. Reginatto, *Ensembles on Configuration Space*,
Fundamental Theories of Physics 184, DOI 10.1007/978-3-319-34166-8_5

This simple relationship suggests that we examine the following question: is it possible to understand the addition of the second term on the right hand side of Eq. (5.1) as due to some property which is characteristic of quantum systems?

Notice that the addition of this term amounts to the *quantization* of the classical system. Thus, this term ought to be related to something that is fundamentally quantum. This is indeed the case. The additional term underlies an *exact uncertainty relation* for quantum systems, which is related to but stronger than the more familiar Heisenberg uncertainty relation. Moreover, the form of this additional term can be uniquely derived from an *exact uncertainty principle*.

We first show how an exact uncertainty relation can be formulated for a quantum system and then describe the quantization procedure based on the exact uncertainty principle. In the next section, we consider a quantum system and show that it is possible to decompose the momentum into the sum of classical and nonclassical components and that the nonclassical component satisfies an exact uncertainty relation. In Sect. 5.3 we invert this approach: we start from a classical ensemble, assume that it is subject to random momentum fluctuations which scale inversely with uncertainty in position, and derive the Schrödinger equation from this exact uncertainty principle. This quantization procedure is not restricted to particles but can also be used to derive bosonic field equations, as shown in Sect. 5.4. To illustrate the approach, we apply the procedure to the electromagnetic and gravitational fields, and show in the latter case that all operator-ordering ambiguities are removed by the approach.

We will substantially follow the exposition of Ref. [1] in Sect. 5.3, and the exposition of Ref. [2] in Sect. 5.4.

5.2   An Exact Uncertainty Relation

One of the fundamental distinctions between classical and quantum systems concerns the class of states which are allowed: in the case of classical systems, it is possible to have states in which both the position and momentum uncertainties are arbitrarily small, while in the case of quantum systems there is a limitation imposed by the Heisenberg uncertainty relation. Indeed, the uncertainty principle is generally considered to be a fundamental conceptual tool for understanding differences between classical and quantum mechanics. As first argued by Heisenberg in 1927 [3], the fact that quantum states do not admit simultaneously precise values of conjugate observables, such as position and momentum, does not necessarily imply an incompleteness of the theory, but rather is consistent with not being able to simultaneously determine such observables experimentally to an arbitrary accuracy. It might be asked whether this "measure of freedom" from classical concepts can be formulated more precisely. The answer is, surprisingly, yes—and, as a consequence, the Heisenberg inequality $\Delta x \Delta p \geq \hbar/2$ can be replaced by an exact *equality*, valid for all states [4, 5].

To obtain this equality, consider that the position and momentum observables of a classical system can be measured simultaneously, to an arbitrary accuracy. For a quantum system we therefore define the *classical component of the quantum momentum* to be that observable which is *closest* to the quantum momentum observable, under the constraint of being comeasurable with the position of the system.

More formally, for the case in which the state of a one-dimensional quantum system is described by the wavefunction $\psi(x) = \sqrt{P}e^{iS/\hbar}$, the classical component p_{cl}^ψ of the momentum is *defined* by the properties

$$[\hat{x}, \hat{p}_{cl}^\psi] = 0, \qquad \langle\psi|(\hat{p} - \hat{p}_{cl}^\psi)^2|\psi\rangle = \text{minimum}. \tag{5.2}$$

where $\hat{x}$ and $\hat{p}$ are the standard quantum mechanical position and momentum operators. The first property implies that $\hat{p}_{cl}^\psi$ has the form

$$\hat{p}_{cl}^\psi = \int dx\,|x\rangle\langle x|\,p_{cl}^\psi(x), \tag{5.3}$$

which in combination with the second property leads to the unique solution [5]

$$p_{cl}^\psi(x) = \frac{\hbar}{2i}\left[\frac{1}{\psi}\frac{\partial\psi}{\partial x} - \frac{1}{\psi^*}\frac{\partial\psi^*}{\partial x}\right] = \frac{\partial S}{\partial x}. \tag{5.4}$$

Thus $p_{cl}^\psi(x)$ provides the best possible estimate of momentum for a state $\psi(x)$ consistent with the position measurement result x.

Having a classical momentum component, it is natural to define the corresponding *nonclassical* component of the momentum, $\hat{p}_{nc}^\psi$, via the decomposition

$$\hat{p} = \hat{p}_{cl}^\psi + \hat{p}_{nc}^\psi. \tag{5.5}$$

The average of $\hat{p}_{nc}^\psi$ is zero for the state ψ, and hence $\hat{p}$ may be thought of as comprising a *nonclassical fluctuation* about a classical average. It is the nonclassical component which is responsible for the commutation relation $[\hat{x}, \hat{p}] = i\hbar$. One has the related decomposition [4, 5]

$$(\Delta p)^2 = (\Delta p_{cl}^\psi)^2 + (\Delta p_{nc}^\psi)^2 \tag{5.6}$$

of the momentum variance into classical and nonclassical components, and there is a similar decomposition of the kinetic energy. The magnitude of the nonclassical momentum fluctuation, Δp_{nc}^ψ, provides a natural measure for that "degree of freedom from the limitations of classical concepts" referred to by Heisenberg [3]. Note that this magnitude can be operationally determined from the statistics of x and p, via equations (5.4) and (5.6). It is remarkable that Δp_{nc}^ψ satisfies an *exact* uncertainty relation [4, 5],

$$\delta x\,\Delta p_{nc}^\psi = \hbar/2, \tag{5.7}$$

where δx denotes a classical measure of position uncertainty, called the "Fisher length," defined via

$$\delta x = \left[\int dx\, P \left(\frac{1}{P} \frac{\partial P}{\partial x} \right)^2 \right]^{-1/2}. \tag{5.8}$$

for probability density $P(x)$. There is thus a *precise* connection between the statistics of complementary observables.

The momentum of a quantum state may be decomposed into classical and nonclassical components

- The classical component $p_{cl}^{\psi}(x)$ provides the best estimate of momentum for the state $\psi(x)$ consistent with the position measurement result x.
- The nonclassical component p_{nc}^{ψ} satisfies the exact uncertainty relation $\delta x \Delta p_{nc}^{\psi} \equiv \hbar/2$.

While we have carried out the derivation for the particular case of a one-dimensional non-relativistic quantum particle, exact uncertainty relations may be generalised and/or applied to, for example, density operators, higher dimensions, energy bounds, photon number and phase, entanglement, optimal estimates, weak values and joint measurements [5–7], illustrating their general applicability.

Recalling that $\langle p_{nc}^{\psi} \rangle = 0$, the exact uncertainty relation and (5.8) imply that the intrinsically 'quantum' term in the ensemble Hamiltonian $\mathscr{H}_Q$ in Eq. (5.1) is equal to $(p_{nc}^{\psi})^2/(2m)$, i.e., to a *nonclassical kinetic energy*. This feature of the exact uncertainty relation for quantum systems raises the possibility of whether a suitable 'exact uncertainty principle' can be used as a basis for *deriving* the Schrödinger equation. This is indeed so, as will be shown in the next section.

5.3   Derivation of the Schrödinger Equation

If regarded as merely asserting a physical limit on the degree to which classical concepts can be applied, the Heisenberg uncertainty principle is not sufficiently restrictive in content to supply a means for moving from classical mechanics to quantum mechanics.[1] Thus Landau and Lifschitz write that [8]

> This principle in itself does not suffice as a basis on which to construct a new mechanics of particles.

---

[1] We substantially follow the exposition of Ref. [1] in this section.

In particular, uncertainty relations expressed as imprecise inequalities are not enough to pin down the essence of what is nonclassical about quantum mechanics.

However, an *exact* form of the uncertainty principle is in fact strong enough to allow for a derivation of the equations of motion of a quantum ensemble. More precisely, if we assume that a classical ensemble is subject to random momentum fluctuations, *where the strength of these fluctuations is precisely determined by and scales inversely with uncertainty in position* (as characterised by the position probability density), then the resulting modified equations of motion are equivalent to the Schrödinger equation given in Eq. (5.7) [1]. This will be the main result of this section.

In any axiomatic-type construction of quantum mechanics one must first choose a classical starting point, to be generalised or modified appropriately. The starting point here is a statistical one—the classical motion of an ensemble of particles on configuration space.

5.3.1 Classical Mechanics

We have seen in Sect. 1.2 that the equations of motion for a classical, non-relativistic particle can be derived from the ensemble Hamiltonian

$$\mathcal{H}_C = \int d^n x \, P \left(\frac{1}{2m} |\nabla S|^2 + V \right),$$

(5.9)

with equations of motion

$$\frac{\partial P}{\partial t} = \{P, \mathcal{H}_C\} = \frac{\delta \mathcal{H}_C}{\delta S}, \qquad \frac{\partial S}{\partial t} = \{S, \mathcal{H}_C\} = -\frac{\delta \mathcal{H}_C}{\delta P}.$$

(5.10)

They are the Hamilton–Jacobi equation,

$$\frac{\partial S}{\partial t} + \frac{1}{2m} |\nabla S|^2 + V = 0,$$

(5.11)

and the continuity equation,

$$\frac{\partial P}{\partial t} + \nabla \cdot \left(P \frac{1}{m} \nabla S \right) = 0.$$

(5.12)

Equations (5.11) and (5.12) completely determine the motion of the classical ensemble.

In standard Hamilton–Jacobi theory, there is an additional assumption: that the velocity field $\mathbf{v}(x, t)$ that describes the motion of the particle is related to $S(x, t)$ by $\mathbf{v} = \frac{1}{m} \nabla S$ and therefore that the momentum of the particle is always given by

$$\mathbf{p} = \nabla S. \tag{5.13}$$

This allows for the notion of individual, classical deterministic particle trajectories.

5.3.2 Nonclassical Momentum Fluctuations

We now introduce momentum fluctuations: we consider the possibility that the classical ensemble Hamiltonian $\mathcal{H}_C$ is not quite right because ∇S is actually an average momentum and one has in addition a fluctuation $\mathbf{N}$ about ∇S. Then the physical momentum is

$$\mathbf{p} = \nabla S + \mathbf{N}. \tag{5.14}$$

No particular underlying physical model will be assumed for the momentum fluctuation $\mathbf{N}$. Indeed, one could instead regard the fluctuations as fundamentally non-analyzable, being introduced as a simple device to remove the notion of individual particle trajectories.

The momentum fluctuation $\mathbf{N}$ may conceivably depend on position. We introduce the following notation: the average over such fluctuations for a given quantity A at point x will be denoted by $\overline{A}$, while the average over fluctuations *and* position will be denoted by $< A >$. One hence has the general relation $< A > = \int d^n x \, P \, \overline{A}$. A physically very reasonable *randomness* assumption for the momentum fluctuation $\mathbf{N}$ is that it vanishes on average everywhere, i.e., $\overline{\mathbf{N}} \equiv 0$. However, here only two weaker assumptions will be made:

$$< \mathbf{N} > = 0, \qquad < \nabla S \cdot \mathbf{N} > = 0. \tag{5.15}$$

The first of these states that the fluctuations are unbiased, and the second that the fluctuations are linearly uncorrelated with the average momentum ∇S.

It follows that when the momentum fluctuations are significant, they may be taken into account by replacing the kinetic energy term $\frac{1}{2m}|\nabla S|^2$ in the ensemble Hamiltonian by $\frac{1}{2m}|\nabla S + \mathbf{N}|^2$ and averaging over the fluctuations, yielding the modified ensemble Hamiltonian

$$\begin{aligned}
\mathcal{H} &= \int d^n x \, P \left(\frac{1}{2m} \overline{|\nabla S + \mathbf{N}|^2} + V \right) \\
&= \int d^n x \, P \left(\frac{1}{2m} |\nabla S|^2 + V \right) + \frac{1}{2m} \int d^n x \, P \overline{|\mathbf{N}|^2} \\
&\equiv \mathcal{H}_C + \frac{1}{2m} < (\Delta N)^2 >
\end{aligned} \tag{5.16}$$

where ΔN is the average rms momentum fluctuation, given by $< \mathbf{N} \cdot \mathbf{N} >^{1/2}$. Thus the consequence of taking into consideration the momentum fluctuations is to add a positive term to the ensemble Hamiltonian, arising from the additional kinetic energy due to the fluctuations.

5.3.3 Exact Uncertainty Principle

How can we estimate the magnitude of this additional term, if we don't know anything else about the system except the probability density P and the average momentum ∇S? To estimate the magnitude of the momentum spread, we will assume that an *exact uncertainty principle* holds, in the sense that the strength of the momentum fluctuations at a given time are inversely correlated with uncertainty in position, where the uncertainty of position is characterized by P. Clearly, this assumption is an additional hypothesis that is *independent* of classical mechanics.

To make this assumption precise, consider the general case of an n-dimensional space and a one-parameter family of probability distributions (which we label with a parameter $k > 0$), related by a rescaling of variables

$$P(x) \rightarrow P_k(x) \equiv k^n P(kx). \tag{5.17}$$

These transformations preserve the normalization,

$$\int d^n x \, P(x) \rightarrow \int d^n x \, k^n P(kx) = \int d^n y \, P(y) \tag{5.18}$$

where we have introduced the change of variables $y = kx$. We also have

$$|\nabla P(x)|^2 \rightarrow k^{2n+2} |\nabla_y P(y)|^2, \quad \mathbf{x} \cdot \nabla P(x) \rightarrow k^n \mathbf{y} \cdot \nabla_y P(y). \tag{5.19}$$

Under such a transformation, any direct measurement of position uncertainty δx such as the rms uncertainty Δx changes according to the rule

$$\delta x \rightarrow \delta x_k \equiv \frac{1}{k} (\delta x) . \tag{5.20}$$

Thus probability densities with different values of k represent physical systems that *only differ in how well we know the location of the particle*, since the shape of the probability densities are the same except for the rescaling. The exact uncertainty principle that we want to make use of corresponds roughly to the assumption that such a scaling of position by a factor $1/k$ scales the momentum fluctuation by a factor k. More precisely:

Exact uncertainty principle: The nonclassical momentum fluctuation ΔN is determined by the uncertainty in position, where the latter is characterized by the probability density P, such that

$$\Delta N \to k \Delta N \tag{5.21}$$

under k transformations.

It follows that the uncertainty product $\delta x \, \Delta N$ is thus preserved under k transformations, for *any* direct measure of position uncertainty δx.

To apply the exact uncertainty principle, note first that for the Hamiltonian formalism to be applicable to $\mathcal{H}$, the additional term in Eq. (5.16) must be an integral over a density that is a function of $\mathbf{x}$, P and S and their derivatives. Moreover, since $(\Delta N)^2$ is determined solely by position fluctuations (where the latter are characterized by P), then this additional term is in fact independent of S. Finally, to get the simplest possible modification of the classical ensemble Hamiltonian which is consistent with the exact uncertainty principle, we will search for a modified ensemble Hamiltonian that does not contain second and higher order derivatives of P. Hence, the additional term in the Hamiltonian of Eq. (5.16) can be written in the form

$$\int d^n x \, (\Delta N)^2 = \int d^n x \, P f(\mathbf{x}, P, \mathbf{x} \cdot \nabla P, |\nabla P|^2). \tag{5.22}$$

The exact uncertainty principle requires f to transform under k transformation as follows,

$$\int d^n x \, P(x) f \left[\mathbf{x}, P(x), \mathbf{x} \cdot \nabla P(x), |\nabla P(x)|^2) \right] \tag{5.23}$$

$$\to \int d^n y \, P(y) f(k^{-1}\mathbf{y}, k^n P(y), k^n \mathbf{y} \cdot \nabla_y P(y), k^{2n+2}|\nabla_y P(y)|^2) \tag{5.24}$$

$$\equiv k^2 \int d^n y \, P(y) f(\mathbf{y}, P(y), \mathbf{y} \cdot \nabla_y P(y), |\nabla_y P(y)|^2). \tag{5.25}$$

This leads to the homogeneity condition

$$f(k^{-1}\mathbf{x}, k^n u, k^n v, k^{2n+2}w) = k^2 f(\mathbf{x}, u, v, w) \tag{5.26}$$

where we have introduced the more compact notation

$$u = P, \quad v = \mathbf{x} \cdot \nabla P, \quad w = |\nabla P|^2. \tag{5.27}$$

From this requirement we derive the first order partial differential equation

$$-\sum_{i=1}^{n} x_i \frac{\partial f}{\partial x_i} + nu\frac{\partial f}{\partial u} + nv\frac{\partial f}{\partial v} + (2n+2)w\frac{\partial f}{\partial w} = 2f. \tag{5.28}$$

The problem of finding the general integral of such an equation is equivalent to the problem of finding the general integral of a system of ordinary differential equations [9], in our case given by

$$-\frac{dx_1}{x_1} = \cdots = -\frac{dx_n}{x_n} = \frac{du}{nu} = \frac{dv}{nv} = \frac{dw}{(2n+2)w} = \frac{df}{2f}. \tag{5.29}$$

This system of ordinary differential equations has $(n + 3)$ independent integrals, which can be chosen as

$$u^{-1}w^{1/2}x_i = \text{constant}, \tag{5.30}$$
$$u^{-1}v = \text{constant}, \tag{5.31}$$
$$u^{2/n}|\mathbf{x}|^2 = \text{constant}, \tag{5.32}$$
$$u^2w^{-1}f = \text{constant}, \tag{5.33}$$

and this implies that the general solution of Eq. (5.28) is of the form

$$f = \left(u^{-2}w\right) g \left(u^{-1}w^{1/2}\mathbf{x}, u^{-1}v, u^{2/n}|\mathbf{x}|^2\right), \tag{5.34}$$

where g is an arbitrary function.

5.3.4 Independent Subsystems

To determine f completely, we need one further condition. We will require subsystem independence (see Sect. 3.2), which is equivalent to the condition that the extra term in the ensemble Hamiltonian $\mathscr{H}$ decomposes into additive subsystem contributions whenever the system is composed of independent subsystems. It is also equivalent to the condition that the momentum fluctuations N_1 and N_2 are linearly uncorrelated for two such subsystems, and hence can equivalently be interpreted as a further randomness assumption for the momentum fluctuations.

To investigate the requirements of subsystem independence, it will be sufficient to consider the case where we have a system consisting of two uncorrelated particles of mass m that do not interact, one particle described by a set of coordinates $\mathbf{x}_1$ and the other by $\mathbf{x}_2$. Then, P is given by Eq. (3.1),

$$P(\mathbf{x}_1, \mathbf{x}_2) = P_1(\mathbf{x}_1)P_2(\mathbf{x}_2), \tag{5.35}$$

which leads immediately to

$$u = u_1 u_2, \tag{5.36}$$
$$v' \equiv u^{-1} v = u_1^{-1} v_1 + u_2^{-1} v_2 = v_1' + v_2', \tag{5.37}$$
$$w' \equiv u^{-2} w = u_1^{-2} w_1 + u_2^{-2} w_2 = w_1' + w_2', \tag{5.38}$$

where the subscripts 1 and 2 refer to quantities corresponding to the subsystems 1 and 2, respectively.

From Eqs. (5.16) and (5.22), subsystem independence requires

$$Pf = P_1 P_2 (f_1 + f_2). \tag{5.39}$$

Using Eq. (5.34), we find that

$$f = \left(w_1' + w_2' \right) g \left(\sqrt{w_1' + w_2'} \, \mathbf{x}, \ v_1' + v_2', \ (u_1 u_2)^{2/n} \, |\mathbf{x}|^2 \right), \tag{5.40}$$

where $\mathbf{x} = (\mathbf{x}_1, \mathbf{x}_2)$. From Eq. (5.39), this form of f must decompose into the sum of a function of u_1, v_1', w_1' and $\mathbf{x}_1$, and a function of u_2, v_2', w_2' and $\mathbf{x}_2$. Since the factor that multiplies g and the second and third arguments of g are such functions (with respect to w', v', and $\mathbf{x}$ respectively), these terms cannot be mixed by the functional form of g. Taking into consideration that the first argument of g is given by $\sqrt{w_1' + w_2'} \, \mathbf{x}$, we find that g must be of the form

$$g (\mathbf{a}, b, c) = C + g_0(\mathbf{a}) + b g_1(\mathbf{a}) + c g_2(\mathbf{a}), \tag{5.41}$$

where C is a constant, and the functions g_j must satisfy the condition

$$g_j(\lambda \mathbf{a}) = \lambda^{-2} g_j(\mathbf{a}), \ j = 0, 1, 2, \tag{5.42}$$

to allow cancellation of the factor $w_1' + w_2'$ that multiplies g.

Hence f has the general form

$$f = C \left(w_1' + w_2' \right) + g_0(\mathbf{x}_1, \mathbf{x}_2) + \left(v_1' + v_2' \right) g_1(\mathbf{x}_1, \mathbf{x}_2)$$
$$+ (u_1 u_2)^{2/n} \left(|\mathbf{x}_1|^2 + |\mathbf{x}_2|^2 \right) g_2(\mathbf{x}_1, \mathbf{x}_2). \tag{5.43}$$

The independence condition Eq. (5.39) places strong requirements on the g_j. First, g_0 is required to be a sum of a function of $\mathbf{x}_1$ and a function of $\mathbf{x}_2$. Hence it only represents a classical additive potential term (satisfying the homogeneity condition of Eq. (5.42) above), and will be ignored as having no nonclassical role (it can be absorbed into the classical potential V in the Lagrangian). Second, to avoid subsystem cross terms, g_1 must be constant. But then the homogeneity condition of Eq. (5.42) can only be satisfied by the choice $g_1 = 0$. Third, cross terms in u_1 and u_2 can only be avoided by choosing $g_2 = 0$.

Thus the form of f reduces to $C\left(w_1' + w_2'\right)$, which from Eq. (5.39) is to be identified with the sum of f_1 and f_2, thus yielding the general form

$$f = Cw' = \frac{C}{P^2}|\nabla P|^2 \tag{5.44}$$

where C is a universal constant.

5.3.5 Equations of Motion

The modified ensemble Hamiltonian follows from Eqs. (5.16), (5.22) and (5.44) as

$$\mathscr{H} = \int dx\, P\left(\frac{1}{2m}|\nabla S|^2 + \frac{C}{2m}\frac{1}{P^2}|\nabla P|^2 + V\right). \tag{5.45}$$

Variation with respect to S leads again to Eq. (5.12), while variation with respect to P leads to

$$\frac{\partial S}{\partial t} + \frac{1}{2m}|\nabla S|^2 + \frac{C}{2m}\left[\frac{1}{P^2}|\nabla P|^2 - \frac{2}{P}\nabla^2 P\right] + V = 0. \tag{5.46}$$

Equations (5.12) and (5.46) are identical to the Schrödinger equation,

$$i\hbar\frac{\partial \psi}{\partial t} = -\frac{\hbar^2}{2m}\nabla^2\psi + V\psi, \tag{5.47}$$

provided we introduce the wave function $\psi = \sqrt{P}e^{iS/\hbar}$ and the constant C is set equal to $C = (\hbar/2)^2$.

Schrödinger equation from an exact uncertainty principle

- The exact uncertainty relation is sufficiently strong to provide the basis for moving from classical mechanics to quantum mechanics.
- Quantization is achieved via nonclassical momentum fluctuations having a strength which scales inversely with uncertainty in position.

We have derived the Schrödinger equation based on three main assumptions: an exact uncertainty principle, additivity, and an action principle with derivatives of up to first order in P. One can show [10] that discarding the first of these assumptions allows for an additional additive term proportional to the configuration space entropy

$$R[P] := - \int dx \, P \log P. \qquad (5.48)$$

Adding this entropic term to the quantum ensemble Hamiltonian leads to the logarithmic Schrödinger equation, a nonlinear modification of quantum mechanics first studied by Bialynicki-Birula and Mycielski [11].

It is worth remarking that the configuration space entropy satisfies all the requirements for observables except for homogeneity (see Sect. 2.2). However, the latter can be regained by replacing the entropy functional by a relative entropy, i.e., by replacing $\log P$ in Eq. (5.48) by $\log (P/Q)$, for some fixed reference distribution Q. Further, if the set of classical ensemble observables, defined in Sect. 2.3.2, is supplemented by this entropy, then one still has closure under the Poisson bracket.

It is of interest to note that a formally related 'hydrodynamic' approach to quantization has since been given by Ván and Tülöp, in which the role of the exact uncertainty principle is replaced by a 'mass invariance' principle [12]. More recently, Rudnicki has considered modifying the exact uncertainty principle to yield corresponding modified exact uncertainty relations compatible with generalised Heisenberg uncertainty relations proposed in the literature, and leading to nonlinear Schrödinger equations [13].

5.3.6 Further Remarks on the Exact Uncertainty Relation

The Schrödinger equation has been derived above using an exact uncertainty principle to fix the strength of random momentum fluctuation in terms of the uncertainty in position. Note that no specific measure of position uncertainty was assumed; it was required only that the momentum fluctuations scale inversely with position uncertainty under k transformations. However, having obtained a unique form, Eq. (5.44), for the function f in (5.22) we return now to the exact uncertainty *relation* relating position and momentum uncertainties that we introduced in Sect. 5.3.

For simplicity, we only consider the case of one dimension. Recall the definition of the "Fisher length" in Eq. (5.8):

$$\delta x = \left[\int dx \, P \left(\frac{1}{P} \frac{\partial P}{\partial x} \right)^2 \right]^{-1/2}. \qquad (5.49)$$

For the case of a Gaussian probability density with standard deviation σ one has $\delta x = \sigma$. More generally, this measure has units of position, scales appropriately with x ($\delta y = \lambda \delta x$ for $y = \lambda x$), and vanishes in the limit that P approaches a delta function. Hence it represents a direct measure of uncertainty for position. We refer to δx as the "Fisher length" of the probability density P due to its connection with the "Fisher information" of statistical estimation theory [14].

From Eqs. (5.22) and (5.44) it follows that

$$\delta x \, \Delta N = \sqrt{C} = \frac{\hbar}{2}. \tag{5.50}$$

Thus, the exact uncertainty principle leads to an exact uncertainty relation between position and momentum, of the same form as the quantum exact uncertainty relation in Eq. (5.7). In particular, the nonclassical momentum fluctuation ΔN may be identified with the rms deviation of the nonclassical momentum operator in Eq. (5.5).

The usual Heisenberg uncertainty relation can be derived from this exact uncertainty relation. From the Cramer-Rao inequality of statistical estimation theory [15] one has $\Delta x \geq \delta x$, while the randomness assumptions in Eq. (5.15) imply

$$(\Delta p)^2 = \text{Var}(\partial S/\partial x + N) = \text{Var}(\partial S/\partial x) + (\Delta N)^2 \geq (\Delta N)^2, \tag{5.51}$$

and hence it follows immediately from Eq. (5.50) that $\Delta x \, \Delta p \geq \hbar/2$.

5.4 Derivation of Bosonic Field Equations

The exact uncertainty principle can be successfully generalized to derive the equations of motion for bosonic fields with Hamiltonians quadratic in the field momenta, including scalar, electromagnetic, and gravitational fields.[2] The field quantization procedure is extremely *minimalist* in nature: unlike canonical quantization, it does not use nor make any assumptions about the existence of operators, Hilbert spaces, complex amplitudes, inner products, linearity, superposition, or the like.

It is remarkable that the basic underlying concept, the addition of "nonclassical" momentum fluctuations to a classical ensemble, carries through from quantum particles to quantum fields, without creating conceptual difficulties (although significant technical generalizations are needed). The approach based on the exact uncertainty principle is thus conceptually very simple, being based on the core notion of statistical uncertainty, intrinsic to any interpretation of quantum theory. This logical consistency and range of applicability is a further strength of the quantization procedure.

As a bonus, the exact uncertainty approach further implies a unique operator ordering for the functional Schrödinger equation associated with the quantum ensemble, something which the canonical quantization procedure is unable to do.

---

[2]This section, and associated Appendices 1 and 2 of this chapter, substantially follow the exposition of Ref. [2].

5.4.1   Classical Ensembles of Fields

We consider a real multicomponent classical field $\phi \equiv (\phi^a)$ with conjugate momentum density $\pi \equiv (\pi^a)$ and Hamiltonian functional $\tilde{H}_C[\phi, \pi, t]$. For example, ϕ may denote the electromagnetic field $A \equiv (A^\mu)$, or some collection of interacting fields labeled by the index a. Spatial coordinates will be denoted by x (irrespective of dimension), and the values of field components ϕ^a and π^b at position x will be denoted by ϕ_x^a and π_x^b respectively.

We restrict to fields for which the associated Hamiltonian functional is *quadratic* in the momentum field density, i.e., of the form

$$\tilde{H}_C[\phi, \pi, t] = \sum_{a,b} \int dx \, \{K_x^{ab}[\phi]\pi_x^a \pi_x^b + V[\phi]\}. \tag{5.52}$$

Here $K_x^{ab}[\phi] = K_x^{ba}[\phi]$ is a kinetic factor coupling components of the momentum density, and $V[\phi]$ is some potential energy functional. Note that cross terms of the form $\pi_x^a \pi_{x'}^b$ with $x \neq x'$ are not permitted in local field theories, and hence are not considered here.

The most direct way of introducing ensembles on configuration space for the fields ϕ is via the Hamilton–Jacobi formulation of the field theory. Given a Hamiltonian functional, it is straightforward to write down the equation of motion for an individual classical field as a Hamilton–Jacobi functional equation,

$$\frac{\partial S}{\partial t} + \tilde{H}_C[f, \delta S/\delta \phi, t] = 0, \tag{5.53}$$

where $S[\phi]$ denotes the Hamilton–Jacobi functional, and $\delta/\delta\phi$ denotes the functional derivative with respect to ϕ. In Appendix 1 of this chapter we discuss in detail how to get from the Hamiltonian formalism for fields to the Hamilton–Jacobi functional equation, and we provide the mathematical tools that are necessary for the formulation of the Hamilton–Jacobi theory for fields. Further information on functional derivatives and functional integrals is provided in Appendix A of this book. Readers who are not familiar with this material are encouraged to consult both Appendices.

The description of an ensemble of such fields requires some additional mathematical structure: a probability density functional $P[\phi]$. The equation of motion for $P[\phi]$ corresponds to the conservation of probability, i.e., to the continuity equation

$$\frac{\partial P}{\partial t} + \sum_a \int dx \, \frac{\delta}{\delta \phi_x^a} \left(P \left. \frac{\delta \tilde{H}_C}{\delta \pi_x^a} \right|_{\pi = \delta S/\delta \phi} \right) = 0. \tag{5.54}$$

This equation is derived in Appendix 1 of this chapter.

Equations (5.53) and (5.54) describe the motion of the ensemble completely, in terms of the two functionals P and S. These equations of motion can be put in the Hamiltonian form

$$\frac{\partial P}{\partial t} = \frac{\Delta \mathcal{H}_C}{\Delta S}, \qquad \frac{\partial S}{\partial t} = -\frac{\Delta \mathcal{H}_C}{\Delta P}, \qquad (5.55)$$

where $\mathcal{H}_C$ denotes the ensemble Hamiltonian given by the functional integral

$$\mathcal{H}_C[P, S, t] := \langle \tilde{H}_C \rangle = \int D\phi \, P \tilde{H}_C[\phi, \delta S/\delta\phi, t], \qquad (5.56)$$

and P and S are regarded as canonically conjugate functionals. Variational derivatives of functional integrals, such as $\Delta \mathcal{H}_C/\Delta S$, are discussed in Appendix A of this book, including for the example of a classical scalar field. Note that Eq. (5.56) implies that $\mathcal{H}_C$ typically corresponds to the average energy of the ensemble.

5.4.2 Momentum Fluctuations and Quantum Ensembles

Our approach to modifying the classical ensemble Hamiltonian, $\mathcal{H}_C[P, S, t]$ of Eq. (5.56), to derive equations of motion for a quantum ensemble of fields is again based on a single ingredient: the addition of nonclassical fluctuations to the momentum density, with the magnitude of the fluctuations determined by the uncertainty in the field. This exact uncertainty approach leads to equations of motion equivalent to those of a bosonic field, with the interpretational advantage of an intuitive statistical picture for quantum field ensembles, and the technical advantage of a unique operator ordering for the associated functional Schrödinger equation.

The assumption that we make is similar to the one that we made in the case of particles. Suppose then that $\delta S/\delta\phi$ is an *average* momentum density associated with the field ϕ, in the sense that the true momentum density is given by

$$\pi = \delta S/\delta\phi + N, \qquad (5.57)$$

where N is a fluctuation field that vanishes on the average for any given field ϕ. No specific underlying model for N is assumed or necessary: in the approach to be followed, one may in fact interpret the "source" of the fluctuations as the field uncertainty itself. Thus the main effect of the fluctuation field is to remove any deterministic connection between ϕ and π.

Similar to the case of particles, we consider the possibility that the momentum fluctuations may depend on the field ϕ. We use the same notation as before: the average over such fluctuations for a given quantity $A[\phi, N]$ will be denoted by $\overline{A}[\phi]$, and the average over fluctuations *and* the field by $\langle A \rangle$. Thus $\overline{N} \equiv 0$ by assumption, and in general $\langle A \rangle = \int D\phi \, P[\phi] \, \overline{A}[\phi]$. Assuming a quadratic dependence on momentum density as per Eq. (5.52), it follows that when the fluctuations are significant the classical ensemble Hamiltonian $\mathscr{H}_C = \langle \tilde{H}_C[\phi, \delta S/\delta \phi, t] \rangle$ in Eq. (5.56) should be replaced by

$$
\begin{aligned}
\mathscr{H} &= \langle \tilde{H}_C[\phi, \delta S/\delta \phi + N, t] \rangle \\
&= \sum_{a,b} \int D\phi \int dx\, P\, K_x^{ab} \overline{(\delta S/\delta \phi_x^a + N_x^a)(\delta S/\delta \phi_x^b + N_x^b)} + \langle V \rangle \\
&= \mathscr{H}_C + \sum_{a,b} \int D\phi \int dx\, P\, K_x^{ab}\, \overline{N_x^a N_x^b}.
\end{aligned}
\tag{5.58}
$$

Thus the momentum fluctuations lead to an additional nonclassical term in the ensemble Hamiltonian, specified by the *covariance matrix* $\mathrm{Cov}_x(N)$ of the fluctuations at position x, where

$$
[\mathrm{Cov}_x(N)]^{ab} := \overline{N_x^a N_x^b}.
\tag{5.59}
$$

To get the simplest possible modification of the classical ensemble Hamiltonian which is consistent with the exact uncertainty principle, we will search for a modified ensemble Hamiltonian that does not contain second and higher order functional derivatives of P and S. Then,

$$
\mathrm{Cov}_x(N) = \alpha(P, \delta P/\delta \phi_x, S, \delta S/\delta \phi_x, \phi_x, t)
\tag{5.60}
$$

for some symmetric matrix function α. Note that in principle one could also allow the covariance matrix to depend on auxiliary fields and functionals; however, the third assumption below immediately removes such a possibility. Given the functional form of Eq. (5.60), the covariance matrix is uniquely determined, up to a multiplicative constant, by three assumptions:

(1) Independence: Consider the case in which the ensemble comprises two independent non-interacting subensembles 1 and 2, with a factorisable probability density functional $P[\phi^{(1)}, \phi^{(2)}] = P_1[\phi^{(1)}]P_2[\phi^{(2)}]$. Then any dependence of the corresponding subensemble fluctuations $N^{(1)}$ and $N^{(2)}$ on P only enters via the corresponding probability densities P_1 and P_2 respectively. Thus

$$
\mathrm{Cov}_x(N^{(1)})\big|_{P_1 P_2} = \mathrm{Cov}_x(N^{(1)})\big|_{P_1}, \quad \mathrm{Cov}_x(N^{(2)})\big|_{P_1 P_2} = \mathrm{Cov}_x(N^{(2)})\big|_{P_2}
\tag{5.61}
$$

for such an ensemble. Note that this assumption implies that the ensemble Hamiltonian $\mathscr{H}$ in Eq. (5.58) is *additive* for independent non-interacting ensembles.

(2) Invariance: The covariance matrix transforms correctly under linear canonical transformations of the field components. Thus, noting that $\phi \to \Lambda^{-1}\phi$, $\pi \to \Lambda^T \pi$ is a canonical transformation for any invertible matrix Λ with transpose Λ^T, which preserves the quadratic form of $\tilde{H}_C$ in Eq. (5.52) and leaves S invariant (since $\delta/\delta\phi \to \Lambda^T \delta/\delta\phi$), one has from Eq. (5.57) that $N \to \Lambda^T N$, and hence that

$$
\mathrm{Cov}_x(N) \to \Lambda^T \mathrm{Cov}_x(N)\Lambda \quad \text{for} \quad \phi \to \Lambda^{-1}\phi.
\tag{5.62}
$$

Note that for single-component fields this reduces to a scaling relation for the variance of the fluctuations at each point x.

(3) Exact uncertainty principle: The uncertainty of the momentum density fluctuations at any given position and time, as characterised by the covariance matrix of the fluctuations, is specified by the field uncertainty at that position and time. Thus, since the field uncertainty is completely determined by the probability density functional P, it follows that $\text{Cov}_x(N)$ cannot depend on S, nor explicitly on t.

The first two assumptions, independence and invariance, are natural on physical grounds, and hence relatively unconstraining (note that invariance replaces the role of k-transformations for particles in Sect. 5.3.3). In contrast, the third assumption is of a special character: it postulates an exact connection between the nonclassical momentum uncertainty and the field uncertainty. Remarkably, these assumptions lead directly to the equations of motion of a bosonic quantum field, as shown by the following Theorem and Corollary.

Theorem 1 *The above assumptions of causality, independence, invariance, and exact uncertainty imply that*

$$\overline{N_x^a N_x^b} = \frac{C}{P^2} \frac{\delta P}{\delta \phi_x^a} \frac{\delta P}{\delta \phi_x^b}, \tag{5.63}$$

where C is a positive universal constant.

The theorem thus yields a unique form for the additional term in Eq. (5.58), up to a multiplicative constant C. The classical equations of motion for the ensemble are recovered in the limit of small fluctuations, i.e., in the limit $C \to 0$. Note that one cannot make the identification $N_x^a \sim (\delta P/\delta\phi_x^a)/P$ from Eq. (5.63), as this is inconsistent with the fundamental property $\overline{N_x^a} = 0$. The proof of the theorem is given in Appendix 2 of this chapter, and is substantially different from (and stronger than) the proofs for the analogous theorem for quantum particles [1, 10] (see also Sect. 5.4), which rely heavily on a "scalar" form which does not carry over in a natural manner to general fields.

The main result of this section is the following Corollary:

Corollary 1 *The equations of motion corresponding to the ensemble Hamiltonian $\mathscr{H}$ can be expressed as the single functional Schrödinger equation*

$$i\hbar\frac{\partial \Psi}{\partial t} = \tilde{H}_C[\phi, -i\hbar\delta/\delta\phi, t]\Psi = -\hbar^2 \left(\sum_{a,b} \int dx \frac{\delta}{\delta\phi_x^a} K_x^{ab} \frac{\delta}{\delta\phi_x^b}\right)\Psi + V\Psi, \tag{5.64}$$

where

$$\hbar := 2\sqrt{C}, \qquad \Psi := \sqrt{P}e^{iS/\hbar}. \tag{5.65}$$

The proof of the Corollary is given in Appendix 2 of this chapter. Equation (5.64) may be recognised as the functional Schrödinger equation for a quantum bosonic field [16, 17], and hence the goal of deriving this equation, via an exact uncertainty principle for nonclassical momentum fluctuations acting on a classical ensemble, has been achieved. Note that the exact uncertainty approach specifies a *unique* operator ordering, $(\delta/\delta\phi_x^a) K_x^{ab} (\delta/\delta\phi_x^b)$, for the functional derivative operators in Eq. (5.64). Thus there is no ambiguity in the ordering for cases where K_x^{ab} depends on the field ϕ, in contrast to traditional approaches (e.g., the Wheeler-DeWitt equation, discussed in Sect. 5.4.4 below). The above results generalise straightforwardly to complex classical fields.

The ensemble of fields corresponding to an ensemble Hamiltonian $\mathscr{H}$ will be called the *quantum ensemble* corresponding to $\mathscr{H}_C$. Note from Eqs. (5.63) and (5.65) that the role of Planck's constant is to fix the relative scale of the nonclassical fluctuations. It is remarkable that the assumptions of independence, invariance and exact uncertainty lead to a *linear operator* equation.

In certain classical field theories, in addition to the equations of motion for the classical ensemble there are constraint equations for P and/or S. For example, each member of an ensemble of electromagnetic fields may have the Lorentz gauge imposed (e.g., see Sect. 5.4.3.1 below). As a guiding principle, we will require that the corresponding quantum ensemble be subject to the same constraint equations for P and/or S. This will ensure a meaningful classical-quantum correspondence for the results of field measurements. However, consistency of the quantum equations of motion with a given set of constraints is not guaranteed by the above Theorem and Corollary, and so must be checked independently for each case.

Bosonic field equations from an exact uncertainty principle

- The basic underlying concept, the addition of "nonclassical" momentum fluctuations to a classical ensemble, carries through from quantum particles to quantum fields, without creating conceptual difficulties.
- The exact uncertainty approach implies a unique operator ordering for the functional Schrödinger equation associated with the quantum ensemble.

5.4.3   Example: Electromagnetic Field

Our first example concerns the electromagnetic field. We consider formulations corresponding to the Lorentz and radiation gauges.

5.4.3.1   Lorentz Gauge

The electromagnetic field is described, up to gauge invariance, by a 4-component field A^μ. In the Lorentz gauge all physical fields satisfy $\partial_\mu A^\mu \equiv 0$, and the classical equations of motion in vacuum are given by $\partial^\nu \partial_\nu A^\mu = 0$. These follow, for example, from the Hamiltonian [17]

$$\tilde{H}_{GB}[A, \pi] = (1/2) \int dx \, \eta_{\mu\nu} \, (\pi^\mu \pi^\nu - \nabla A^\mu \cdot \nabla A^\nu), \qquad (5.66)$$

where $\eta_{\mu\nu}$ denotes the metric in Minkowski space, π^μ denotes the conjugate momentum density, and ∇ denotes the spatial derivative. Here $\tilde{H}_{GB}$ corresponds to the gauge-breaking Lagrangian $\tilde{L} = -(1/2) \int dx \, A^{\mu,\nu} A_{\mu,\nu}$, and is seen to have the quadratic form of Eq. (5.52) with $K_x^{\mu\nu} \equiv \eta_{\mu\nu}/2$.

The exact uncertainty approach implies, via the Corollary of the previous section, that the evolution of a *quantum* ensemble of electromagnetic fields is described by the functional Schrödinger equation

$$i\hbar \frac{\partial \Psi}{\partial t} = \tilde{H}_{GB}[A, -i\hbar(\delta/\delta A^\mu)]\Psi, \qquad (5.67)$$

in agreement with the Gupta-Bleuler formalism [17].

Further, note that the probability of a member of the classical ensemble not satisfying the Lorentz gauge condition $\partial_\mu A^\mu \equiv 0$ is zero by assumption, i.e., the Lorentz gauge is equivalent to the condition that the product $(\partial_\mu A^\mu) P[A^\mu]$ vanishes for all physical fields. For the quantum ensemble to satisfy this condition, as per the guiding principle discussed at the end of Sect. 5.4.2 above, one equivalently requires, noting Eq. (5.65), that

$$(\partial_\mu A^\mu)\Psi[A^\mu] = 0. \qquad (5.68)$$

As is well known, this constraint, if initially satisfied, is satisfied for all times [18] (as is the weaker constraint that only the 4-divergence of the positive frequency part of the field vanishes [17]). Hence the evolution of the quantum ensemble is consistent with the Lorentz gauge.

5.4.3.2 Radiation Gauge

It is well known that one can also obtain the classical equations of motion for the electromagnetic field via an alternative Hamiltonian, obtained by exploiting the degree of freedom left by the Lorentz gauge to remove a dynamical coordinate (corresponding to the longitudinal polarisation). In particular, since $\partial_\mu A^\mu$ is invariant under $A^\mu \rightarrow A^\mu + \partial^\mu \chi$ for any function χ satisfying $\partial^\nu \partial_\nu \chi = 0$, one may completely fix the gauge in a given Lorentz frame by choosing χ such that $A^0 = 0$. One thus obtains, writing $A^\mu \equiv (A^0, \mathbf{A})$, the radiation gauge $A^0 = 0$, $\nabla \cdot \mathbf{A} = 0$.

The classical equations of motion, $\partial^\nu \partial_\nu \mathbf{A} = 0$, follow from the Hamiltonian

$$\tilde{H}_R[\mathbf{A}, \pi] = \frac{1}{2} \int dx \left(\frac{\pi \cdot \pi}{\varepsilon_0} + \frac{|\nabla \times \mathbf{A}|^2}{\mu_0} \right), \qquad (5.69)$$

where π denotes the conjugate momentum density. Here $\tilde{H}_R$ corresponds to the standard Lagrangian $\tilde{L} = -(1/4\mu_0) \int dx\, F^{\mu\nu} F_{\mu\nu}$, with $A^0 \equiv 0$.

This Hamiltonian has the quadratic form of Eq. (5.52), and hence the exact uncertainty approach yields the corresponding functional Schrödinger equation

$$i\hbar \frac{\partial \Psi}{\partial t} = \tilde{H}_R[\mathbf{A}, -i\hbar(\delta/\delta\mathbf{A})]\Psi \qquad (5.70)$$

for a quantum ensemble of electromagnetic fields in the radiation gauge (this is also the form of the functional Schrödinger equation obtained via the Schwinger–Tomonaga formalism [19]).

For the electric field we have

$$\mathbf{E} = -\frac{\partial \mathbf{A}}{\partial t} = -\frac{\delta H_R}{\delta \pi} = -\frac{\pi}{\varepsilon_0}, \qquad (5.71)$$

therefore $\mathbf{E}$ is directly proportional to the classical momentum density π. Fluctuations of the momentum density thus correspond to fluctuations of the electric field $\mathbf{E}$. Further, the constraint $\nabla \cdot \mathbf{A} = 0$ implies there is a one-to-one relation between $\mathbf{A}$ and the magnetic field $\mathbf{B} = \nabla \times \mathbf{A}$ (up to an additive constant). Uncertainty in the vector potential thus corresponds to uncertainty in the magnetic field $\mathbf{B}$. Hence, in the radiation gauge, the exact uncertainty approach corresponds to adding nonclassical fluctuations to the electric field components of an ensemble of electromagnetic fields, with the fluctuation strength determined by the uncertainty in the magnetic field components.

5.4.4 Example: Gravitational Field

In our second example, we derive the Wheeler-DeWitt equation with a unique operator ordering from a classical ensemble of gravitational fields (see also Chap. 10 where classical ensembles of gravitational fields are investigated further).

5.4.4.1 Ensembles of Classical Gravitational Fields

The gravitational field is described, up to arbitrary coordinate transformations, by the metric tensor $g_{\mu\nu}$. The line element may be decomposed as [20]

$$ds^2 = g_{\mu\nu}dx^\mu dx^\nu = -(N^2 - N_i N^i)dt^2 + 2N_i dx^i dt + h_{ij}dx^i dx^j, \qquad (5.72)$$

in terms of the lapse function N, the shift vector N^i, and the spatial 3-metric h_{ij}, with $N_i = g_{ij}N^j$. The equations of motion are the Einstein field equations, which follow from the Hamiltonian functional [20]

$$\tilde{H}[h_{ij}, \pi^{ij}, N, N_i] = \int dx \left\{ N \left[G_{ijkl}[h_{ij}]\pi^{ij}\pi^{kl} - \sqrt{h}\,^{(3)}R[h_{ij}] \right] - 2\,N_i D_j \pi^{ij} \right\},$$
(5.73)

where π^{ij} denotes the momentum density conjugate to h_{ij}, D_j denotes the covariant 3-derivative, h is the determinant of h_{ij}, $^{(3)}R[h_{ij}]$ is the curvature scalar corresponding to h_{ij}, and

$$G_{ijkl}[h_{ij}] = \frac{1}{2\sqrt{h}}(h_{ik}h_{jl} + h_{il}h_{jk} - h_{ij}h_{kl})$$
(5.74)

is the (inverse) DeWitt metric.

The Hamiltonian functional $\tilde{H}$ is the one that corresponds to the standard Lagrangian for gravity, $\tilde{L} = \int dx \sqrt{g}\, R[g]$, where the momenta π^0 and π^i conjugate to N and N_i respectively vanish identically. However, the lack of dependence of $\tilde{H}$ on π^0 and π^i is consistently maintained only if the rates of change of these momenta also vanish; i.e., noting Eq. (5.83) of Appendix 1, only if the constraints [20]

$$\frac{\delta \tilde{H}}{\delta N} = G_{ijkl}\pi^{ij}\pi^{kl} - \sqrt{h}\,^{(3)}R = 0, \qquad \frac{\delta \tilde{H}}{\delta N_i} = -2D_j\,\pi^{ij} = 0,$$
(5.75)

are satisfied. Thus the dynamics of the field is independent of N and N_i, so that these functions may be fixed arbitrarily. Moreover, these constraints immediately yield $\tilde{H} = 0$ in Eq. (5.73), and hence the system is static, with no explicit time dependence.

It follows that, in the Hamilton–Jacobi formulation of the equations of motion (see Appendix 1 of this chapter), S is independent of N, N_i and t. Noting that $\pi^{ij} \equiv \delta S/\delta h_{ij}$ in this formulation, Eq. (5.75) therefore yield the corresponding constraints

$$\frac{\delta S}{\delta N} = \frac{\delta S}{\delta N_i} = \frac{\partial S}{\partial t} = 0, \quad D_j \left(\frac{\delta S}{\delta h_{ij}} \right) = 0,$$
(5.76)

for S. A given functional $F[h_{ij}]$ of the 3-metric is invariant under spatial coordinate transformations if and only if $D_j(\delta F/\delta h_{ij}) = 0$ [21], and hence the fourth constraint in Eq. (5.76) is equivalent to the invariance of S under such transformations. This fourth constraint moreover implies that the second term in Eq. (5.73) may be dropped from the Hamiltonian, yielding the reduced Hamiltonian

$$\tilde{H}_G[h_{ij}, \pi^{ij}, N] = \int dx\, N \left[G_{ijkl}[h_{ij}]\pi^{ij}\pi^{kl} - \sqrt{h}\,^{(3)}R[h_{ij}] \right],$$
(5.77)

as the basis for the Hamiltonian-Jacobi formulation [21, 22].

For an *ensemble* of classical gravitational fields, the independence of the dynamics with respect to N, N_i and t implies that members of the ensemble are distinguishable only by their corresponding 3-metric h_{ij}. Moreover, it is natural to impose the additional geometric requirement that the ensemble is invariant under spatial coordinate transformations. One therefore has the constraints

$$\frac{\delta P}{\delta N} = \frac{\delta P}{\delta N_i} = \frac{\partial P}{\partial t} = 0, \quad D_j \left(\frac{\delta P}{\delta h_{ij}} \right) = 0, \tag{5.78}$$

for the corresponding probability density functional $P[h_{ij}]$, analogous to Eq. (5.76). The first two constraints imply that ensemble averages only involve integration over h_{ij}.

5.4.4.2   Quantum Ensembles and Operator-Ordering

The Hamiltonian $\tilde{H}_G$ in Eq. (5.77) has the quadratic form of Eq. (5.52). Hence the exact uncertainty approach is applicable, and immediately leads to the functional Schrödinger equation

$$i\hbar \frac{\partial \Psi}{\partial t} = \tilde{H}_G[h_{ij}, -i\hbar(\delta/\delta h_{ij}), N]\Psi \tag{5.79}$$

for a *quantum* ensemble of gravitational fields, as per the Corollary of Sect. 5.4.2.

As discussed at the end of Sect. 5.4.2, we follow the guiding principle that all constraints imposed on the classical ensemble should be carried over to corresponding constraints on the quantum ensemble. Thus, from Eqs. (5.76) and (5.78) we require that P and S, and hence Ψ in Eq. (5.65), should be independent of N, N_i and t and invariant under spatial coordinate transformations, i.e.,

$$\frac{\delta \Psi}{\delta N} = \frac{\delta \Psi}{\delta N_i} = \frac{\partial \Psi}{\partial t} = 0, \quad D_j \left(\frac{\delta \Psi}{\delta h_{ij}} \right) = 0. \tag{5.80}$$

Applying the first and third of these constraints to Eq. (5.79) immediately yields the reduced functional Schrödinger equation

$$- \hbar^2 \frac{\delta}{\delta h_{ij}} G_{ijkl}[h_{ij}] \frac{\delta}{\delta h_{kl}} \Psi - \sqrt{h} \, {}^{(3)}R[h_{ij}]\Psi = 0, \tag{5.81}$$

which may be recognised as the Wheeler-DeWitt equation in the metric representation [20].

A notable feature of Eq. (5.81) is that the Wheeler-DeWitt equation has not only been derived from an exact uncertainty principle: it has, as a consequence of Eq. (5.64), been derived with a *precisely* defined operator ordering (with G_{ijkl} sandwiched between the two functional derivatives). Thus the exact uncertainty approach does not admit ambiguity in this respect, unlike the standard approach [20]. Such removal of ambiguity is essential to making definite physical predictions, and hence may be regarded as an advantage of the exact uncertainty approach.

For example, Kontoleon and Wiltshire [23] have pointed out that Vilenkin's prediction of inflation in minisuperspace, from a corresponding Wheeler-DeWitt equation with "tunneling" boundary conditions [24], depends critically upon the operator ordering used. In particular, considering the class of orderings defined by an integer power p, with corresponding Wheeler-DeWitt equation for a Friedmann-Robertson-Walker metric coupled to a scalar field ϕ [24]

$$\left[\frac{\partial^2}{\partial a^2} + \frac{p}{a} \frac{\partial}{\partial a} - \frac{1}{a^2} \frac{\partial^2}{\partial \phi^2} - U(a, \phi) \right] \Psi = 0, \tag{5.82}$$

Kontoleon and Wiltshire show that Vilenkin's approach fails for orderings with $p \geq 1$ [23]. Moreover, they suggest that the only natural ordering is in fact the "Laplacian" ordering corresponding to $p = 1$, which has been justified on geometric grounds by Hawking and Page [25].

However, noting that the relevant Hamiltonian functional in Eq. (2.7) of Ref. [24] is quadratic in the momentum densities of the metric and the scalar field, the exact uncertainty approach may be applied and yields the Wheeler-DeWitt equation corresponding to $p = -1$ in Eq. (5.82). Hence the criticism in Ref. [23] is avoided. One also has the nice feature that the associated Wheeler-DeWitt equation can be exactly solved for this "exact uncertainty" ordering [24].

A certain degree of ambiguity remains, which derives from the need to introduce some sort of regularisation scheme to remove divergences arising from the product of two functional derivatives acting at the same point in the Wheeler-DeWitt equation. Such considerations, however, do not play a role in the example that we have just discussed, which concerns minisuperspace quantisation involving a finite number of degrees of freedom. It is important to distinguish this regularisation problem from the far more difficult one associated with the requirement of Dirac consistency; i.e., the need to find a choice of operator ordering *and* regularisation scheme that will permit mapping the classical Poisson bracket algebra of constraints to an algebra of operators within the context of the Dirac quantisation of canonical gravity [26]. Our approach is based on the Hamilton–Jacobi formulation of classical gravity and, as shown by Bergmann [27], the functional form of the Hamilton–Jacobi functional S is already invariant under the action of the group generated by the constraints.

Finally, we point out that a similar approach may be applied to the Ashtekar formalism for gravity [28], where again the Hamiltonian is quadratic in the field momentum density.

Appendix 1: Hamilton–Jacobi Ensembles

The salient aspects of the Hamilton–Jacobi formulation of classical field theory [29] are collected here, with particular attention to the origin of the associated continuity equation for ensembles of classical fields.

Two classical fields ϕ, π are canonically conjugate if there is a Hamiltonian functional $\tilde{H}[\phi, \pi, t]$ such that

$$\frac{\partial \phi}{\partial t} = \frac{\delta \tilde{H}}{\delta \pi}, \qquad \frac{\partial \pi}{\partial t} = -\frac{\delta \tilde{H}}{\delta \phi}. \qquad (5.83)$$

These equations follow from the action principle $\delta A = 0$, with the action functional $A = \int dt \, [-\tilde{H} + \int dx \, \pi_x(\partial \phi_x / \partial t)]$. The rate of change of an arbitrary functional $G[\phi, \pi, t]$ follows from

$$\frac{dG}{dt} = \frac{\partial G}{\partial t} + \int dx \left(\frac{\delta G}{\delta \phi_x} \frac{\partial \phi_x}{\partial t} + \frac{\delta G}{\delta \pi_x} \frac{\partial \pi_x}{\partial t} \right) \qquad (5.84)$$

and Eq. (5.83) as

$$\frac{dG}{dt} = \frac{\partial G}{\partial t} + \int dx \left(\frac{\delta G}{\delta \phi_x} \frac{\delta \tilde{H}}{\delta \pi_x} - \frac{\delta G}{\delta \pi_x} \frac{\delta \tilde{H}}{\delta \phi_x} \right) =: \frac{\partial G}{\partial t} + \{G, H\}, \qquad (5.85)$$

where $\{\ ,\ \}$ is a generalised Poisson bracket.

A canonical transformation maps ϕ, π and $\tilde{H}$ to ϕ', π' and $\tilde{H}'$, such that the equations of motion for the latter retain the canonical form of Eq. (5.83). Equating the variations of the corresponding actions A and A' to zero, it follows that all physical trajectories must satisfy

$$- \tilde{H} + \int dx \, \pi_x \frac{\partial \phi_x}{\partial t} = -\tilde{H}' + \int dx \, \pi_x' \frac{\partial \phi_x'}{\partial t} + \frac{dF}{dt} \qquad (5.86)$$

for some generating functional F. Now, any two of the fields ϕ, π, ϕ', π' determine the remaining two fields for a given canonical transformation. Choosing ϕ and π' as the two independent fields, defining the new generating functional $G[\phi, \pi', t] = F + \int dx \, \phi_x' \pi_x'$, and using Eq. (5.84), then yields

$$\tilde{H}' = \tilde{H} + \frac{\partial G}{\partial t} + \int dx \left[\frac{\partial \phi_x}{\partial t} \left(\frac{\delta G}{\delta \phi_x} - \pi_x \right) + \frac{\partial \pi_x'}{\partial t} \left(\frac{\delta G}{\delta \pi_x'} - \phi_x' \right) \right] \qquad (5.87)$$

for all physical trajectories. The terms in round brackets therefore vanish identically, yielding the generating relations

$$\tilde{H}' = \tilde{H} + \frac{\partial G}{\partial t}, \qquad \pi = \frac{\delta G}{\delta \phi}, \qquad \phi' = \frac{\delta G}{\delta \pi'}. \qquad (5.88)$$

A canonical transformation is thus completely specified by the associated generating functional G.

To obtain the *Hamilton–Jacobi* formulation of the equations of motion, consider a canonical transformation to fields ϕ', π' which are time-independent (e.g., to the fields

ϕ and π at some fixed time t_0). From Eq. (5.83) one may choose the corresponding Hamiltonian $\tilde{H}' \equiv 0$ without loss of generality, and hence using the integration by parts formula

$$\int D\phi \, P \frac{\delta F}{\delta \phi} = -\int D\phi \, \frac{\delta P}{\delta \phi} F \qquad (5.89)$$

(see Appendix A.2 of this book), the momentum density and the associated generating functional S are specified by the functional equations

$$\pi = \frac{\delta S}{\delta \phi}, \quad \frac{\partial S}{\partial t} + \tilde{H}[\phi, \delta S/\delta \phi, t] = 0. \qquad (5.90)$$

The latter is the desired Hamilton–Jacobi equation. Solving this equation for S is equivalent to solving Eqs. (5.83) for ϕ and π.

Note that along a physical trajectory one has $\pi' \equiv$ constant, and hence from Eqs. (5.84) and (5.92) that

$$\frac{dS}{dt} = \frac{\partial S}{\partial t} + \int dx \frac{\delta S}{\delta \phi_x} \frac{\partial \phi_x}{\partial t} = -\tilde{H} + \int dx \, \pi_x \frac{\partial \phi_x}{\partial t} = \frac{dA}{dt}. \qquad (5.91)$$

Thus the Hamilton–Jacobi functional S is equal to the action functional A, up to an additive constant. This relation underlies the connection between the derivation of the Hamilton–Jacobi equation from a particular type of canonical transformation, as above, and the derivation from a particular type of variation of the action, as per the Schwinger–Tomonaga formalism [30].

The Hamilton–Jacobi formulation has the interesting feature that once S is specified, the momentum density is determined by the relation $\pi = \delta S/\delta \phi$, i.e., it is a functional of ϕ. Thus, unlike the Hamiltonian formulation of Eq. (5.83), an *ensemble* of fields is specified by a probability density functional $P[\phi]$, not by a phase space density functional $\rho[\phi, \pi]$.

In either case, the equation of motion for the probability density corresponds to the conservation of probability, i.e., to a continuity equation of the form

$$\frac{\partial P}{\partial t} + \int dx \frac{\delta}{\delta \phi_x} [P V_x] = 0. \qquad (5.92)$$

For example, in the Hamiltonian formulation the associated continuity equation for $\rho[\phi, \pi]$ is

$$\frac{\partial \rho}{\partial t} + \int dx \left[\frac{\delta}{\delta \phi_x} \left(\rho \frac{\partial \phi_x}{\partial t} \right) + \frac{\delta}{\delta \pi_x} \left(\rho \frac{\partial \pi_x}{\partial t} \right) \right] = 0, \qquad (5.93)$$

which reduces to the Liouville equation $\partial \rho/\partial t = \{H, \rho\}$ via Eq. (5.83).

Similarly, in the Hamilton–Jacobi formulation, the rate of change of the field ϕ follows from Eqs. (5.83) and (5.92) as the functional

$$V_x[\phi] = \frac{\partial \phi_x}{\partial t} = \left. \frac{\delta \tilde{H}}{\delta \pi_x} \right|_{\pi = \delta S/\delta \phi} \tag{5.94}$$

and hence the associated continuity equation for an ensemble of fields described by $P[f]$ follows via Eq. (5.92) as

$$\frac{\partial P}{\partial t} + \int dx \, \frac{\delta}{\delta f_x} \left[\left. P \frac{\delta \tilde{H}}{\delta g_x} \right|_{g=\delta S/\delta f} \right]. \tag{5.95}$$

Equations (5.92) and (5.95) generalise immediately to multicomponent fields.

Appendix 2: Proofs of the Theorem and Corollary

Proof of Theorem (Eq. 5.63): From the causality and exact uncertainty assumptions in Sect. 5.4.2, one has $\text{Cov}_x(N) = \alpha(P, \delta P/\delta \phi_x, \phi_x)$. To avoid issues of regularisation, it is convenient to consider a position-dependent canonical transformation, $\phi_x \to \Lambda_x^{-1}\phi_x$, such that $A[\Lambda] := \exp[\int dx \, \ln |\det \Lambda_x|]$ is finite. Then the probability density functional P and the measure $D\phi$ transform as $P \to AP$ and $D\phi \to A^{-1}D\phi$ respectively, and so the invariance assumption in Sect. 3 requires that

$$\alpha(AP, A\Lambda_x^T u, \Lambda_x^{-1}w) \equiv \Lambda_x^T \alpha(P, u, w)\Lambda_x, \tag{5.96}$$

where u^a and w^a denote the vectors $\delta P/\delta \phi_x^a$ and ϕ_x^a respectively, for a given value of x. Since Λ_x can remain the same at a given point x while varying elsewhere, this homogeneity condition must hold for A and Λ_x independently. Thus, choosing Λ_x to be the identity matrix at some point x, one has $\alpha(AP, Au, w) = \alpha(P, u, w)$ for all A, implying that α can involve P only via the combination $v := u/P$.

The above homogeneity condition for α therefore reduces to

$$\alpha(\Lambda^T v, \Lambda^{-1}w) = \Lambda^T \alpha(v, w)\Lambda. \tag{5.97}$$

Note that this equation is linear, and invariant under multiplication of α by any function of the scalar $J := v^T w$. Moreover, it may easily be checked that if σ and τ are solutions, then so are $\sigma\tau^{-1}\sigma$ and $\tau\sigma^{-1}\tau$. Choosing the two independent solutions $\sigma = vv^T$, $\tau = (ww^T)^{-1}$, it follows that the general solution has the form

$$\alpha(v, w) = \beta(J)vv^T + \gamma(J)(ww^T)^{-1} \tag{5.98}$$

for arbitrary functions β and γ.

For $P = P_1 P_2$ one finds $v = (v_1, v_2)$, $w = (w_1, w_2)$, where the subscripts label corresponding subensemble quantities, and hence the independence assumption in Sect. 5.4.2, reduces to the requirements

$$\beta(J_1 + J_2)v_1 v_1^T + \gamma(J_1 + J_2)(w_1 w_1^T)^{-1} = \beta_1(J_1)v_1 v_1^T + \gamma_1(J_1)(w_1 w_1^T)^{-1}, \quad (5.99)$$

$$\beta(J_1 + J_2)v_2 v_2^T + \gamma(J_1 + J_2)(w_2 w_2^T)^{-1} = \beta_2(J_2)v_2 v_2^T + \gamma_2(J_2)(w_2 w_2^T)^{-1}, \quad (5.100)$$

for the respective subensemble covariance matrices. Thus $\beta = \beta_1 = \beta_2 = C$, $\gamma = \gamma_1 = \gamma_2 = D$ for universal (i.e., system-independent) constants C and D, yielding the general form

$$[\mathrm{Cov}_x(N)]^{ab} = \frac{C}{P^2} \frac{\delta P}{\delta \phi_x^a} \frac{\delta P}{\delta \phi_x^b} + D W_x^{ab}[\phi] \qquad (5.101)$$

for the fluctuation covariance matrix, where $W_x[\phi]$ denotes the inverse of the matrix with ab-coefficient $\phi_x^a \phi_x^b$.

Since $W_x[\phi]$ is purely a functional of ϕ, it merely contributes a classical additive potential term to the ensemble Hamiltonian of Eq. (5.58). It thus has no nonclassical role, and can be absorbed directly into the classical potential $\langle V \rangle$ (indeed, for fields with more than one component this term is singularly ill-defined, and hence can be discarded on physical grounds). Thus we may take $D = 0$ without loss of generality. Finally, the positivity of C follows from the positivity of the covariance matrix $\mathrm{Cov}_x(N)$, and the theorem is proved.

Proof of Corollary (Eq. 5.64) First, the equations of motion corresponding to the ensemble Hamiltonian $\mathcal{H}$ follow via the theorem and Eq. (5.55) as: (a) the continuity equation, Eq. (5.54), as before (since the additional term does not depend on S), which from Eq. (5.52) has the explicit form

$$\frac{\partial P}{\partial t} + 2 \sum_{a,b} \int dx \, \frac{\delta}{\delta \phi_x^a} \left(P K_x^{ab} \frac{\delta S}{\delta \phi_x^b} \right) = 0; \qquad (5.102)$$

and (b) the modified Hamilton–Jacobi equation

$$\frac{\partial S}{\partial t} = -\frac{\Delta \mathcal{H}}{\Delta P} = -H[\phi, \delta S/\delta \phi, t] - \frac{\Delta(\mathcal{H} - \mathcal{H}_c)}{\Delta P}. \qquad (5.103)$$

Calculating the last term, this simplifies to

$$\frac{\partial S}{\partial t} + H[\phi, \delta S/\delta \phi, t] - \frac{4C}{P^{1/2}} \sum_{a,b} \int dx \left(K_x^{ab} \frac{\delta^2 P^{1/2}}{\delta \phi_x^a \delta \phi_x^b} + \frac{\delta K_x^{ab}}{\delta \phi_x^a} \frac{\delta P^{1/2}}{\delta \phi_x^b} \right) = 0. \qquad (5.104)$$

Second, writing $\Psi = P^{1/2}e^{iS/\hbar}$, multiplying each side of Eq. (5.64) on the left by Ψ^{-1}, and expanding, gives a complex equation for P and S. The imaginary part is just the continuity equation, Eq. (5.102), and the real part is the modified Hamilton–Jacobi equation, Eq. (5.104) above, providing that one identifies C with $\hbar^2/4$.

References

1. Hall, M.J.W., Reginatto, M.: Schrödinger equation from an exact uncertainty principle. J. Phys A **35**, 3289–3303 (2002)
2. Hall, M.J.W., Kumar, K., Reginatto, M.: Bosonic field equations from an exact uncertainty principle. J. Phys A **36**, 9779–9794 (2003)
3. Heisenberg, W.: Über den anschaulichen Inhalt der quantentheoretischen Kinematik und Mechanik. Z. Phys. **43**, 172–198 (1927). English translation in: Wheeler, J.A., Zurek, W.H. (eds.) Quantum theory and measurement, pp. 62–84. Princeton University Press, Princeton (1983)
4. Hall, M.J.W.: Quantum properties of classical Fisher information. Phys. Rev. A **62**, 012107 (2000)
5. Hall, M.J.W.: Exact uncertainty relations. Phys. Rev. A **64**, 052103 (2001)
6. Johansen, L.M.: What is the value of an observable between pre- and postselection? Phys. Lett. A **322**, 298–300 (2004)
7. Hall, M.J.W.: Prior information: how to circumvent the standard joint-measurement uncertainty relation. Phys. Rev. A **69**, 052113 (2004)
8. Landau, L.D., Lifschitz, E.M.: Quantum mechanics. Pergamon Press, Oxford (1977)
9. Schuh, J.F.: Mathematical tools for modern physics. Philips Technical Library, Eindehoven (1968)
10. Hall, M.J.W., Reginatto, M.: Quantum mechanics from a Heisenberg-type equality. Fortschr. Phys. **50**, 646–651 (2002). Reprinted in: Papenfuss, D., Lüst, D., Schleich W.P. (eds.) 100 Years Werner Heisenberg: Works and Impact, pp. 217–222. Wiley, Berlin (2002)
11. Bialynicki-Birula, I., Mycielski, J.: Nonlinear wave mechanics. Ann. Phys. **100**, 62–93 (1976)
12. Ván, P., Fülöp, T.: Weakly non-local fluid mechanics: the Schrödinger equation. Proc. R. Soc. A **462**, 541–557 (2006)
13. Rudnicki, L.: Nonlinear Schrödinger equation from generalized exact uncertainty principle. http://arxiv.org/abs/1602.05004v1 (2016)
14. Fisher, R.A.: Theory of statistical estimation. Proc. Camb. Philos. Soc. **22**, 700–725 (1925)
15. Cox, D.R., Hinkley, D.V.: Theoretical Statistics. Chapman and Hall, London (1974)
16. Brown, L.S.: Quantum Field Theory. Cambridge University Press, Cambridge (1992)
17. Schweber, S.S.: An Introduction to Relativistic Quantum Field Theory. Row, Peterson and Co., New York (1961)
18. Dirac, P.A.M.: Lectures on Quantum Field Theory. Academic Press, New York (1966)
19. Wheeler, J.A.: Superspace. In: Gilbert, R.P., Newton, R.G. (eds.) Analytical Methods in Mathematical Physics, pp. 335–378. Gordon and Breach, New York (1970)
20. Kiefer, C.: Quantum Gravity. Oxford University Press, Oxford (2012)
21. Gerlach, U.H.: Derivation of the the Einstein equations from the semiclassical approximation to quantum geometrodynamics. Phys. Rev. **177**, 1929–1941 (1969)
22. Peres, A.: On Cauchy's problem in general relativity – II. Nuovo Cim. **XXVI**, 53–62 (1962)
23. Kontoleon, N., Wiltshire, D.L.: Operator ordering and consistency of the wave function of the Universe. Phys. Rev. D **59**, 063513 (1999)
24. Vilenkin, A.: Boundary conditions in quantum cosmology. Phys. Rev. D **33**, 3560–3569 (1986)
25. Hawking, S.W., Page, D.N.: Operator ordering and the flatness of the universe. Nucl. Phys. B **264**, 185–196 (1986)

26. Tsamis, N.C., Woodard, R.P.: The factor-ordering problem must be regulated. Phys. Rev. D **36**, 3641–3650 (1987)
27. Bergmann, P.G.: Hamilton-Jacobi and Schrödinger theory in theories with first-class Hamiltonian constraints. Phys. Rev. **144**, 1078–1080 (1966)
28. Ashtekar, A.: New variables for classical and quantum gravity. Phys. Rev. Lett **57**, 2244–2247 (1986)
29. Goldstein, H.: Classical Mechanics. Addison-Wesley, New York (1950)
30. Roman, P.: Introduction to Quantum Field Theory. Wiley, New York (1969)

Chapter 6
The Geometry of Ensembles on Configuration Space

Abstract A description of ensembles on configuration space incorporates at least two geometrical structures which arise in a natural way: a metric structure, which derives from the natural geometry associated with a space of probabilities, and a symplectic structure, which derives from the symplectic geometry associated with a Hamiltonian description of motion. We show that these two geometrical structures give rise to a Kähler geometry. We first consider probabilities P and introduce the information metric. This leads to information geometry, a Riemannian geometry defined on the space of probabilities. We then bring in dynamics via a Hamiltonian formalism defined on a phase space with canonically conjugate coordinates P and S. This leads to more geometrical structure, a symplectic geometry defined on this phase space. The next step is to extend the information metric, which is defined over the space of probabilities only, to a metric over the full phase space. This requires satisfying certain conditions which ensure the compatibility of the metric and symplectic structures. These conditions are equivalent to requiring that the space have a Kähler structure. In this way, we are led to a Kähler geometry. This rich geometrical structure allows for a reconstruction of the geometric formulation of quantum theory. One may associate a Hilbert space with the Kähler space and this leads to the standard version of quantum theory. Thus the theory of ensembles on configuration space permits a geometric derivation of quantum theory.

6.1 Introduction

Geometrical methods have proven quite fruitful in physics, leading sometimes to novel quantisation methods. For example, Hamiltonian mechanics allows for a formulation in terms of symplectic geometry and this geometric formulation leads in turn to geometric quantization. In this chapter, we consider the geometry of ensembles on configuration space and show that there is a rich geometrical structure that allows for a reconstruction of the geometric formulation of quantum theory first introduced by Kibble [1]. Thus, in addition to the derivation of quantum theory from an exact uncertainty principle which was discussed in the previous chapter, the theory of ensembles on configuration space permits also a geometric derivation [2–4].

© Springer International Publishing Switzerland 2016
M.J.W. Hall and M. Reginatto, *Ensembles on Configuration Space*,
Fundamental Theories of Physics 184, DOI 10.1007/978-3-319-34166-8_6

The basic elements of this geometrical reconstruction of quantum theory are the natural metric on the space of probabilities (information geometry), the description of dynamics using a Hamiltonian formalism (symplectic geometry), and requirements of consistency (Kähler geometry). The procedure, which can be carried out for both continuous and discrete systems, has a number of remarkable features: the complex structure appears by requiring consistency between metric and symplectic structures; wave functions arise as the natural complex coordinates of the Kähler space; and time evolution is described by a one-parameter group of unitary transformations. One may associate a Hilbert space with the Kähler space, which leads to the standard version of quantum theory.

In describing the geometry of ensembles with a continuous configuration space, we substantially follow the exposition of Ref. [2].[1] For the case of a discrete configuration space, we substantially follow the exposition of Refs. [3][2] and [4].

6.2 Information Metric, Symplectic Structure and Kähler Geometry

Probabilities play a fundamental role in the theory of ensembles on configuration space. Spaces of probabilities have a natural Riemannian geometry called *information geometry*. We show in this section that there is also a natural geometry associated with ensembles on configuration space that is more general than Riemannian geometry. It is a *Kähler geometry*, which brings together metric, symplectic and complex structures in a harmonious way.

6.2.1 Information Geometry

We introduce information geometry for probabilities defined over both discrete and continuous configuration spaces.

We consider first a system with a discrete configuration space (see also Chap. 1). We make the assumption that the configuration of the system is subject to uncertainty and the state of the system is described by a probability $P = (P^1, \ldots, P^n)$, where n is the number of states. The probability that the system is in state i is P^i, where P^i satisfies $P^i \geq 0$ and $\sum_i P^i = 1$. The use of a superscript, rather than a subscript as in Chap. 1, is convenient for contractions with a metric tensor further below.

[1]Reproduced with permission from: Reginatto, M., Hall, M.J.W.: AIP Conf. Proc. **1443**, 96–103 (2012). Copyright 2012, AIP Publishing LLC.
[2]Reproduced with permission from: Reginatto, M., Hall, M.J.W.: AIP Conf. Proc. **1553**, 246–253 (2013). Copyright 2013, AIP Publishing LLC.

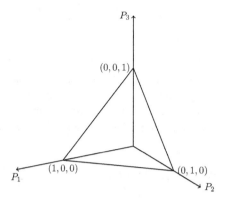

Fig. 6.1 The simplex S_2

Probability distributions can be visualized as points in the space corresponding to the simplex $S_{n-1} = \{P \in \mathbf{R}_n^+ : \sum P^i = 1\}$. As an example, in Fig. 6.1 we show the representation of the simplex S_2.

There is a natural line element in this space, given by

$$ds^2 = G_{ij}\, dP^i\, dP^j = \frac{\alpha}{2P^i}\, \delta_{ij}\, dP^i\, dP^j, \tag{6.1}$$

where α is a constant and repeated indices are to be summed over as per the usual convention. The metric G_{ij} that appears in Eq. (6.1) is known as the *information metric*,

$$G_{ij} = \frac{\alpha}{2P^i}\, \delta_{ij}. \tag{6.2}$$

The value of α can not be determined a priori; it is usually set to $1/2$, but it will be convenient not to follow this convention. Instead, α will be treated as a parameter. The line element of Eq. (6.1) defines a distance on the space of probabilities. This distance seems to have been introduced into statistics by Bhattacharyya [5, 6] as a way of providing a measure of divergence for multinomial probabilities [7]. Wootters [8] refers to this distance as the "statistical distance".

To derive the statistical distance from the metric G_{ij}, consider two points in probability space, P_A and P_B, joined by a curve $P^i(t)$, $0 \le t \le 1$, and write the expression for the length l of the curve in the form

$$l = \int_0^1 dt \sqrt{G_{ij} \frac{dP^i(t)}{dt} \frac{dP^j(t)}{dt}}. \tag{6.3}$$

The statistical distance is defined as the shortest distance between P_A and P_B. To compute the statistical distance, it is convenient to do the change of coordinates $X^i = \sqrt{P^i}$. Then

$$l = \sqrt{2\alpha} \int_0^1 dt \sqrt{\sum_{i=1}^n \left[\frac{dX^i(t)}{dt}\right]^2} . \tag{6.4}$$

Since the curve $P(t)$ is assumed to lie in the probability space, it must satisfy the condition $\sum_{i=1}^n P^i(t) = \sum_{i=1}^n [X^i(t)]^2 = 1$; that is, the curve must lie on a unit n-dimensional sphere in the X space. The shortest distance on the n-dimensional sphere is equal to the angle between the unit vectors X_A and X_B. This leads immediately to

$$d(P_A, P_B) = \sqrt{2\alpha} \cos^{-1} \left(\sum_{i=1}^n X_A^i X_B^i\right) = \sqrt{2\alpha} \cos^{-1} \left(\sum_{i=1}^n \sqrt{P_A^i} \sqrt{P_B^i}\right). \tag{6.5}$$

The case of a continuous configuration space has to be treated in a slightly different manner. Consider an n-dimensional continuous configuration space, with coordinates $x \equiv \{x^1, \ldots, x^n\}$ and probability densities $P(x)$ satisfying $P(x) \geq 0$ and $\int d^n x P(x) = 1$.

If we consider the action of the translation group T on $P(x)$, $T : P(x) \to P(x+\theta)$, there is a natural metric γ_{jk} on the space of parameters θ: the Fisher–Rao metric [9],

$$\gamma_{jk} = \frac{\alpha}{2} \int d^n x \frac{1}{P(x+\theta)} \frac{\partial P(x+\theta)}{\partial \theta^j} \frac{\partial P(x+\theta)}{\partial \theta^k}, \tag{6.6}$$

where α is a constant. The line element $d\sigma^2 = \gamma_{jk} \Delta^j \Delta^k$ (where $|\Delta^k| << 1$) defines a distance between the two probability distributions $P(x+\theta)$ and $P(x+\theta+\Delta)$.

It will be convenient to consider another form of the metric. In particular, using the equality $\partial P(x+\theta)/\partial \theta^j = \partial P(x+\theta)/\partial x^j$, and making the change of integration variables $x \to x+\theta$, the metric becomes proportional to the Fisher information matrix,

$$\gamma_{jk} = \frac{\alpha}{2} \int d^n x \frac{1}{P(x)} \frac{\partial P(x)}{\partial x^j} \frac{\partial P(x)}{\partial x^k}. \tag{6.7}$$

It follows that the line element $d\sigma^2 = \gamma_{jk} \Delta^j \Delta^k$ induces a line element in the space of probability densities,

$$ds^2 = \frac{\alpha}{2} \int d^n x \frac{1}{P_x} \delta P_x \delta P_x = \int d^n x \, d^n x' \, g_{PP}(x, x') \, \delta P_x \, \delta P_{x'}, \tag{6.8}$$

where we have introduced the notation $P_x = P(x)$, $\delta P_x \equiv \frac{\partial P(x)}{\partial x^j} \Delta^j$. The line element of Eq. (6.8) was introduced by Jeffreys [10]. The induced metric g_{PP} is diagonal, and given by

$$g_{PP}(x, x') = \frac{\alpha}{2P_x} \delta(x - x'). \tag{6.9}$$

This metric is the generalization of the metric of Eq. (6.2) to the case of a continuum configuration space.

The information metric is the natural metric on the space of probabilities. This metric defines a geometrical structure known as information geometry.

6.2.2 Uniqueness of the Information Metric via Markov Mappings

It has been shown by Čencov that the information metric, Eq. (6.2), is the only metric that is invariant under a family of probabilistically natural mappings known as *congruent embeddings by a Markov mapping* [11]. A simpler proof, which also makes use of these mappings, was given later by Campbell [12]. Thus the metric of Eq. (6.2) is unique. The proofs of Čencov and Campbell assume a discrete probability space, therefore they do not apply directly to the metric of Eq. (6.9). However, in the case of physical systems, one can always carry out a discretization of the continuous configuration space and approximate it by a discrete configuration space. The continuum case is recovered in the limit of an arbitrarily fine discretization. Therefore, if we restrict to the applications that we consider in this monograph, we may also assume uniqueness of the metric of Eq. (6.9).

We give a brief description of these mappings because later we will have to extend such mappings from the space of probabilities P^i to the complete phase space with coordinates P^i, S^i. We follow the presentation of Ref. [12]. A Markov mapping is a particular type of linear transformation between a simplex S_{m-1} and a simplex S_{n-1} (with $m \leq n$) which preserves the probability; i.e., $\sum_{a=1}^{m} P^a = \sum_{b=1}^{n} \tilde{P}^b = 1$. For $m = n$, the mapping is just a permutation of the components P^i, but for $m < n$, the mapping relates spaces of different dimensions.

A Markov mapping may be constructed in the following way. Let $A = \{A_1, \ldots, A_m\}$ be a partition of the set $\{1, 2, \ldots, n\}$ into disjoint sets. Associate a probability vector $Q_{(a)} = (q_{a1}, \ldots, q_{an})$ to each of the A_a, where the q_{ab} satisfy

$$q_{ab} = 0 \text{ if } b \notin A_a, \qquad q_{ab} > 0 \text{ if } b \in A_a, \qquad \sum_{b=1}^{n} q_{ab} = 1. \qquad (6.10)$$

The probability vector $Q_{(a)}$ is therefore concentrated on A_a. Note that the $m \times n$ matrix Q with elements q_{ab} has the following properties: Each column has precisely one non-zero element and each row sums to one.

Define mappings between $f : S_{m-1} \to S_{n-1}$ and $g : S_{n-1} \to S_{m-1}$ by

$$\tilde{P}^b = \sum_{a=1}^{m} P^a q_{ab},$$

$$P^a = \sum_{b \in A_a} \tilde{P}^b, \qquad a \in \{1, 2, \ldots, m\}. \tag{6.11}$$

Following Čencov, the mapping f is known as a congruent embedding of S_{m-1} in S_{n-1} by a Markov mapping. The mapping g, which is also defined in terms of the partition A, has the property that the composition $g \circ f$ is the identity map on S_{m-1}.

To illustrate such a mapping, consider a simple example where $n = m + 1$. Set

$$\mathbf{Q} = \begin{pmatrix} 1 & 0 & \cdots & 0 & 0 & 0 \\ 0 & 1 & \cdots & 0 & 0 & 0 \\ \vdots & \vdots & \ddots & \vdots & \vdots & \vdots \\ 0 & 0 & \cdots & 1 & 0 & 0 \\ 0 & 0 & \cdots & 0 & k & (1-k) \end{pmatrix}, \tag{6.12}$$

with $0 < k < 1$. Then

$$P = (P^1, \ldots, P^m) \to \tilde{P} = (\tilde{P}^1, \ldots, \tilde{P}^m, \tilde{P}^{m+1}) := (P^1, \ldots, kP^m, (1-k)P^m). \tag{6.13}$$

A vector in the tangent space of the simplex transforms in a similar way,

$$V = (V^1, \ldots, V^m) \to \tilde{V} = (\tilde{V}^1, \ldots, \tilde{V}^m, \tilde{V}^{m+1}) := (V^1, \ldots, kV^m, (1-k)V^m). \tag{6.14}$$

To prove uniqueness of the information metric, Čencov [11] and Campbell [12] show that the only metric that preserves the inner product $< A, B >$ of two tangent vectors A, B under a Markov mapping is precisely the information metric. It is straightforward to show that the metric has this property. To see this for the simple example discussed above, simply compute

$$< \tilde{A}, \tilde{B} > = \sum_{i=1}^{m+1} \left\{ \frac{\tilde{A}^i \tilde{B}^i}{\tilde{P}^i} \right\} = \sum_{i=1}^{m-1} \left\{ \frac{A^i B^i}{P^i} \right\} + \frac{kA^m \, kB^m}{kP^m} + \frac{(1-k)A^m \, (k-1)B^m}{(1-k)P^m}$$

$$= \sum_{i=1}^{m} \left\{ \frac{A^i B^i}{P^i} \right\} = < A, B > . \tag{6.15}$$

A proof of uniqueness is, as one would expect, not straightforward. Since the details of the proof are not needed here, we will not discuss it further and refer the reader instead to the proofs of Čencov [11] and Campbell [12].

6.2.3   Dynamics and Symplectic Geometry

As we saw in Chap. 1, the time evolution of the P^i in the discrete case may be described using a Hamiltonian formalism. To do this, we introduce additional coordinates S^i which are canonically conjugate to the P^i and a corresponding Poisson bracket for any two functions $F(P^i, S^i)$ and $G(P^i, S^i)$,

$$\{F, G\} = \sum_i \left(\frac{\partial F}{\partial P^i} \frac{\partial G}{\partial S^i} - \frac{\partial F}{\partial S^i} \frac{\partial G}{\partial P^i} \right). \tag{6.16}$$

The new important insight that we will need here is that the Poisson bracket can be rewritten geometrically as

$$\{F, G\} = (\partial F/\partial P, \ \partial F/\partial S) \, \Omega \left(\begin{array}{c} \partial G/\partial P \\ \partial G/\partial S \end{array} \right), \tag{6.17}$$

where Ω is the symplectic form, given in this case by

$$\Omega = \left(\begin{array}{cc} 0 & 1 \\ -1 & 0 \end{array} \right), \tag{6.18}$$

where 1 is the $n \times n$ unit matrix. We thus have a *symplectic structure* and a corresponding *symplectic geometry*. The equations of motion for P^i and S^i are given by $\partial P^i/\partial t = \{P^i, \mathcal{H}\}$, $\partial S^i/\partial t = \{S^i, \mathcal{H}\}$ where $\mathcal{H}(P^i, S^i)$ is the ensemble Hamiltonian that generates time translations.

The $2n$-dimensional phase space with coordinates P^i and S^i has a richer structure than the n-dimensional space of probabilities P^i; in particular, as we discussed in Chap. 2, one may introduce the notion of *observables*, which are functions $F(P, S)$ of the coordinates, together with an algebra of observables defined in terms of the Poisson brackets of these functions. We also saw that normalization of P implies gauge invariance of the theory under $S^i \rightarrow S^i + c$, where c is a constant.

The generalization to the continuous case is straightforward. In this case, the Poisson bracket for any two functionals $F[P, S]$ and $G[P, S]$ is given by

$$\{F, G\} = \int d^n x \left\{ \frac{\delta F}{\delta P} \frac{\delta G}{\delta S} - \frac{\delta F}{\delta S} \frac{\delta G}{\delta P} \right\}. \tag{6.19}$$

The Poisson bracket can be rewritten geometrically as

$$\{F, G\} = \int d^n x \, d^n x' \, (\delta P_x \ \delta S_x) \, \Omega(x, x') \left(\begin{array}{c} \delta P_{x'} \\ \delta S_{x'} \end{array} \right), \tag{6.20}$$

where $\Omega(x, x')$ is the corresponding symplectic form, given in this case by

$$\Omega(x, x') = \left(\begin{array}{cc} 0 & 1 \\ -1 & 0 \end{array} \right) \delta(x - x'). \tag{6.21}$$

Thus in the continuum case we also have a symplectic structure and a corresponding symplectic geometry. The equations of motion for P and S are now given by $\partial P/\partial t =$

$\{P, \mathcal{H}\}$, $\partial S/\partial t = \{S, \mathcal{H}\}$ where $\mathcal{H}[P, S]$ is the ensemble Hamiltonian that generates time translations.

> The description of dynamics via a Hamiltonian formalism leads to the doubling of the dimensionality of the space (configuration space $\rightarrow$ phase space), a symplectic structure and a corresponding symplectic geometry.

6.2.4 Kähler Geometry

The $2n$-dimensional discrete phase space with coordinates P^i and S^i is an extension of the n-dimensional subspace of probabilities P^i. It is natural to ask the following question: is it possible to extend the metric G_{ij} in Eq. (6.2), which is only defined on the n-dimensional subspace of probabilities P^i, to the full $2n$-dimensional phase space of the P^i and S^i? This can be done, but certain conditions which ensure the compatibility of the metric and symplectic structures have to be satisfied. These conditions are equivalent to requiring that the space have a Kähler structure (for a proof, see Appendix 1 of this chapter).

A Kähler structure brings together metric, symplectic and complex structures in a harmonious way. To define such a space, introduce a complex structure $J^a_{\ b}$ and impose the following conditions [13],

$$\Omega_{ab} = g_{ac} J^c_{\ b}, \tag{6.22}$$

$$J^a_{\ c} g_{ab} J^b_{\ d} = g_{cd}, \tag{6.23}$$

$$J^a_{\ b} J^b_{\ c} = -\delta^a_{\ c}. \tag{6.24}$$

Equation (6.22) is a compatibility equation between the symplectic structure Ω_{ab} and the metric g_{ab}, Eq. (6.23) is the condition that the metric should be Hermitian, and Eq. (6.24) is the condition that $J^a_{\ b}$ should be a complex structure.

The metric over the n-dimensional subspace of probabilities is the information metric, Eq. (6.2). Then, the metric over the full space will be of the form

$$g_{ab} = \begin{pmatrix} G & E \\ E^T & F \end{pmatrix}, \tag{6.25}$$

where $G = \mathrm{diag}(\alpha/2P^i)$, and E and F are $n \times n$ matrices that still need to be determined.

A simple matrix calculation using the Kähler conditions and the expression for Ω_{ab}, Eq. (6.18), leads to general forms for the metric g_{ab} and the complex structure $J^a_{\ b}$,

$$g_{ab} = \begin{pmatrix} G & A^T \\ A & (1+A^2)G^{-1} \end{pmatrix}, \quad J^a_{\ b} = \begin{pmatrix} A & (1+A^2)G^{-1} \\ -G & -GAG^{-1} \end{pmatrix}. \tag{6.26}$$

where the $n \times n$ matrix A satisfies $GAG^{-1} = A^T$ but is otherwise arbitrary.

In the case of a continuous configuration space we can derive a similar result, although we need to proceed in a somewhat different manner. We saw before that the metric over the subspace of probabilities is diagonal and given by $g_{PP}(x, x') = (\alpha/2P_x)\delta(x - x')$. We assume that the full metric g_{ab} is also diagonal; that is, of the form $g_{ab}(x, x') = g_{ab}(x)\delta(x - x')$. This assumption corresponds to a locality assumption. Then g_{ab} is a real, symmetric matrix of the form

$$g_{ab} = \begin{pmatrix} \frac{\alpha}{2P_x} & g_{PS} \\ g_{SP} & g_{SS} \end{pmatrix} \delta(x - x'). \tag{6.27}$$

The elements $g_{PS} = g_{SP}$ and g_{SS} still need to be determined.

Since $\Omega_{ab}(x, x')$ in Eq. (6.21) is also diagonal, the Kähler conditions imply that $J^c_{\ b}(x, x')$ is diagonal; i.e. $J^c_{\ b}(x, x') = J^c_{\ b}(x)\delta(x - x')$. Using the Kähler conditions and the expression for $\Omega_{ab}(x, x')$ derived from Eq. (6.21), it is possible to show that the metric and the complex structure take the form

$$g_{ab} = \begin{pmatrix} \alpha/2P_x & A_x \\ A_x & (2P_x/\alpha)(1+A_x^2) \end{pmatrix} \delta(x - x'), \tag{6.28}$$

$$J^a_{\ b} = \begin{pmatrix} A_x & (2P_x/\alpha)(1+A_x^2) \\ -\alpha/2P_x & -A_x \end{pmatrix} \delta(x - x'), \tag{6.29}$$

which depend on an arbitrary functional A_x.

> The information metric can be extended to a metric over the full phase space but this requires satisfying conditions which ensure the compatibility of the metric and symplectic structures. This leads to a Kähler structure and a corresponding Kähler geometry.

6.2.5 On the Geometry of Ensembles on Configuration Space

It may be useful to make some general remarks now regarding the approach and the results obtained so far.

The starting point is a space of probabilities defined over a configuration space. The first step of the procedure is to introduce the information metric, the natural metric on the space of probabilities. This leads to the most basic geometrical structure of the construction, *information geometry*. The second step is to consider dynamics and to introduce an action principle to derive equations of motion. This is done using a Hamiltonian formalism: introduce coordinates S canonically conjugate to the P, a Poisson bracket structure, and an ensemble Hamiltonian. This leads to additional geometrical structure, a *symplectic structure*. The third step is to extend the metric structure of information geometry, to define a geometry over the full space of the P and S. This can not be done in an arbitrary way. Consistency between the metric tensor and the symplectic form leads to a *Kähler geometry*.

Notice that the construction is very general: it applies to any system that is described probabilistically and which admits equations of motion that can be derived from an action principle. In particular, the construction does not require any assumptions regarding classical or quantum physics.

The few assumptions that enter into the analysis lead to the beautiful result that *the natural geometry of ensembles on configuration space is a Kähler geometry*.

6.2.6 Uniqueness of the Kähler Metric via Generalized Markov Mappings

We first consider the discrete case. The Kähler conditions impose strong restrictions on the form of the metric g_{ab}, Eq. (6.26), but that they do not determine the metric uniquely because it depends on a matrix **A** which is to some extent arbitrary. Additional requirements are therefore needed to determine the form of g_{ab}.

As discussed before, Markov mappings play a crucial role in the proof of uniqueness of the information metric, Eq. (6.2). In this section, it will be shown that the form of **A** can be determined by requiring invariance of the metric g_{ab} under a particular type of canonical transformation which extends the notion of a Markov mapping to the phase space with coordinates P^i and S^i. The simplest way to introduce a generalization of a Markov mapping is to treat it as a point transformation (i.e., a transformation of the P^i) and to use the well known fact that a point transformation can always be extended to a canonical transformation.

Consider first the simple case of Markov mappings between spaces of the same dimension; i.e., where $n = m$. In this case, the Markov mappings are just permutations of the components P^i and their extension to canonical transformations is trivial: carry out simultaneous permutations of the P^i and S^i. These transformations have two important properties: they are linear and they do not mix the P^i and the S^i.

The case that is non-trivial is the case where $n > m$. Here there are some subtle issues that need to be addressed, foremost that such canonical transformations will map spaces of different dimensions. This, however, is not a fundamental difficulty because there is a well developed theory of "nonclassical canonical transformations" which extends the concept of canonical transformations to allow for mappings of phase spaces of different dimensions [14, 15]. Although the canonical transformation that we derive in this section may be formulated within this formalism, a simpler approach is possible and therefore there will be no need to make use of the full theory of "nonclassical canonical transformations". The simpler approach followed here consists of extending the dimensionality of the space in a trivial way and imposing constraints.

To illustrate the transformation, it will be sufficient to work out the generalization for the case of the simple example of a Markov mapping that we discussed before, where the dimensionality of the space of probabilities is increased by one; i.e., $n = m + 1$. More general cases can be derived by considering a series of successive transformations that are either permutations or which increase the dimensionality of the space by one at each step.

Consider then a system with states described by the coordinates $P^j, S^j, j = 1, \ldots m$, and a Hamiltonian $H(P^j, S^j)$ which describes the dynamics.

As a first step, increase the dimensionality of the space in a trivial way by adding coordinates P^{m+1}, S^{m+1} to the phase space. This increase in the dimensionality does not change the Hamiltonian. Therefore, the time evolution of the system remains the same and the additional coordinates P^{m+1}, S^{m+1} are *constants of the motion*. Consider now a point transformation relating old and new coordinates $P^k, \tilde{P}^k$ which satisfies the following relations,

$$
\begin{aligned}
P^i &= \tilde{P}^i, \\
P^m &= \tilde{P}^m + \tilde{P}^{m+1}, \\
P^{m+1} &= (1-k)\tilde{P}^m - k\tilde{P}^{m+1} \approx 0.
\end{aligned}
\tag{6.30}
$$

where $i = 1, \ldots, m-1$, and the symbol "$\approx$" is used to indicate a weak equality (i.e., a constraint in the sense of an equality of numerical values, not of functions of the phase space coordinates). Using the last two relations of Eq. (6.30), it is straightforward to show that $\tilde{P}^m$ and $\tilde{P}^{m+1}$ satisfy the constraints

$$
\begin{aligned}
\tilde{P}^m &\approx kP^m, \\
\tilde{P}^{m+1} &\approx (1-k)P^m,
\end{aligned}
\tag{6.31}
$$

which agree with Eq. (6.13). Therefore, the relations defined in Eq. (6.30) are equivalent to a Markov mapping. Notice that these constraints are preserved because P^{m+1} is a constant of the motion.

The second step is to extend this point transformation to a canonical transformation. In analogy to the case discussed above in which $n = m$, we will look for a linear canonical transformation which does not mix the P^i and the S^i. Notice that these two conditions lead to a *unique* canonical transformation (up to an additive constant which is unimportant). To define the canonical transformation, introduce the generating function

$$
K = \sum_{i=1}^{m-1} \left\{ \tilde{P}^i S^i \right\} + (\tilde{P}^m + \tilde{P}^{m+1}) S^m + [(1-k)\tilde{P}^m - k\tilde{P}^{m+1}] S^{m+1}.
\tag{6.32}
$$

Derive the canonical transformation from the generating function in the standard way; i.e., $P^k = \partial K/\partial S^k$ and $\tilde{S}^k = \partial K/\partial \tilde{P}^k$. This leads to the following equations,

$$P^i = \tilde{P}^i, \quad P^m = \tilde{P}^m + \tilde{P}^{m+1}, \quad\quad P^{m+1} = (1-k)\tilde{P}^m - k\tilde{P}^{m+1},$$
$$S^i = \tilde{S}^i, \quad S^m = k\tilde{S}^m + (1-k)\tilde{S}^{m+1}, \quad S^{m+1} = \tilde{S}^m - \tilde{S}^{m+1}, \qquad (6.33)$$

and

$$\tilde{P}^i = P^i, \quad \tilde{P}^m = kP^m + P^{m+1}, \quad\quad \tilde{P}^{m+1} = (1-k)P^m - P^{m+1},$$
$$\tilde{S}^i = S^i, \quad \tilde{S}^m = S^m + (1-k)S^{m+1}, \quad \tilde{S}^{m+1} = S^m - kS^{m+1}. \qquad (6.34)$$

where $i = 1, \ldots, m-1$.

We have increased the dimensionality of the phase space by two dimensions and we have two constants of the motion, P^{m+1} and S^{m+1}. As shown in Appendix 2 of this chapter, consistency requires both $P^{m+1} \approx 0$ and $S^{m+1} \approx 0$, which leads to

$$\tilde{P}^m \approx kP^m, \quad \tilde{P}^{m+1} \approx (1-k)P^m,$$
$$\tilde{S}^m \approx S^m, \quad \tilde{S}^{m+1} \approx S^m. \qquad (6.35)$$

It is also shown in Appendix 2 of this chapter that the time evolution is preserved; i.e., the dynamics in the $2(m+1)$-dimensional phase space (with coordinates with tildes) reproduces precisely the dynamics in the $2m$-dimensional phase space (with coordinates without tildes).

Given the canonical transformation, Eq. (6.33), and its inverse, Eq. (6.34), one can examine the restrictions imposed on the matrix $\mathbf{A}$ by the requirement of invariance of the metric g_{ab} under this generalization of a Markov mapping. The calculation is summarized in Appendix 2 of this chapter. The result is that the matrix $\mathbf{A}$ must be proportional to the $n \times n$ unit matrix, $\mathbf{A} = A\mathbf{1}$, where A is a constant. The line element depends on two parameters only, α and A, and it takes the remarkably simple form

$$d\sigma^2 = \sum_k \left[\frac{\alpha}{2P^k}(dP^k)^2 + 2AdP^k dS^k + \frac{2P^k}{\alpha}(1+A^2)(dS^k)^2 \right]. \qquad (6.36)$$

In the continuous case, it is not possible to carry out an analogous proof based on invariance under generalized Markov morphisms because such transformations are no longer well defined. However, based on the results of the finite case and assuming that the continuous configuration space may be discretized and approximated by a finite configuration space, we will restrict to metrics of the form

$$g_{ab} = \begin{pmatrix} \frac{\alpha}{2P_x} & A \\ A & \frac{2P_x}{\alpha}(1+A^2) \end{pmatrix} \delta(x - x'), \qquad (6.37)$$

where A is a constant.

6.2.7 Complex Coordinates and Wave Functions

Up to now, we have made use of real coordinates P^i, S^i. Kähler geometry, however, is best expressed in terms of *complex coordinates*. We carry out a complex transformation that shows that the metric of Eq. (6.36) describes in fact a flat Kähler space.

Set $A = A1$ in Eq. (6.26) and consider first the particular case $A = 0$. The tensors that define the Kähler structure take the form

$$\Omega_{ab} = \begin{pmatrix} 0 & 1 \\ -1 & 0 \end{pmatrix}, \quad g_{ab} = \begin{pmatrix} G & 0 \\ 0 & G^{-1} \end{pmatrix}, \quad J^a_{\ b} = \begin{pmatrix} 0 & G^{-1} \\ -G & 0 \end{pmatrix}. \tag{6.38}$$

Define now the *Madelung transformation*,

$$\psi^k = \sqrt{P^k}\, e^{iS^k/\alpha}, \qquad \bar{\psi}^k = \sqrt{P^k}\, e^{-iS^k/\alpha}. \tag{6.39}$$

A simple calculation shows that the tensors that define the Kähler geometry, expressed in terms of ψ^k, $\bar{\psi}^k$, take the standard form which is characteristic of flat-space [13],

$$\Omega_{ab} = \begin{pmatrix} 0 & i\alpha 1 \\ -i\alpha 1 & 0 \end{pmatrix}, \quad g_{ab} = \begin{pmatrix} 0 & \alpha 1 \\ \alpha 1 & 0 \end{pmatrix}, \quad J^a_{\ b} = \begin{pmatrix} -i1 & 0 \\ 0 & i1 \end{pmatrix}. \tag{6.40}$$

One may conclude that in this case ($A = 0$) there is a natural set of fundamental variables given by ψ^i and $\bar{\psi}^i$. In terms of these variables, the tensors that define the Kähler geometry take their simplest form. If the constant α is set equal to $\hbar$, these fundamental variables are precisely the *wave functions* of quantum mechanics. This is a remarkable result because it is based on geometrical arguments only. *The derivation does not use any assumptions from quantum theory.*

Consider now the more general case $A \neq 0$. The tensors that define the Kähler structure take the form

$$\Omega_{ab} = \begin{pmatrix} 0 & 1 \\ -1 & 0 \end{pmatrix}, \quad g_{ab} = \begin{pmatrix} G & A1 \\ A1 & (1+A^2)G^{-1} \end{pmatrix}, \quad J^a_{\ b} = \begin{pmatrix} A1 & (1+A^2)G^{-1} \\ -G & -A1 \end{pmatrix}. \tag{6.41}$$

In this case, define the *modified Madelung transformation*

$$\phi^k = \sqrt{P^k}\, e^{i\left[S^k/(\alpha\Lambda^{-1}) - \gamma \ln\sqrt{P^k}\right]}, \qquad \bar{\phi}^k = \sqrt{P^k}\, e^{-i\left[S^k/(\alpha\Lambda^{-1}) - \gamma \ln\sqrt{P^k}\right]}, \tag{6.42}$$

where $\Lambda = 1/(1+A^2)$ and $\gamma = -A/(1+A^2)$. Once more, the tensors that define the Kähler geometry, expressed now in terms of ϕ^k, $\bar{\phi}^k$, take the standard form which is characteristic of flat-space,

$$\Omega_{ab} = \begin{pmatrix} 0 & i\left(\frac{\alpha}{\Lambda}\right)1 \\ -i\left(\frac{\alpha}{\Lambda}\right) & 0 \end{pmatrix}, \quad g_{ab} = \begin{pmatrix} 0 & \left(\frac{\alpha}{\Lambda}\right)1 \\ \left(\frac{\alpha}{\Lambda}\right)1 & 0 \end{pmatrix}, \quad J^a_{\ b} = \begin{pmatrix} -i1 & 0 \\ 0 & i1 \end{pmatrix}. \tag{6.43}$$

This shows that the geometry of the Kähler space is the same whether $A = 0$ or $A \neq 0$. In fact, it is possible to map one case to the other using an A-dependent canonical transformation. It is clear then that both cases lead to the *same* theory (provided one sets $\alpha = \hbar$ when $A = 0$ or $\alpha \Lambda^{-1} = \hbar$ when $A \neq 0$), and in the following sections we will set $A = 0$ and use the complex coordinates (wave functions) ψ^i and $\bar{\psi}^i$.

The transformation from the coordinates of Eq. (6.39) to the coordinates of Eq. (6.42) is a particular case of a family of nonlinear gauge transformations introduced by Doebner and Goldin [16] (compare to their Eq. (2.2)). As pointed out by Doebner and Goldin, the theory that results from this particular family of nonlinear gauge transformations is physically equivalent to standard quantum mechanics. Here we arrive at the same conclusion, but now on the basis of the equivalence of the two cases $A = 0$ and $A \neq 0$ via a canonical transformation. One may therefore view the present derivation of the geometric formulation of quantum mechanics as providing a new route to this family of Doebner-Goldin nonlinear gauge transformations.

The continuous case is completely analogous to the discrete case. We consider therefore only the case $A = 0$. The complex coordinate transformation that is required is the Madelung transformation, $\psi = \sqrt{P}e^{iS/\alpha}$, $\bar{\psi} = \sqrt{P}e^{-iS/\alpha}$. In terms of the new variables, the geometrical structures take the standard flat-space form,

$$\Omega_{ab} = \begin{pmatrix} 0 & i\alpha \\ -i\alpha & 0 \end{pmatrix} \delta(x - x'), \quad g_{ab} = \begin{pmatrix} 0 & \alpha \\ \alpha & 0 \end{pmatrix} \delta(x - x'), \quad J^a{}_b = \begin{pmatrix} -i & 0 \\ 0 & i \end{pmatrix} \delta(x - x').$$

(6.44)

Here again, if α is set equal to $\hbar$, the fundamental variables are precisely the wave functions of quantum mechanics.

> The natural geometry of ensembles on configuration space is a Kähler geometry, which brings together metric, symplectic and complex structures in a harmonious way. The natural complex coordinates of the Kähler space are precisely the wave functions of quantum mechanics.

6.2.8   Statistical Distance in the Kähler Space

We now want to introduce a new expression for statistical distance which will be valid in the Kähler space. In going from the discrete n-dimensional space of probabilities P^i to the full $2n$-dimensional phase space of the P^i and S^i, the metric has been extended, and this should be taken in consideration when carrying out the generalization of the statistical distance.

Consider two points ψ_A and ψ_B representing states in this space which are joined by a curve $\psi^i(t), 0 \leq t \leq 1$. The generalization of the expression for the distance l is

$$l = \int_0^1 dt \sqrt{g_{ij} \frac{d\psi^i(t)}{dt} \frac{d\bar{\psi}^j(t)}{dt}} = \sqrt{2\alpha} \int_0^1 dt \left| \left(\frac{d\psi(t)}{dt}, \frac{d\psi(t)}{dt} \right) \right|. \qquad (6.45)$$

where we have introduced the notation $|(\psi, \psi)| = \sqrt{\sum_i \psi^i \bar{\psi}^i}$.

The statistical distance in the Kähler space is defined as the shortest distance computed with Eq. (6.45). Since the curve $P(t)$ is assumed to lie in the probability space, then $\psi(t)$ must satisfy the condition

$$1 = \sum_i \psi^i(t) \bar{\psi}^i(t), \qquad (6.46)$$

that is, the curve must lie on the unit sphere in the $\{\psi(t), \bar{\psi}(t)\}$ space. The shortest distance on the unit sphere is equal to the angle between the unit vectors ψ_A and ψ_B. This leads immediately to the expression for the statistical distance that appears at the end of section III of Wootters' paper [8], provided α is set to the standard value of $1/2$,

$$d(\psi_A, \psi_B) = \sqrt{2\alpha} \cos^{-1} |(\psi_A, \psi_B)|. \qquad (6.47)$$

In this way we obtain a very brief, geometrical derivation of Wootters' expression for the statistical distance in quantum mechanics, which he derived using a completely different argument [8].

6.3 Geometrical Reconstruction of Quantum Mechanics

We have shown that there is a natural Kähler geometry associated with ensembles on configuration space. This rich geometrical structure allows for a reconstruction of quantum mechanics. To derive this result, we look at the group of transformations of the theory and show that the requirement of metric invariance, defined below, leads precisely to the group of unitary transformations. This is an easily formulated condition which requires however a finite value of α because the metric and complex structures become singular when $\alpha \to 0$ (which is equivalent to taking the limit $\hbar \to 0$). We also show that one may associate a complex Hilbert space with the Kähler space. This leads to the standard formulation of quantum mechanics.

6.3.1 Group of Unitary Transformations

To understand the role of the unitary group in the case of a discrete configuration space, one can argue in general terms as follows. Since the Kähler structure includes a symplectic structure, one would expect the group of symplectic transformations, $Sp(2n,R)$, to play a prominent role. However, the group of transformations

of the theory must be a subgroup of Sp(2n,R) because not all symplectic transformation are allowed, as physically meaningful transformations must satisfy additional
requirements. The first requirement is that they preserve the normalization of the
probability. We will impose a second requirement, metric invariance (see below).
These two conditions are quite strong: requiring normalization of the probability
and metric invariance leads to the group of rotations on the 2n-dimensional sphere,
$O(2n, R)$. Unitary transformations are the only symplectic transformations which are
also rotations; i.e., $Sp(2n,R) \cap O(2n, R) = U(n)$ [17]. Therefore, we are led to the
group of unitary transformations. Notice that the first requirement is quite general,
but the second requirement only makes sense for $\alpha > 0$ because the metric becomes
singular when $\alpha \to 0$. We make some remarks concerning this limit at the end of
this section.

We now show explicitly that the evolution generated by any ensemble Hamiltonian $\mathcal{H}$ is described by a one-parameter group of unitary transformations. It will be
convenient to work with the complex coordinates ψ^k, $\bar{\psi}^k$ (that is, we set set $A = 0$;
see the discussion in Sect. 6.2.7). We introduce the two requirements that we mentioned above: that the transformations generated by $\mathcal{H}$ preserve the normalization
of the probability, $\sum_i P^i = \sum_i \psi^i \bar{\psi}^i = 1$, and that the metric be form invariant
under those transformations; i.e., that the transformations preserve the line element
$d\sigma^2 = 2\alpha \sum_j d\bar{\psi}^j d\psi^j$ of the Kähler space. Requiring normalization of the probability and metric invariance leads to the group of rotations on the 2n-dimensional
sphere. Such rotations are linear with respect to ψ^j and $\bar{\psi}^j$. For an infinitesimal
transformation, it follows that

$$\frac{\partial \psi^j}{\partial t} = -i\frac{\partial \mathcal{H}}{\partial \bar{\psi}^j}, \quad \frac{\partial \bar{\psi}^j}{\partial t} = i\frac{\partial \mathcal{H}}{\partial \psi^j}, \tag{6.48}$$

are linear in ψ and $\bar{\psi}$. Then $\mathcal{H}$ must be of the form

$$\mathcal{H} = E(t) + \sum_{j,k} \left[M_{jk} \bar{\psi}^j \psi^k + N_{jk} \psi^j \psi^k + \bar{N}_{jk} \bar{\psi}^j \bar{\psi}^k \right] \tag{6.49}$$

where $E(t)$ is a arbitrary function of time, M is Hermitian, and N is symmetric.

The third and final requirement is to consider rotations on the 2n-dimensional
sphere that are compatible with the equations of motion. Conservation of probability
requires that the ensemble Hamiltonian be invariant (up to an additive constant) under
$S^j \to S^j + c$, since to first-order [18]

$$0 = \epsilon \sum_j \frac{\partial P^j}{\partial t} = \epsilon \sum_j \frac{\partial \mathcal{H}}{\partial S^j} = \mathcal{H}(P, S + \epsilon) - \mathcal{H}(P, S), \tag{6.50}$$

as we discussed in Sect. 1.4. This condition, when written in terms of complex coordinates, is equivalent to invariance of the ensemble Hamiltonian under $\psi \to \psi e^{ic}$.

Using the notation $Q := \sum N_{jk} \psi^j \psi^k$, Eq. (6.49) leads to the equality

$$\left[Q e^{2ic} + \bar{Q} e^{-2ic} \right] = 0 \tag{6.51}$$

which must be valid for all c. Differentiating with respect to c gives the additional equality

$$2i \left[Q e^{2ic} - \bar{Q} e^{-2ic} \right] = 0. \tag{6.52}$$

Combining these two expressions leads to $Q = 0$. Since this must hold for all ψ, it follows that $N_{jk} \equiv 0$, i.e., the ensemble Hamiltonian has the Hermitian form

$$\mathcal{H} = E(t) + \sum_{j,k} M_{jk} \bar{\psi}^j \psi^k \tag{6.53}$$

as desired.

The same conclusion holds for the case of a continuous configuration space. We consider again a canonical transformation generated by some ensemble Hamiltonian $\mathcal{H}$. Requiring normalization of the probability and metric invariance leads again to the group of rotations, and such rotations are linear with respect to ψ and $\bar{\psi}$. Therefore, we restrict to the group of transformations that preserve the linearity of these equations,

$$\varphi(y) = \int d^n x \, K(x, y) \psi(x), \qquad \bar{\varphi}(y) = \int d^n x \, \bar{K}(x, y) \bar{\psi}(x), \tag{6.54}$$

with $\{\varphi, \bar{\varphi}'\} = \{\psi, \bar{\psi}'\} = \delta(y - y')$, as the transformation is canonical. A simple calculation leads to the following condition,

$$\{\varphi, \bar{\varphi}'\} = \int d^n x'' \left[\frac{\delta \varphi}{\delta \psi''} \frac{\delta \bar{\varphi}'}{\delta \bar{\psi}''} - \frac{\delta \varphi}{\delta \bar{\psi}''} \frac{\delta \bar{\varphi}''}{\delta \psi''} \right]$$

$$= \int d^n x'' \left[K(x'', y) \bar{K}(x'', y') \right] = \delta(y - y'). \tag{6.55}$$

This is precisely the condition for the transformation to be unitary.

We end this section with a few remarks concerning the limit $\alpha \to 0$, which corresponds to the limit $\hbar \to 0$ in quantum mechanics. In this case, neither the Madelung transformation, Eq. (6.39), nor the modified Madelung transformation, Eq. (6.42), are well defined so we have to consider the description in terms of real coordinates rather than complex coordinates (wavefunctions). Of the three tensors that describe the Kähler structure, Eq. (6.38), only the symplectic structure Ω_{ab} remains well defined: both the metric g_{ab} and the complex structure $J^a_{\ b}$ become singular when $\alpha \to 0$.

Therefore, in the limit $\alpha \to 0$, the metric g_{ab} and complex structure $J^a_{\ b}$ are no longer available and the Kähler structure is lost. The only mathematical structure that remains is the symplectic structure, which can be used to define equations of motion for the canonically conjugate variables P^i, S^i given an appropriate ensemble

Hamiltonian. But this means that the argument concerning the role of the unitary group given in this section no longer applies as $\alpha \to 0$ because without a metric one can no longer require metric invariance.

6.3.2   Hilbert Space Formulation

There is a standard construction that associates a complex Hilbert space with any Kähler space. Given two complex vectors ψ^i and φ^i, define the Dirac product by [1]

$$
\begin{aligned}
\langle \psi | \varphi \rangle &= \frac{1}{2\alpha} \sum_i \left\{ (\psi^i, \bar{\psi}^i) \cdot [g + i\Omega] \cdot \begin{pmatrix} \varphi^i \\ \bar{\varphi}^i \end{pmatrix} \right\} \\
&= \frac{1}{2} \sum_i \left\{ (\psi^i, \bar{\psi}^i) \left[\begin{pmatrix} 0 & 1 \\ 1 & 0 \end{pmatrix} + i \begin{pmatrix} 0 & i1 \\ -i1 & 0 \end{pmatrix} \right] \begin{pmatrix} \varphi^i \\ \bar{\varphi}^i \end{pmatrix} \right\} \\
&= \sum_i \bar{\psi}^i \varphi^i.
\end{aligned}
\tag{6.56}
$$

An analogous result is valid in the continuous case. Given two complex functions ϕ and φ, define the Dirac product by [1]

$$
\begin{aligned}
\langle \phi | \varphi \rangle &= \frac{1}{2} \int d^n x \left\{ (\phi(x^\mu), \phi^*(x^\mu)) \cdot [g + i\Omega] \cdot \begin{pmatrix} \varphi(x^\mu) \\ \varphi^*(x^\mu) \end{pmatrix} \right\} \\
&= \frac{1}{2} \int d^n x \left\{ (\phi(x^\mu), \phi^*(x^\mu)) \left[\begin{pmatrix} 0 & 1 \\ 1 & 0 \end{pmatrix} + i \begin{pmatrix} 0 & i \\ -i & 0 \end{pmatrix} \right] \begin{pmatrix} \varphi(x^\mu) \\ \varphi^*(x^\mu) \end{pmatrix} \right\} \\
&= \int d^n x \, \phi^*(x^\mu) \varphi(x^\mu).
\end{aligned}
\tag{6.57}
$$

In this way one arrives at the *Hilbert space formulation of quantum mechanics*.

The Kähler geometry associated with ensembles on configuration space allows for a reconstruction of the geometric formulation of quantum theory. One may associate a Hilbert space with the Kähler space and this leads to the standard version of quantum mechanics.

6.4   Information Geometry and Quantum Mechanics

We have shown that the geometry of ensembles on configuration space is a Kähler geometry, that this rich geometrical structure allows for a reconstruction of the geometric formulation of quantum theory, and that one may associate a Hilbert space with the Kähler space, obtaining in this way the standard formulation of quantum theory.

However, the analysis may also be interpreted as *a derivation of quantum theory from information geometry*, as the starting point of the analysis is very basic: a space of probabilities and the information metric. The derivation has a number of interesting features, in the discrete case given by:

- The doubling of the dimensionality of the space (i.e., $\{P^i\} \rightarrow \{P^i, S^i\}$) and a symplectic structure from dynamical considerations
- The complex structure that is characteristic of quantum mechanics from requiring consistency between the metric and symplectic structures
- Wave functions ψ^i as the natural complex coordinates of the Kähler space
- The representation of a particular case of a family of Doebner-Goldin nonlinear gauge transformations in terms of canonical transformations
- Time evolution described by a one-parameter group of unitary transformations
- The Hilbert space formulation of quantum mechanics expressed in terms of geometrical quantities associated with the Kähler space

Mehrafarin [19] and Goyal [20, 21] have developed reconstructions of discrete quantum theory using information-geometrical approaches which provide alternatives to the one presented here. Their approaches are based on assumptions that are very different from the ones that we have used. Perhaps one of the main differences between their reconstructions and ours is the handling of *dynamics*, which plays a crucial role in our derivation: as we have seen in this chapter, the use of an action principle to describe the dynamics of probabilities leads in a natural way to much of the geometrical structure that is needed for a geometrical formulation of quantum theory.

Appendix 1: Symplectic Geometry, Compatibility Conditions, and Kähler Structure

We consider a finite space, but similar relations hold for infinite dimensional spaces. A symplectic vector space is a vector space V that is equipped with a bilinear form $\Omega : V \times V \rightarrow R$ that is [22]:

a. Skew-symmetric: $\Omega(u, v) = -\Omega(v, u)$ for all $u, v \in V$,

b. Non-degenerate: if $\Omega(u, v) = 0$ for all $v \in V$, then $u = 0$.

The standard space is $\Re^{2n}$, and typically Ω is chosen to be the matrix

$$\Omega_{ab} = \begin{pmatrix} 0 & 1 \\ -1 & 0 \end{pmatrix}, \tag{6.58}$$

where 1 is the unit matrix in n dimensions.

Consider the dual space V^*. The symplectic structure can be identified with an element of $V^* \times V^*$, so that $\Omega(u, v) = \Omega_{ab}u^a v^b$. Since the spaces V and V^* are isomorphic, there is a Ω^{ab} that is the dual of Ω_{ab}. This Ω^{ab} can be identified with an element of $V \times V$. The convention is to set [22]

$$\Omega^{ab} = -(\Omega^{-1})^{ab} = \begin{pmatrix} 0 & 1 \\ -1 & 0 \end{pmatrix}, \qquad (6.59)$$

so that $\Omega^{ac}\Omega_{cb} = -\delta^a_b$.

We assume there is a metric in the space, $g_{ab} = g_{ba}$, and a corresponding inverse metric g^{ab} with $g_{ab}g^{bc} = \delta_{ac}$ (indices are raised and lowered with g_{ab} and g^{ab}). The metric also defines a map $V \to V^*$ to the dual space in an obvious way. Therefore, the space has two linear operators that induce maps $V \to V^*$, Ω_{ab} and g_{ab}. They will be related by an equation of the form $\Omega_{ab} = g_{ac}J^c_{\ b}$ for some choice of linear operator $J^c_{\ b}$. This is Eq. (6.22), the first of the Kähler conditions.

The relations $\Omega^{ac}\Omega_{cb} = -\delta^a_b$ and $\Omega_{ab} = g_{ac}J^c_{\ b}$ lead to the condition $J^a_{\ s}J^s_{\ c} = -\delta^a_c$, that is, that $J^a_{\ b}$ is a complex structure. This is Eq. (6.24), the third of the Kähler conditions.

Finally, $\Omega_{ab} = g_{ac}J^c_{\ b}$ and $J^a_{\ s}J^s_{\ c} = -\delta^a_c$, together with the symmetries $-\Omega_{cb} = \Omega_{bc}$ and $g_{ba} = g_{ab}$, lead to $g_{cd} = J^a_{\ c}g_{ab}J^b_{\ d}$ which is Eq. (6.23), the second of the Kähler conditions.

This shows that consistency requirements imply that a space with both symplectic and metric structures must have a Kähler structure.

Appendix 2: Generalized Markov Mappings

Constants of the Motion and Dynamics

To derive the canonical transformation that generalizes the Markov mapping of Eq. (6.13), the dimensionality of the original phase space was increased by two in a trivial way. This led to two constants of the motion, P^{m+1} and S^{m+1}. P^{m+1} was set to $P^{m+1} \approx 0$, with corresponding constraints for the $\tilde{P}^k$ of the form

$$\tilde{P}^m \approx kP^m,$$
$$\tilde{P}^{m+1} \approx (1-k)P^m. \qquad (6.60)$$

These are precisely the conditions that are needed to get a generalization of the Markov mapping of Eq. (6.13). When $k = 1/2$, $\tilde{P}^m \approx \tilde{P}^{m+1}$, which is expected because in this case there should be invariance under the re-labeling $m \leftrightarrow m+1$. To fix the value of S^{m+1}, notice that $S^{m+1} \approx c$ leads to constraints for the $\tilde{S}^k$ of the form

$$\tilde{S}^m \approx S^m + (1-k)c,$$
$$\tilde{S}^{m+1} \approx S^m - kc. \qquad (6.61)$$

We can argue once more that there should be invariance under the re-labeling $m \leftrightarrow m + 1$ in the case when $k = 1/2$. But this can only be satisfied if $c = 0$. We can conclude therefore that the constants of the motion must satisfy

$$P^{m+1} = (1 - k)\tilde{P}^m - k\tilde{P}^{m+1} \approx 0,$$
$$S^{m+1} = \tilde{S}^m - \tilde{S}^{m+1} \approx 0. \tag{6.62}$$

The corresponding constraints for the unprimed coordinates are of the form

$$\tilde{P}^m \approx kP^m, \quad \tilde{P}^{m+1} \approx (1 - k)P^m,$$
$$\tilde{S}^m \approx S^m, \quad \tilde{S}^{m+1} \approx S^m. \tag{6.63}$$

We now need to check that the dynamics on the $2(n+1)$-dimensional phase space (with coordinates with tildes) reproduces precisely the dynamics on the original $2n$-dimensional phase space (with coordinates without tildes). Using Eqs. (6.33–6.34), one can show that

$$\frac{\partial \tilde{P}^i}{\partial t} = \frac{\partial H}{\partial S^i}, \quad \frac{\partial \tilde{P}^m}{\partial t} = \frac{\partial H}{\partial S^m}k, \quad \frac{\partial \tilde{P}^{m+1}}{\partial t} = \frac{\partial H}{\partial S^m}(1 - k),$$
$$\frac{\partial \tilde{S}^i}{\partial t} = -\frac{\partial H}{\partial P^i}, \quad \frac{\partial \tilde{S}^m}{\partial t} = -\frac{\partial H}{\partial P^m}, \quad \frac{\partial \tilde{S}^{m+1}}{\partial t} = -\frac{\partial H}{\partial P^m}. \tag{6.64}$$

These equations lead to

$$\frac{\partial P^i}{\partial t} = \frac{\partial \tilde{P}^i}{\partial t} = \frac{\partial H}{\partial S^i}, \quad \frac{\partial P^m}{\partial t} = \frac{\partial \tilde{P}^m}{\partial t} + \frac{\partial \tilde{P}^{m+1}}{\partial t} = \frac{\partial H}{\partial S^m}$$
$$\frac{\partial S^i}{\partial t} = \frac{\partial \tilde{S}^i}{\partial t} = -\frac{\partial H}{\partial P^i}, \quad \frac{\partial S^m}{\partial t} = k\frac{\partial \tilde{S}^m}{\partial t} + (1 - k)\frac{\partial \tilde{S}^{m+1}}{\partial t} = -\frac{\partial H}{\partial P^m}, \tag{6.65}$$

which are the correct equations of motion for the original space.

Invariance of the Kähler Metric Under Generalized Markov Mappings

The metric of the Kähler space is given by

$$g_{ab} = \begin{pmatrix} G & A^T \\ A & (1 + A^2)G^{-1} \end{pmatrix}, \tag{6.66}$$

where $G = \text{diag}(\alpha/2P^i)$ and the $n \times n$ matrix A satisfies $GAG^{-1} = A^T$. For the calculations in this Appendix it is convenient to introduce the matrix B with matrix elements given by

$$B_{jk} = \sqrt{P_j/P_k}\, A_{jk}. \tag{6.67}$$

It is straightforward to show that $\mathbf{B}$ is a symmetric matrix, $B_{jk} = B_{kj}$.

To restrict the form of $\mathbf{B}$, it will be sufficient to consider the invariance of the metric under the particular generalized Markov mapping which corresponds to the inverse canonical transformation given by Eq. (6.34). After taking into consideration the constraints, Eq. (6.35), the generalized Markov mapping can be written in the form

$$P = (P^i, P^m) \rightarrow \tilde{P} = (\tilde{P}^i, \tilde{P}^m, \tilde{P}^{m+1}) := (P^i, kP^m, (1-k)P^m),$$
$$S = (S^i, S^m) \rightarrow \tilde{S} = (\tilde{S}^i, \tilde{S}^m, \tilde{S}^{m+1}) := (S^i, S^m, S^m), \tag{6.68}$$

where $i = 1, \ldots, m-1$.

As a first step, look at the contribution to the line element $d\sigma^2$ from the mixed terms $dP^k dS^k$. In terms of the coordinates without tildes,

$$d\sigma^2 = \sum_{i=1}^{m-1} \left\{ B_{ii} dP^i dS^i + \sqrt{\frac{P^i}{P^m}} B_{im} dP^i dS^m + \sqrt{\frac{P^m}{P^i}} B_{mi} dP^m dS^i \right\}$$
$$+ B_{mm} dP^m dS^m. \tag{6.69}$$

There is a corresponding expression for the coordinates with tildes, and with the help of Eq. (6.68) it can be rewritten in terms of coordinates without tildes. This leads to

$$d\sigma^2 = \sum_{i=1}^{m-1} \left\{ \tilde{B}_{ii} dP^i dS^i + \left[\sqrt{\frac{P^i}{kP^m}} \tilde{B}_{im} + \sqrt{\frac{P^i}{(1-k)P^m}} \tilde{B}_{i(m+1)} \right] dP^i dS^m \right\}$$
$$+ \sum_{i=1}^{m-1} \left\{ \left[\sqrt{\frac{k^3 P^m}{P^i}} \tilde{B}_{mi} + \sqrt{\frac{(1-k)^3 P^m}{P^i}} \tilde{B}_{(m+1)i} \right] dP^m dS^i \right\}$$
$$+ \left[k\tilde{B}_{mm} + \sqrt{\frac{k^3}{1-k}} \tilde{B}_{m(m+1)} \right.$$
$$\left. + \sqrt{\frac{(1-k)^3}{k}} \tilde{B}_{(m+1)m} + (1-k)\tilde{B}_{(m+1)(m+1)} \right] dP^m dS^m. \tag{6.70}$$

Equate terms in Eqs. (6.69) and (6.70) proportional to the same $dP^a dS^b$, where $a, b = 1, \ldots, m$. This leads to the four relations

$$B_{ii} = \tilde{B}_{ii},$$

$$B_{im} = \sqrt{\frac{1}{k}} \tilde{B}_{im} + \sqrt{\frac{1}{(1-k)}} \tilde{B}_{i(m+1)},$$

$$B_{mi} = \sqrt{k^3}\tilde{B}_{mi} + \sqrt{(1-k)^3}\tilde{B}_{(m+1)i},$$

$$B_{mm} = k\tilde{B}_{mm} + \sqrt{\frac{k^3}{1-k}}\tilde{B}_{m(m+1)} + \sqrt{\frac{(1-k)^3}{k}}\tilde{B}_{(m+1)m}$$

$$+ (1-k)\tilde{B}_{(m+1)(m+1)}. \tag{6.71}$$

Since the matrix B is symmetric, $B_{im} = B_{mi}$, which leads to

$$\sqrt{\frac{1}{k}}\tilde{B}_{im} + \sqrt{\frac{1}{(1-k)}}\tilde{B}_{i(m+1)} = \sqrt{k^3}\tilde{B}_{mi} + \sqrt{(1-k)^3}\tilde{B}_{(m+1)i}. \tag{6.72}$$

By symmetry, $\tilde{B}_{im} = \tilde{B}_{i(m+1)}$ at $k = 1/2$, but this relation can only be satisfied if $\tilde{B}_{im} = \tilde{B}_{i(m+1)} = 0$, which in turn implies $B_{im} = B_{mi} = 0$. Since B_{im} and B_{mi} are independent of k, it follows that they must always be zero. This shows that the off-diagonal elements of the matrix B are zero.

Now look at the contribution to the line element $d\sigma^2$ from terms proportional to $dP^a dP^a$ and $dS^a dS^a$. The terms proportional to $dP^a dP^a$ give the two relations

$$B_{ii} = \tilde{B}_{ii},$$
$$B_{mm} = k\tilde{B}_{mm} + (1-k)\tilde{B}_{(m+1)(m+1)}, \tag{6.73}$$

while the terms proportional to $dS^a dS^a$ give the two relations

$$1 + B_{ii}^2 = 1 + \tilde{B}_{ii}^2,$$
$$1 + B_{mm}^2 = k(1 + \tilde{B}_{mm}^2) + (1-k)\left(1 + \tilde{B}_{(m+1)(m+1)}^2\right). \tag{6.74}$$

Combining Eqs. (6.73) and (6.74) leads to

$$B_{ii} = \tilde{B}_{ii},$$
$$B_{mm} = \tilde{B}_{mm} = \tilde{B}_{(m+1)(m+1)}. \tag{6.75}$$

Notice that this result is valid for arbitrary values of k. Since there is nothing special about the particular labels m and $(m+1)$, *all* the diagonal elements of the matrices B and $\tilde{B}$ must be equal. Then

$$B = B1_{m \times m},$$
$$\tilde{B} = B1_{(m+1) \times (m+1)}, \tag{6.76}$$

where $1_{n \times n}$ is the $n \times n$ unit matrix and B still has to be determined.

To carry out this last step, use the relations

$$B(P, S) = B(P^i, P^m, S^i, S^m),$$
$$\tilde{B}(\tilde{P}, \tilde{S}) = B(P(\tilde{P}), S(\tilde{S})) = B(\tilde{P}^i, \tilde{P}^m + \tilde{P}^{m+1}, \tilde{S}^i, k\tilde{S}^m + (1 - k)\tilde{S}^{m+1}).$$

$$(6.77)$$

The functional form of $B(P, S)$ must be the same as the functional form of $\tilde{B}(\tilde{P}, \tilde{S})$, and these expressions must be both invariant under permutations and independent of k. The only functional form that seems to satisfy all these conditions appears to be $B(P, S) = B(\sum_i P^i)$. But $\sum_i P^i = 1$, therefore one can conclude that B is a *constant* and the matrix

$$\mathsf{B} = B\mathbf{1} \tag{6.78}$$

is a constant matrix proportional to the unit matrix. This in turn implies that

$$\mathsf{A} = A\mathbf{1} \tag{6.79}$$

where A is a constant.

References

1. Kibble, T.W.B.: Geometrization of quantum mechanics. Commun. Math. Phys. **65**, 189–201 (1979)
2. Reginatto, M., Hall, M.J.W.: Quantum theory from the geometry of evolving probabilities. In: Goyal, P., Giffin, A., Knuth, K.H., Vrscay, E. (eds.) Bayesian Inference and Maximum Entropy Methods in Science and Engineering, 31st International Workshop on Bayesian Inference and Maximum Entropy Methods in Science and Engineering, Waterloo, Canada, 10–15 July 2011, AIP Conference Proceedings, vol. 1443, American Institute of Physics, Melville, New York (2012)
3. Reginatto, M., Hall, M.J.W.: Information geometry, dynamics and discrete quantum mechanics. In: von Toussaint, U. (ed.) Bayesian Inference and Maximum Entropy Methods in Science and Engineering, 32nd International Workshop on Bayesian Inference and Maximum Entropy Methods in Science and Engineering, Garching, Germany, 15–20 July 2012. AIP Conference Proceedings, vol. 1553, American Institute of Physics, Melville, New York (2013)
4. Reginatto, M.: From probabilities to wave functions: a derivation of the geometric formulation of quantum theory from information geometry. J. Phys. Conf. Ser. **538**, 012018 (2014)
5. Bhattacharyya, A.: On a measure of divergence between two statistical populations defined by their probability distributions. Bull. Calcutta Math. Soc. **35**, 99–109 (1943)
6. Bhattacharyya, A.: On a measure of divergence between two multinomial populations. Sankhya **7**, 401–406 (1946)
7. Good, I.J.: Invariant distance in statistics: some miscellaneous comments. J. Stat. Comput. Simul. **36**, 179–186 (1990)
8. Wootters, W.K.: Statistical distance and Hilbert space. Phys. Rev. D. **23**, 357–362 (1981)
9. Rao, C.R.: Differential metrics in probability spaces. In: Amari, S.-I., Barndorff-Nielsen, O.E., Kass, R.E., Lauritzen, S.L., Rao, C.R. (eds.) Differential Geometry in Statistical Inference, pp. 217–240. Institute of Mathematical Statistics, Hayward (1987)
10. Jeffreys, H.: An invariant form for the prior probability in estimation problems. Proc. Roy. Soc. A **186**, 453–461 (1946)
11. Čencov, N.N.: Statistical Decision Rules and Optimal Inference. Translations of Mathematical Monographs. American Mathematical Society, Providence (1981)

12. Campbell, L.L.: An extended Čencov characterization of the information metric. Proc. Am. Math. Soc. **98**, 135–141 (1986)
13. Goldberg, S.I.: Curvature and Homology. Dover Publications, New York (1982)
14. Scheifele, G.: On nonclassical canonical systems. Celest. Mech. **2**, 296–310 (1970)
15. Kurcheeva, I.V.: Kustaanheimo-Stiefel regularization and nonclassical canonical transformations. Celest. Mech. **15**, 353–365 (1977)
16. Doebner, H.-D., Goldin, G.A.: Introducing nonlinear gauge transformations in a family of nonlinear Schrdinger equations. Phys. Rev. A **54**, 3764–3771 (1996)
17. Arnold, V.I.: Mathematical Methods of Classical Mechanics. Springer, Berlin (1978)
18. Hall, M.J.W.: Superselection from canonical constraints. J. Phys. A **37**, 7799–7811 (2004)
19. Mehrafarin, M.: Quantum mechanics from two physical postulates. Int. J. Theor. Phys. **44**, 429–442 (2005)
20. Goyal, P.: Information-geometric reconstruction of quantum theory. Phys. Rev. A **78**, 052120 (2008)
21. Goyal, P.: From information geometry to quantum theory. New J. Phys. **12**, 023012 (2010)
22. Stewart, J.: Advanced General Relativity. Cambridge University Press, Cambridge (1990)

Chapter 7
Local Representations of Rotations on Discrete Configuration Spaces

Abstract A spin-half system may be characterised as having a set of two-valued observables which generate infinitesimal rotations in three dimensions. This abstract formulation can be given a concrete realization using ensembles on configuration space. We derive very general probabilistic models for ensembles that consist of one and two spin-half systems. In the case of a single spin-half system, there are two main requirements that need to be satisfied: the configuration space must be a discrete set, labelling the outcomes of two-valued spin observables, and these observables must provide a representation of $so(3)$. These two requirements are sufficient to lead to a model which is equivalent to the quantum theory of a single qubit. The case of a pair of spin-half systems is more complicated, in that additional physical requirements concerning locality and subsystem independence must also be taken into account, and now the observables must provide a representation of $so(3) \oplus so(3)$. We show in this case that, in addition to a model equivalent to the quantum theory of a pair of qubits, it may also be possible to have non-quantum local models.

7.1 Introduction

Systems that consist of a pair of spin-half quantum particles have proven extremely useful for elucidating the structure of quantum mechanics. This became apparent after Bohm [1] provided a reformulation of the EPR paradox [2] in terms of a pair of spin-half systems. This model was later used by Bell [3] to derive his famous theorem. Since then, innumerable papers have been devoted to this example. Such systems are simple enough to be tractable, yet have enough complexity to display many of the most puzzling features of quantum mechanics.

In this chapter, we look into alternative possible physical theories of spin-half systems. The quantum mechanical formulation is shown to be a particular case of a more general family of theories.

A spin-half system may be characterised as having a set of two-valued observables which generate infinitesimal rotations in three dimensions. This abstract formulation can be given a concrete realization using ensembles on configuration space, as we show in the next section. We call such a system a 'rotational bit', or 'robit', to

© Springer International Publishing Switzerland 2016

M.J.W. Hall and M. Reginatto, *Ensembles on Configuration Space*,

Fundamental Theories of Physics 184, DOI 10.1007/978-3-319-34166-8_7

distinguish it from the standard quantum qubit. We have seen in previous chapters that the formulation of ensembles on configuration space is more general than quantum mechanics, in the sense that it allows us to describe not only quantum mechanical systems but also classical and hybrid systems. Therefore, it is reasonable to expect robits to be more general than qubits. Single robits however are strictly equivalent to single qubits (see Chap. 2), as reviewed in Sect. 7.2 below.

The case of a pair of robits is more involved. In Sect. 7.3 we construct a model of a pair of robits which is very general while satisfying the following requirements motivated by basic physical principles:

- A measurement of 'spin' in any direction, for either system, gives a result $\pm 1/2$ (we choose units in which $\hbar = 1$)
- The corresponding spin observables correspond to a representation of the Lie algebra $so(3) \oplus so(3)$, and generate local rotations of the respective systems
- No signalling is possible via a rotation of either system (evolution locality)
- No signalling is possible via a measurement of either system (update locality)
- Initially independent systems remain independent under local rotations

We show that in this case, in addition to a theory which is equivalent to the quantum theory of a pair of qubits, it may also be possible to have non-quantum local models, and we discuss the conditions under which such models can provide viable alternatives to pairs of quantum mechanical qubits. Implications of the results are discussed in Sect. 7.4.

7.2 One Robit

We have already developed the theory of a single robit in Sect. 2.5.2, but as it is needed for the analysis of the two robit system, we provide a summary of the relevant results.

We define the probability distribution P in terms of the possible measurement outcomes of spin in the z-direction, which may be labelled by $\pm 1/2$. Thus, we may write $P \equiv \{P_+, P_-\}$, where P_α denotes the probability of measuring spin value $\alpha/2$ in the z-direction. The canonically conjugate quantities are labelled $S \equiv \{S_+, S_-\}$. The requirement that the set of spin observables form a representation of $so(3)$ under the Poisson bracket, thus generating spatial rotations, imply that the observable corresponding to a spin measurement in unit direction $\mathbf{n}$ has the form $L \cdot \mathbf{n}$, where $L = (L_1, L_2, L_3)$ satisfies the $so(3)$ Lie algebra,

$$\{L_j, L_k\} = \varepsilon_{jkl} L_l, \tag{7.1}$$

for $j, k = 1, 2, 3$.

The fundamental variables for a two-level system are $\{P_+, P_-, S_+, S_-\}$, and so the phase space is four-dimensional. However, since $\sum P_k = 1$ is a quantity that is conserved, it is possible to describe the system in a reduced phase space. To do this, introduce coordinates

$$
\begin{aligned}
q_0 &= (P_+ + P_-)/2, \\
q_1 &= (P_+ - P_-)/2, \\
p_0 &= S_+ + S_-, \\
p_1 &= S_+ - S_-.
\end{aligned}
\tag{7.2}
$$

It is easy to check that this transformation is a canonical transformation. Since the P_k are probabilities, q_0 is fixed to $q_0 = 1/2$. As the conservation of probability requires that observables are invariant under $S \to S + $ constant (see Sects. 1.4.1 and 2.2), they are therefore independent of p_0. Thus, the true degrees of freedom are q_1 and p_1.

The most general representation of $so(3)$ on this two-dimensional reduced phase space is given by

$$
L_1 = \sqrt{\left(\frac{1}{2}\right)^2 - q_1^2} \, \cos(p_1) = \sqrt{P_+ P_-} \, \cos(S_+ - S_-),
\tag{7.3}
$$

$$
L_2 = -\sqrt{\left(\frac{1}{2}\right)^2 - q_1^2} \, \sin(p_1) = -\sqrt{P_+ P_-} \, \sin(S_+ - S_-),
\tag{7.4}
$$

$$
L_3 = q_1 = (P_+ - P_-)/2.
\tag{7.5}
$$

up to a canonical transformation (recall that we choose units in which $\hbar = 1$). The proof is given in Sect. 2.5.2

It is instructive to compare Eqs. (7.3–7.5) to the corresponding results for the case of a quantum mechanical qubit. The operators $\hat{L}_k$ for spin-half satisfy the commutation relations $[\hat{L}_j, \hat{L}_k] = i\varepsilon_{jkl}\hat{L}_l$ and may be represented in terms of the Pauli matrices according to [4]

$$
\hat{L}_1 = \frac{1}{2}\begin{pmatrix} 0 & 1 \\ 1 & 0 \end{pmatrix}, \quad \hat{L}_2 = \frac{1}{2}\begin{pmatrix} 0 & -i \\ i & 0 \end{pmatrix}, \quad \hat{L}_3 = \frac{1}{2}\begin{pmatrix} 1 & 0 \\ 0 & -1 \end{pmatrix}.
\tag{7.6}
$$

If we now introduce the two-component wave function

$$
|\psi> = \begin{pmatrix} \psi_+ \\ \psi_- \end{pmatrix} = \begin{pmatrix} \sqrt{P_+}\, e^{iS_+} \\ \sqrt{P_-}\, e^{iS_-} \end{pmatrix},
\tag{7.7}
$$

we find that the expectation values of the quantum operators $\hat{L}_k$ are in agreement with Eqs. (7.3–7.5),

$$
< \psi|\hat{L}_k|\psi > = L_k.
\tag{7.8}
$$

Furthermore, all quantum mechanical expectation values $< \hat{A} >$, $< \hat{B} >$ can be expressed as observables $A(q_1, p_1)$, $B(q_1, p_1)$ in the reduced configuration space and, as we saw in Sect. 2.3.3, the Poisson bracket for the observables on configuration space is isomorphic to the usual commutator on Hilbert space. Thus

> Single robits are equivalent to quantum mechanical qubits.

Finally, we look at the how *Casimir functions* [5] are represented in the reduced phase space of a single robit. These are defined as the functions $F^c(q_1, p_1)$ which satisfy $\{F^c, L_k\} = 0$. The computation is straightforward, with the result that the only solutions are those where the functions F^c are *constants*.

The *Casimir operator* is a Hilbert space operator defined for quantum mechanical systems with symmetry and it is given in this case by $\hat{C} = \sum_k \hat{L}_k^2 = \frac{3}{4}\hat{1}$, where $\hat{1}$ is the unit matrix [4]. The corresponding function on the reduced phase is then given by $C = < \psi|\hat{C}|\psi > = \frac{3}{4}$; i.e., it is the particular Casimir function with $F^c = \frac{3}{4}$. Notice that

$$C \neq \sum_{k=1}^{3} L_k^2 = \frac{1}{4}. \tag{7.9}$$

Thus we see that the algebraic relation which exists between the Casimir operator $\hat{C}$ and the operators $\hat{L}_k$ does not hold for the functions C and L_k. This is because two (commuting) operators have a natural product algebra but there is no natural product algebra available for arbitrary observables in the theory of ensembles on configuration space, only the algebra defined by the Poisson bracket (see also the remarks in Sect. 9.2.2 concerning product algebras).

7.3 Two Robits

We now consider the case of two robits. The formulation is similar to that of a single robit, except that now we consider a four-level system and are interested in representations of $so(3) \oplus so(3)$ in the corresponding phase space. Furthermore, additional requirements such as locality and subsystem independence need to be taken into consideration, as we discuss below.

We define the probability distribution P in terms of the possible measurement outcomes of spins in the z-direction for each of the particles, so that P has the form $P \equiv \{P_{++}, P_{+-}, P_{-+}, P_{--}\}$. The canonically conjugate quantities are then $S \equiv \{S_{++}, S_{+-}, S_{-+}, S_{--}\}$.

7.3.1 Reduced Phase Space for Two Robits

The phase space for a two-robit system is eight-dimensional, with coordinates $\{P_{\alpha\beta}, S_{\alpha\beta}\}$ where $\alpha, \beta \in \{+, -\}$. However, since $\sum P_{\alpha\beta} = 1$ is a quantity that is conserved, it is possible to describe the system in a reduced phase space. To do this, define new coordinates

$$q_0 = (P_{++} + P_{+-} + P_{-+} + P_{--})/2, \tag{7.10}$$
$$q_1 = (P_{++} + P_{+-} - P_{-+} - P_{--})/2, \tag{7.11}$$
$$q_2 = (P_{++} - P_{+-} + P_{-+} - P_{--})/2, \tag{7.12}$$
$$q_3 = (P_{++} - P_{+-} - P_{-+} + P_{--})/2, \tag{7.13}$$
$$p_0 = (S_{++} + S_{+-} + S_{-+} + S_{--})/2, \tag{7.14}$$
$$p_1 = (S_{++} + S_{+-} - S_{-+} - S_{--})/2, \tag{7.15}$$
$$p_2 = (S_{++} - S_{+-} + S_{-+} - S_{--})/2, \tag{7.16}$$
$$p_3 = (S_{++} - S_{+-} - S_{-+} + S_{--})/2. \tag{7.17}$$

It is easy to check that this transformation is a canonical transformation. The inverse transformation is given by

$$P_{++} = (q_0 + q_1 + q_2 + q_3)/2, \tag{7.18}$$
$$P_{+-} = (q_0 + q_1 - q_2 - q_3)/2, \tag{7.19}$$
$$P_{-+} = (q_0 - q_1 + q_2 - q_3)/2, \tag{7.20}$$
$$P_{--} = (q_0 - q_1 - q_2 + q_3)/2, \tag{7.21}$$

with corresponding expressions for $S_{\alpha\beta}$ in terms of the p_j.

Since the $P_{\alpha\beta}$ are probabilities, q_0 is fixed to $q_0 = 1/2$. An argument analogous to the one given in the previous section shows that we can describe the two-robits system in the reduced phase space with coordinates $\{q_1, q_2, q_3, p_1, p_2, p_3\}$ and the constraint $q_0 = 1/2$.

Before moving on to the next section, we introduce some quantities which we need for our discussion of representations of $so(3) \oplus so(3)$. The marginal probabilities for spin up or down in the z-direction, for each of the robits, are given by

$$P_{\pm}^{(1)} := P_{\pm+} + P_{\pm-} = (q_0 \pm q_1) = (1/2 \pm q_1), \tag{7.22}$$
$$P_{\pm}^{(2)} := P_{+\pm} + P_{-\pm} = (q_0 \pm q_2) = (1/2 \pm q_2). \tag{7.23}$$

Further, the conditional probability $P_{\alpha|\beta}^{(1)}$, for spin α in the z-direction for robit 1 given spin β in the z-direction for robit 2, as well as the similarly defined conditional probability $P_{\alpha|\beta}^{(2)}$, follow from classical probability theory as

$$P_{\alpha|\beta}^{(1)} = \frac{P_{\alpha\beta}}{P_{\beta}^{(2)}}, \qquad P_{\alpha|\beta}^{(2)} = \frac{P_{\beta\alpha}}{P_{\beta}^{(1)}} \tag{7.24}$$

7.3.2   Two-Robit System with $so(3) \oplus so(3)$ Symmetry

We now look for representations of $so(3) \oplus so(3)$ in the reduced phase space. These are given in terms of six generators M_j, N_k, which satisfy

$$\{M_j, M_k\} = \varepsilon_{jkl} M_l, \quad \{N_j, N_k\} = \varepsilon_{jkl} N_l, \quad \{M_j, N_k\} = 0, \tag{7.25}$$

for $j, k, l = 1, 2, 3$. The interpretation of the generators is analogous to that discussed in Sect. 7.2. In particular, the observable corresponding to a spin measurement of robit 1 in unit direction $\mathbf{n}$ is of the form $M \cdot \mathbf{n}$ and the corresponding observable for robit 2 is of the form $N \cdot \mathbf{n}$.

The identification of generators with expectation values (see Sect. 2.2) immediately fixes the forms of M_3 and N_3. In particular, the average value of spin measurements in the z-direction may be calculated directly from the probability distribution,

$$M_3 = s \left(P_+^{(1)} - P_-^{(1)} \right) = q_1, \quad N_3 = s \left(P_+^{(2)} - P_-^{(2)} \right) = q_2, \tag{7.26}$$

where $s = 1/2$ for spin-1/2 particles (recall we have chosen units in which $\hbar = 1$). Now the task is to derive the remaining generators. To do this, we will impose certain physical requirements.

7.3.2.1   Constraints from Locality

There are a wide variety of possible choices for M_j and N_k satisfying Eqs. (7.25) and (7.26). However, a physically meaningful choice must satisfy at a minimum the following *locality* requirements:

1. no signalling is possible via a rotation of either system (evolution locality),
2. no signalling is possible via a measurement of either system (update locality).

The first requirement, evolution locality, is in fact guaranteed by the property $\{M_j, N_k\} = 0$ in Eq. (7.25), keeping in mind that the generators have the interpretation of expectation values. Note that this property of the Lie algebra is an instance of the strong separability property discussed in Sect. 3.3.3.

The second requirement, update locality, is satisfied if the generators M_j and N_k are of the form

$$M_j = P_+^{(2)} L_j(q_+^{(1)}, p_+^{(1)}) + P_-^{(2)} L_j(q_-^{(1)}, p_-^{(1)}) := M_j^+ + M_j^-, \tag{7.27}$$

$$N_k = P_+^{(1)} L_k(q_+^{(2)}, p_+^{(2)}) + P_-^{(1)} L_k(q_-^{(2)}, p_-^{(2)}) := N_k^+ + N_k^-, \tag{7.28}$$

where the L_k are of the same form as the single-robit generators given in Eqs. (7.3–7.5), and the $q_\pm^{(1)}, q_\pm^{(2)}, p_\pm^{(1)}, p_\pm^{(2)}$, are appropriate functions of the reduced phase space coordinates. The generators $M_j^\pm, N_k^\pm$ may be given the following interpretation:

the M_j^+, M_j^- correspond to $so(3)$ generators associated with the first robit conditioned on the second robit being spin-up and spin-down respectively, while the N_k^+, N_k^- correspond to $so(3)$ generators associated with the second robit conditioned on the first robit being spin-up and spin-down respectively. Note that we do *not* impose the stronger requirement that the M_j and N_k are given by the formula for extending local observables to larger configuration spaces in Eq. (3.13), as we wish to make a minimal number of physically-based assumptions.

We now look for a concrete realization of all the quantities that appear in Eqs. (7.27) and (7.28). We first derive expressions for $M_3^\pm$, $N_3^\pm$. To put M_3 and N_3 in a form that corresponds to the right hand side of Eqs. (7.27) and (7.28), we use Eqs. (7.22–7.24) and write

$$M_3 = q_1 = P_+^{(2)} \frac{P_{+|+} - P_{-|+}}{2} + P_-^{(2)} \frac{P_{+|-} - P_{-|-}}{2} = \frac{q_1 + q_3}{2} + \frac{q_1 - q_3}{2}, \quad (7.29)$$

$$N_3 = q_2 = P_+^{(1)} \frac{P_{+|+} - P_{+|-}}{2} + P_-^{(1)} \frac{P_{-|+} - P_{-|-}}{2} = \frac{q_2 + q_3}{2} + \frac{q_2 - q_3}{2}, \quad (7.30)$$

which leads immediately to

$$M_3^\pm = \frac{q_1 \pm q_3}{2}, \qquad N_3^\pm = \frac{q_2 \pm q_3}{2}. \quad (7.31)$$

Thus the generators $M_3^\pm$ act in the subspaces spanned by $\{(q_1 \pm q_3)/2, p_1 \pm p_3\}$ and the generators $N_3^\pm$ act in the subspaces spanned by $\{(q_2 \pm q_3)/2, p_2 \pm p_3\}$. The forms of the $q_\alpha^{(A)}$ in Eqs. (7.27) and (7.28) follow from Eqs. (7.29) and (7.30) as

$$q_\pm^{(1)} = \frac{M_3^\pm}{P_\pm^{(2)}} = \frac{q_1 \pm q_3}{1 \pm 2q_2}, \qquad q_\pm^{(2)} = \frac{N_3^\pm}{P_\pm^{(1)}} = \frac{q_2 \pm q_3}{1 \pm 2q_1}. \quad (7.32)$$

We can now derive the remaining generators using the same computations that we carried out in Sect. 2.5.2. This leads immediately to the relatively simple expressions

$$M_1^\pm = \sqrt{c_M^\pm - \left(\frac{q_1 \pm q_3}{2}\right)^2} \cos((p_1 \pm p_3) + b_M^\pm) \quad (7.33)$$

$$M_2^\pm = -\sqrt{c_M^\pm - \left(\frac{q_1 \pm q_3}{2}\right)^2} \sin((p_1 \pm p_3) + b_M^\pm) \quad (7.34)$$

$$M_3^\pm = \frac{q_1 \pm q_3}{2}, \quad (7.35)$$

$$N_1^\pm = \sqrt{c_N^\pm - \left(\frac{q_2 \pm q_3}{2}\right)^2} \cos((p_2 \pm p_3) + b_N^\pm) \quad (7.36)$$

$$N_2^\pm = -\sqrt{c_N^\pm - \left(\frac{q_2 \pm q_3}{2}\right)^2}\ \sin((p_2 \pm p_3) + b_N^\pm) \tag{7.37}$$

$$N_3^\pm = \frac{q_1 \pm q_3}{2}, \tag{7.38}$$

where $c_M^\pm$, $c_N^\pm$ follow from Eqs. (7.27), (7.28), (7.32–7.38) as

$$c_M^\pm = \left(\frac{1/2 \pm q_2}{2}\right)^2, \qquad c_N^\pm = \left(\frac{1/2 \pm q_1}{2}\right)^2, \tag{7.39}$$

and $b_M^\pm$ and $b_N^\pm$ still need to be determined.

Notice that there is more freedom in the choice of functions $b_M^\pm$ and $b_N^\pm$ than there is in the case of a single robit because the phase space is now six-dimensional. Thus we can set $b_M^\pm = b_M^\pm(q_1 \pm q_3, p_1 \mp p_3, q_2, p_2)$ and $b_N^\pm = b_M^\pm(q_2 \pm q_3, p_2 \mp p_3, q_1, p_1)$ and the $M_j^\pm$ and $N_K^\pm$ still provide representations of $so(3)$. One must keep in mind however that $b_M^\pm$ and $b_N^\pm$ are related via the symmetry $M_j \leftrightarrow N_j$ under the relabeling $(q_1, p_1) \leftrightarrow (q_2, p_2)$, thus $b_M^\pm$ and $b_N^\pm$ must be the *same* function of their respective arguments.

We now show that there are restrictions on $b_M^\pm$ and $b_N^\pm$ which are a consequence of the requirements $\{M_j^+, M_k^-\} = \{N_j^+, N_k^-\} = 0$, which follow from the first two equalities in Eq. (7.25) together with the relation

$$\{M_j, M_k\} = \{M_j^+ + M_j^-, M_k^+ + M_k^-\} = \varepsilon_{jkl} M_l + \{M_j^+, M_k^-\} + \{M_j^-, M_k^+\} \tag{7.40}$$

and the corresponding relation for $\{N_j, N_k\}$. Consistency between the first two equalities in Eqs. (7.25) and (7.40) leads immediately to $\{M_j^+, M_k^-\} = 0$. A similar argument involving N_j and N_k leads to $\{N_j^+, N_k^-\} = 0$. We derive the restrictions on $b_M^\pm$ and $b_N^\pm$ from particular cases of $\{M_j^+, M_k^-\} = \{N_j^+, N_k^-\} = 0$. A short calculation shows that $\{M_1^\pm, M_3^\mp\} = 0$ implies that $b_M^\pm$ cannot be a function of $p_1 \mp p_3$. Furthermore, we have $\{M_1, N_3\} = 0$ from evolution locality, which implies that $b_M^\pm$ cannot be a function of p_2. This leads to $b_M^\pm = b_M^\pm(q_1 \pm q_3, q_2)$ and a similar argument forces $b_N^\pm = b_N^\pm(q_2 \pm q_3, q_1)$, with no further restrictions from Poisson brackets of this type.

We will see that subsystem independence (see Sect. 7.3.2.3 below) adds a further constraint, which results in

$$b_M^\pm = b_M^\pm(q_1 \pm q_3, q_2) = b\left(q_\pm^{(1)}\right) = b\left(\frac{q_1 \pm q_3}{1 \pm 2q_2}\right), \tag{7.41}$$

$$b_N^\pm = b_N^\pm(q_2 \pm q_3, q_1) = b\left(q_\pm^{(2)}\right) = b\left(\frac{q_2 \pm q_3}{1 \pm 2q_1}\right), \tag{7.42}$$

for some arbitrary function b (more precisely, subsystem independence leads to the second of the equalities above, see the discussion at the end of Sect. 7.3.2.3). Thus it will be sufficient to consider the functional form of b given by Eqs. (7.41) and (7.42)

Since we derived the generators using the same computations that we carried out in Sect. 2.5.2, each of the four sets of generators $M_j^\pm$, $N_k^\pm$ provide representations of $so(3)$. This can also be checked by carrying out a direct computation. Thus, taking into consideration Eqs. (7.41) and (7.42), the expressions for M_j, N_k are given by

$$M_1 = M_1^+ + M_1^-$$

$$= (1/2 + q_2)\sqrt{\left(\frac{1}{2}\right)^2 - \left(q_+^{(1)}\right)^2} \cos\left(p_1 + p_3 + b\left(q_+^{(1)}\right)\right)$$

$$+ (1/2 - q_2)\sqrt{\left(\frac{1}{2}\right)^2 - \left(q_-^{(1)}\right)^2} \cos\left(p_1 - p_3 + b\left(q_-^{(1)}\right)\right), \qquad (7.43)$$

$$M_2 = M_2^+ + M_2^-$$

$$= -(1/2 + q_2)\sqrt{\left(\frac{1}{2}\right)^2 - \left(q_+^{(1)}\right)^2} \sin\left(p_1 + p_3 + b\left(q_+^{(1)}\right)\right)$$

$$- (1/2 - q_2)\sqrt{\left(\frac{1}{2}\right)^2 - \left(q_-^{(1)}\right)^2} \sin\left(p_1 - p_3 + b\left(q_-^{(1)}\right)\right), \qquad (7.44)$$

$$M_3 = M_3^+ + M_3^-$$

$$= (1/2 + q_2)\, q_+^{(1)} + (1/2 - q_2)\, q_-^{(1)}, \qquad (7.45)$$

and

$$N_1 = N_1^+ + N_1^-$$

$$= (1/2 + q_1)\sqrt{\left(\frac{1}{2}\right)^2 - \left(q_+^{(2)}\right)^2} \cos\left(p_2 + p_3 + b\left(q_+^{(2)}\right)\right)$$

$$+ (1/2 - q_1)\sqrt{\left(\frac{1}{2}\right)^2 - \left(q_-^{(2)}\right)^2} \cos\left(p_2 - p_3 + b\left(q_-^{(2)}\right)\right), \qquad (7.46)$$

$$N_2 = N_2^+ + N_2^-$$

$$= -(1/2 + q_1)\sqrt{\left(\frac{1}{2}\right)^2 - \left(q_+^{(2)}\right)^2} \sin\left(p_2 + p_3 + b\left(q_+^{(2)}\right)\right)$$

$$- (1/2 - q_1)\sqrt{\left(\frac{1}{2}\right)^2 - \left(q_-^{(2)}\right)^2} \sin\left(p_2 - p_3 + b\left(q_-^{(2)}\right)\right), \qquad (7.47)$$

$$N_3 = N_3^+ + N_3^-$$

$$= (1/2 + q_1)\, q_+^{(2)} + (1/2 - q_1)\, q_-^{(2)}. \qquad (7.48)$$

In the next two sections, we show that these expressions provide a representation of $so(3) \oplus so(3)$ which satisfies the constraints imposed by subsystem independence.

7.3.2.2   Representation of $so(3) \oplus so(3)$

The form of the expressions given in Eqs. (7.46–7.48) was motivated by the requirement of update locality. We now check that M_j and N_k satisfy Eq. (7.25) and thus provide a representation of $so(3) \oplus so(3)$, as required.

Since the four sets of generators $M_j^+, M_j^-, N_k^+, N_k^-$ provide representations of the $so(3)$ Lie algebra and $\{M_j^+, M_k^-\} = \{N_j^+, N_k^-\} = 0$, it follows that the two sets of generators $M_j = M_j^+ + M_k^-$ and $N_k = N_j^+ + N_k^-$ also provide representations of the $so(3)$ Lie algebra, as required (see the discussion in Sect. 7.3.2.1 where Eq. (7.40) is introduced). Thus the first two equalities of Eq. (7.25) are satisfied.

It remains to show that the third equality, $\{M_j, N_k\} = 0$, is also satisfied. We do this in three steps.

We first examine the cases of M_3 and N_3. A simple computation shows that $\{M_3, N_k\} = \partial N_k / \partial p_1 = 0$ and $\{M_K, N_3\} = -\partial M_k / \partial p_2 = 0$. Thus $\{M_j, N_k\} = 0$ whenever $M_j = M_3$ or $N_k = N_3$.

We next compute the Poisson bracket $\{M_j, N_j\}$. To do this, consider the canonical transformation which exchanges coordinate labels according to $q_1 \leftrightarrow q_2$, $p_1 \leftrightarrow p_2$ but leaves the other coordinates unchanged. Under this transformation, $M_j \to N_j$, $N_j \to M_j$. But this implies that $\{M_j, N_j\} = \{N_j, M_j\}$, which can be checked by writing out the first Poisson bracket explicitly and carrying out the canonical transformation. But we also have $\{M_j, N_j\} = -\{N_j, M_j\}$ because the Poisson bracket is antisymmetric. Thus, $\{M_j, N_k\} = 0$ whenever $j = k$.

Finally, we examine the remaining cases $\{M_1, N_2\}$ and $\{M_2, N_1\}$. Consider the canonical transformation with $q_1 \to q_1$ and $p_1 \to p_1 + \pi/2$ and the other coordinates unchanged. Under this transformation, $M_1 \to M_2$, $M_2 \to -M_1$ and N_1 and N_2 remain unchanged. Thus $\{M_1, N_2\} \to \{M_2, N_2\} = 0$ and $\{M_2, N_1\} \to \{-M_1, N_1\} = 0$, where the equalities follow from the relation $\{M_j, N_j\} = 0$ proved above. Since the Poisson bracket is invariant under canonical transformations, we must conclude that $\{M_1, N_2\} = \{M_2, N_1\} = 0$

Thus all the equalities of Eq. (7.25) are satisfied, therefore the generators M_j, N_k provide a representation of $so(3) \oplus so(3)$, as required. In particular, evolution locality is satisfied as a consequence of $\{M_j, N_k\} = 0$.

7.3.2.3   Constraints from Subsystem Independence

In the case of two subsystems which are independent (see Sect. 3.2), the joint probability factorises and

$$P_{\alpha\beta} = P_\alpha^{(1)} P_\beta^{(2)}. \qquad (7.49)$$

Such an equation amounts to a constraint on the coordinates q_1, q_2, q_3. For example, for the case P_{++}, Eq. (7.18) together with Eqs. (7.22) and (7.23) and $q_0 = 1/2$ lead to the relation

$$P_{++} = (1/2 + q_1 + q_2 + q_3)/2 = P_+^{(1)} P_+^{(2)} = (1/2 + q_1)(1/2 + q_2) \quad (7.50)$$

whenever Eq. (7.49) holds, from which we derive the constraint

$$C_1 = q_3 - 2q_1 q_2 = 0 \quad (7.51)$$

which is then required when two subsystems are independent. It is straightforward to check that we do not get any further conditions from the other $P_{\alpha\beta}$. Notice that

$$q_\pm^{(1)} \to q_1, \quad q_\pm^{(2)} \to q_2, \quad (7.52)$$

when $C_1 = 0$.

There is an additional requirement which concerns p_3. In the case of subsystem independence, one may introduce individual S_α^1, S_β^2 for each of the subsystems (see Sect. 3.2). They must however satisfy

$$S_{\alpha\beta} \approx S_\alpha^{(1)} + S_\beta^{(2)}, \quad (7.53)$$

where we use the notation $\approx$ to indicate equality up to an additive constant. From the four Eqs. (7.53) one may derive the relations

$$S_+^{(1)} - S_-^{(1)} \approx S_{++} - S_{-+} \approx S_{+-} - S_{--},$$
$$S_+^{(2)} - S_-^{(2)} \approx S_{++} - S_{+-} \approx S_{-+} - S_{--}, \quad (7.54)$$

or, equivalently, expressing the $S_{\alpha\beta}$ in terms of p_1, p_2, p_3,

$$p_1 + p_3 \approx p_1 - p_3,$$
$$p_2 + p_3 \approx p_2 - p_3. \quad (7.55)$$

Keeping in mind that $p_1 \pm p_3$ and $p_2 \pm p_3$ appear in M_j, N_k in the argument of sine and cosine functions, the most general solution of Eqs. (7.55) can be written as the constraint

$$C_2 = p_3 - n\pi = 0, \quad n = 0, \pm 1, \pm 2, \ldots \quad (7.56)$$

We have seen in Sect. 7.3.2.1 that the generators M_j are of the form

$$M_j = P_+^{(2)} L_j(q_+^{(1)}, p_+^{(1)}) + P_-^{(2)} L_j(q_-^{(1)}, p_-^{(1)}), \quad (7.57)$$

with a similar expression for the N_k. When the two subsystems are independent, an additional constraint must be satisfied given that the generators have the interpretation of expectation values: the averages computed for the first subsystem cannot depend on the quantities defining the second subsystem, i.e., on q_2 and p_2, with a similar requirement for the second subsystem. A direct computation shows that this is indeed the case, and that

$$M_j \rightarrow L_j(q_1, p_1), \qquad N_k \rightarrow L_k(q_2, p_2), \tag{7.58}$$

when $C_1 = C_2 = 0$, as required.

Finally, one can show that the constraints $C_1 = C_2 = 0$ are preserved under rotations generated by M_j and N_k, which is also required for subsystem independence. This means that $\{M_j, C_\lambda\} = \{N_k, C_\lambda\} = 0$ for $\lambda = 1, 2$. The proof is presented in the appendix of this chapter.

We end this section by commenting on how subsystem independence plays a role when establishing the functional form of b given in Eqs. (7.41) and (7.42). The point is that $b_M^\pm(q_1 \pm q_3, q_2)$ cannot be an arbitrary function of $q_1 \pm q_3$ and q_2 because then Eq. (7.58) would not be satisfied. The only way for Eq. (7.58) to be valid is to set $b_M^\pm(q_1 \pm q_3, q_2) = b_M^\pm(q_\pm^{(1)})$ because then Eq. (7.52) implies $b_M^\pm(q_\pm^{(1)}) \rightarrow b_M^\pm(q_1)$ under subsystem independence, as required. Similar considerations lead to $b_N^\pm(q_2 \pm q_3, q_1) = b_N^\pm(q_\pm^{(2)})$.

7.3.3 Wave Function Representation and a Condition for Equivalence to Quantum Mechanics

We have shown that the generators M_j and N_k given by Eqs. (7.43–7.48) provide *local* representations of $so(3) \oplus so(3)$ which depend on the function $b(q)$ defined in Eqs. (7.41) and (7.42). As we discuss below, the choice $b(q) = 0$ leads to the theory of a pair of quantum mechanical qubits. The alternative choices with $b(q) \neq 0$ may lead to viable non-quantum local models. However, for each choice $b(q) \neq 0$ it becomes necessary to establish positivity of probability (see Sect. 1.4.2) and one must check as well that the theory is not equivalent to quantum mechanics via some canonical transformation. We will not discuss this issue further here. Instead, we now consider in more detail the important case $b(q) = 0$.

To show equivalence to the quantum mechanics of a pair of qubits when $b(q) = 0$, we go to the wavefunction representation where the states are described by the wavefunction $\psi_{\alpha\beta} = (\psi_{++}, \psi_{+-}, \psi_{-+}, \psi_{-})$ and its complex conjugate $\bar{\psi}_{\alpha\beta}$ rather than by the coordinates q_k, p_k of the reduced phase space or, equivalently, the coordinates $P_{\alpha\beta}$, $S_{\alpha\beta}$. The wavefunction is related to $P_{\alpha\beta}$, $S_{\alpha\beta}$ by the equation

$$\psi_{\alpha\beta} = \sqrt{P_{\alpha\beta}}\, e^{iS_{\alpha\beta}}. \tag{7.59}$$

In quantum mechanics, the generators of $so(3) \oplus so(3)$ are represented by matrix operators

$$\hat{M}_1 = \frac{1}{2} \begin{pmatrix} 0\,0\,1\,0 \\ 0\,0\,0\,1 \\ 1\,0\,0\,0 \\ 0\,1\,0\,0 \end{pmatrix}, \quad \hat{M}_2 = \frac{1}{2} \begin{pmatrix} 0\,\,0\,\,i\,\,0 \\ 0\,\,\,0\,\,\,0\,\,i \\ -i\,\,\,0\,\,0\,\,0 \\ 0\,\,-i\,\,0\,\,0 \end{pmatrix}, \quad \hat{M}_3 = \frac{1}{2} \begin{pmatrix} 1\,0\,\,0\,\,\,0 \\ 0\,1\,\,0\,\,\,0 \\ 0\,0\,-1\,\,\,0 \\ 0\,0\,\,0\,-1 \end{pmatrix},$$

$$(7.60)$$

and

$$\hat{N}_1 = \frac{1}{2} \begin{pmatrix} 0\,1\,0\,0 \\ 1\,0\,0\,0 \\ 0\,0\,0\,1 \\ 0\,0\,1\,0 \end{pmatrix}, \quad \hat{N}_2 = \frac{1}{2} \begin{pmatrix} 0\,\,i\,\,0\,\,0 \\ -i\,0\,\,0\,\,0 \\ 0\,\,0\,\,0\,\,i \\ 0\,\,0\,-i\,\,0 \end{pmatrix}, \quad \hat{N}_3 = \frac{1}{2} \begin{pmatrix} 1\,\,\,0\,\,0\,\,\,0 \\ 0\,-1\,\,0\,\,\,0 \\ 0\,\,\,0\,\,1\,\,\,0 \\ 0\,\,\,0\,\,0\,-1 \end{pmatrix}.$$

$$(7.61)$$

When $b(q) = 0$, the expectation values of the quantum operators $\hat{M}_j$, $\hat{M}_K$ correspond precisely to the generators M_j, N_k, so that we have

$$< \psi|\hat{M}_j|\psi > = M_j, \qquad < \psi|\hat{N}_k|\psi > = N_k. \qquad (7.62)$$

We express the generators M_j, N_k for this case in both the wavefunction and q_j, p_j representations,

$$M_1 = \frac{1}{2}(\psi_{++}^*\psi_{-+} + \psi_{-+}^*\psi_{++} + \psi_{+-}^*\psi_{--} + \psi_{--}^*\psi_{+-})$$

$$= \sqrt{\left(\frac{q_0 + q_2}{2}\right)^2 - \left(\frac{q_1 + q_3}{2}\right)^2}\,\cos((p_1 + p_3))$$

$$+ \sqrt{\left(\frac{q_0 - q_2}{2}\right)^2 - \left(\frac{q_1 - q_3}{2}\right)^2}\,\cos((p_1 - p_3)), \qquad (7.63)$$

$$M_2 = \frac{i}{2}(\psi_{++}^*\psi_{-+} - \psi_{-+}^*\psi_{++} + \psi_{+-}^*\psi_{--} - \psi_{--}^*\psi_{+-})$$

$$= -\sqrt{\left(\frac{q_0 + q_2}{2}\right)^2 - \left(\frac{q_1 + q_3}{2}\right)^2}\,\sin((p_1 + p_3))$$

$$- \sqrt{\left(\frac{q_0 - q_2}{2}\right)^2 - \left(\frac{q_1 - q_3}{2}\right)^2}\,\sin((p_1 - p_3)), \qquad (7.64)$$

$$M_3 = \frac{1}{2}(\psi_{++}^*\psi_{++} + \psi_{+-}^*\psi_{+-} - \psi_{-+}^*\psi_{-+} - \psi_{--}^*\psi_{--})$$

$$= q_1, \qquad (7.65)$$

and

$$N_1 = \frac{1}{2}(\psi^*_{++}\psi_{+-} + \psi^*_{+-}\psi_{++} + \psi^*_{-+}\psi_{--} + \psi^*_{--}\psi_{-+})$$

$$= \sqrt{\left(\frac{q_0 + q_1}{2}\right)^2 - \left(\frac{q_2 + q_3}{2}\right)^2} \cos((p_2 + p_3))$$

$$+ \sqrt{\left(\frac{q_0 - q_1}{2}\right)^2 - \left(\frac{q_2 - q_3}{2}\right)^2} \cos((p_2 - p_3)), \qquad (7.66)$$

$$N_2 = \frac{i}{2}(\psi^*_{++}\psi_{+-} - \psi^*_{+-}\psi_{++} + \psi^*_{-+}\psi_{--} - \psi^*_{--}\psi_{-+})$$

$$= -\sqrt{\left(\frac{q_0 + q_1}{2}\right)^2 M - \left(\frac{q_2 + q_3}{2}\right)^2} \sin((p_2 + p_3))$$

$$- \sqrt{\left(\frac{q_0 - q_1}{2}\right)^2 - \left(\frac{q_2 - q_3}{2}\right)^2} \sin((p_2 - p_3)), \qquad (7.67)$$

$$N_3 = \frac{1}{2}(\psi^*_{++}\psi_{++} - \psi^*_{+-}\psi_{+-} + \psi^*_{-+}\psi_{-+} - \psi^*_{--}\psi_{--})$$

$$= q_2. \qquad (7.68)$$

All quantum mechanical expectation values $< \hat{A} >$, $< \hat{B} >$ can be expressed as observables $A(q_j, p_j)$, $B(q_j, p_j)$ in the reduced configuration space and, as we saw in Sect. 2.3.3, the Poisson bracket for the observables on configuration space is isomorphic to the usual commutator on Hilbert space. Thus *pairs of robits are equivalent to pairs of quantum mechanical qubits when $b(q) = 0$.*

Finally, we look at the how the *Casimir functions* [5] (i.e., the functions F^c which satisfy $\{F^c, M_j\} = \{F^c, N_k\} = 0$) are represented in the reduced phase space of a pair of robits when $b(q) = 0$ and show that they are necessarily constants. To see this, evaluate first $\{F^c, M_3\} = \{F^c, N_3\} = 0$, which leads to $F^c = F^c(q_1, q_2, q_3, p_3)$. Now evaluate $\{F^c, M_1\} = 0$, which leads to

$$\{F^c, M_1\} = \left[\left(\frac{\partial F^c}{\partial q_1} + \frac{\partial F^c}{\partial q_3}\right) M_2^+\right] + \left[\left(\frac{\partial F^c}{\partial q_1} - \frac{\partial F^c}{\partial q_3}\right) M_2^-\right]$$

$$- \left[\frac{\partial F^c}{\partial p_3}\frac{\partial M_1}{\partial q_3}\right] = 0 \qquad (7.69)$$

One can check that each of the terms in square brackets must be equal to zero because they depend on different trigonometrical functions of $p_1 \pm q_3$, which leads to $F^c = F^c(q_2)$. A similar calculation of $\{F^c, N_1\} = 0$ shows that the functions F^c cannot depend on q_2 either, thus they must be constants.

The *Casimir operator* is defined by $\hat{C} = \sum_k \left(\hat{M}_k^2 + \hat{N}_k^2 \right) = \frac{3}{2}\hat{1}$, where $\hat{1}$ is the unit matrix. The corresponding function on the reduced phase is $C = < \psi | \hat{C} | \psi > = \frac{3}{2}$; i.e., it is the particular Casimir function with $F^c = \frac{3}{2}$.

Pairs of robits provide *local* representations of $so(3) \oplus so(3)$ which depend on the function $b(q)$ defined in Eqs. (7.41) and (7.42).

- Equivalence to a pair of quantum mechanical qubits is achieved for the choice $b(q) = 0$
- Alternative choices with $b(q) \neq 0$ may lead to viable non-quantum local models

---

7.4 Discussion

In the previous sections we derived a probabilistic model of an ensemble consisting of a pair of spin 1/2 particles. This was accomplished with a minimum of assumptions: (i) the outcome of a measurement results in a value of spin $s = \pm 1/2$, (ii) the relevant observables provide a representation of $so(3) \oplus so(3)$, (iii) requirements of locality and subsystem independence are satisfied. More importantly, we have shown that a particular case of this theory, obtained by setting $b(q) = 0$ in Eqs. (7.43–7.48), leads to a theory which is equivalent to the quantum theory of two qubits. This is despite the fact that *none* of the conditions listed above were formulated using the standard building blocks of quantum mechanics; e.g., operators that act on states in a complex Hilbert space, unitary transformations, etc.

In is important to emphasize that the starting point for the probabilistic model developed here is the observation that the configuration space of a spin 1/2 particle is *discrete*, corresponding to two possible measurement outcomes. Since we assign probabilities to configurations only, we end up with probabilities over the four-outcome configuration space of the pair of spin 1/2 particles. In contrast, one can easily get into serious difficulties if, rather than following the configuration ensemble approach, one assumes instead a probability model where there is one joint probability distribution for *all* observables. Such a model would go beyond what is justified by the experimental results, and is in conflict with quantum mechanics.

For example, if we restrict to the particular case of singlet states (i.e., the two particles have opposite spins in all directions), we can compare our probabilistic models to the local hidden variable models introduced by Bell [3], which have been the starting point for various derivations of Bell inequalities [6]. Since our models include quantum theory as a special case, it is possible for them to violate Bell inequalities for particular choices of the relative angle between measuring devices.

This is quite remarkable, given the way in which the probabilistic model has been derived using only very basic assumptions, none of which are particular to quantum theory. It has been claimed that the axioms of standard probability theory cannot be valid for quantum systems because Bell's inequality is a theorem in standard probability theory and quantum mechanics does not satisfy it. And yet the models that we have derived here *only* use standard probability theory, proving that the often repeated claim concerning the inapplicability of standard probability theory to quantum systems *cannot* be justified on the basis of Bell's theorem, and is certainly incorrect for the type of systems that we discuss here.

Appendix: Invariance of Subsystem Independence Constraints Under Rotations

We show that the constraints $C_1 = 0$ and $C_2 = 0$ given by Eq. (7.51) and Eq. (7.56) are preserved under local rotations.

We first consider the invariance of C_1. A simple computation shows that we have $\{C_1, M_3\} = \{(q_3 - 2q_1q_2), q_1\} = 0$ and $\{C_1, N_3\} = \{(q_3 - 2q_1q_2), q_2\} = 0$, as required. Next we compute the Poisson bracket with M_1,

$$
\begin{aligned}
\{C_1, M_1\} =&\{(q_3 - 2q_1q_2), M_1\}\\
=&\frac{\partial M_1}{\partial p_3} - 2q_2\frac{\partial M_1}{\partial p_1}\\
=&-(1/2 + q_2)\sqrt{\left(\frac{1}{2}\right)^2 - \left(q_+^{(1)}\right)^2} \sin\left(p_1 + p_3 + b\left(q_+^{(1)}\right)\right)\\
&+ (1/2 - q_2)\sqrt{\left(\frac{1}{2}\right)^2 - \left(q_-^{(1)}\right)^2} \sin\left(p_1 - p_3 + b\left(q_-^{(1)}\right)\right)\\
&+ 2q_2(1/2 + q_2)\sqrt{\left(\frac{1}{2}\right)^2 - \left(q_+^{(1)}\right)^2} \sin\left(p_1 + p_3 + b\left(q_+^{(1)}\right)\right)\\
&+ 2q_2(1/2 - q_2)\sqrt{\left(\frac{1}{2}\right)^2 - \left(q_-^{(1)}\right)^2} \sin\left(p_1 - p_3 + b\left(q_-^{(1)}\right)\right)\\
&\rightarrow -q_2L_2(q_1, p_1) + q_2L_2(q_1, p_1) \quad \text{when} \quad C_1 = C_2 = 0\\
=&0, \tag{7.70}
\end{aligned}
$$

as required. Under the canonical transformation $q_1 \rightarrow q_1$ and $p_1 \rightarrow p_1 + \pi/2$, we have $C_1 \rightarrow C_1$ and $M_1 \rightarrow M_2$, which leads to $\{C_1, M_2\} = 0$. Finally, under $q_1 \leftrightarrow q_2$ and $p_1 \leftrightarrow p_2$, we have $C_1 \rightarrow C_1$ and $M_j \rightarrow N_j$, which leads to $\{C_1, N_1\} = \{C_1, N_2\} = 0$. Thus we conclude that $\{C_1, M_j\} = \{C_1, N_k\} = 0$ when $C_1 = C_2 = 0$.

Now we look at the invariance of C_2. A simple computations shows that we have $\{C_2, M_3\} = \{(p_3 - n\pi), q_1\} = 0$ and $\{C_2, N_3\} = \{(p_3 - n\pi), q_2\} = 0$, as required. Next we compute the Poisson bracket with M_1,

$$
\begin{aligned}
\{C_2, M_1\} = & \{(p_3 - n\pi), M_1\} \\
= & -\frac{\partial M_1}{\partial q_3} \\
= & \frac{q_+^{(1)}}{2\sqrt{(1/2)^2 - \left(q_+^{(1)}\right)^2}} \cos\left(p_1 + p_3 + b\left(q_+^{(1)}\right)\right) \\
& -\frac{q_-^{(1)}}{2\sqrt{(1/2)^2 - \left(q_-^{(1)}\right)^2}} \cos\left(p_1 - p_3 + b\left(q_-^{(1)}\right)\right) \\
& + (1/2 + q_2)\sqrt{(1/2)^2 - \left(q_+^{(1)}\right)^2} \sin\left(p_1 + p_3 + b\left(q_+^{(1)}\right)\right) \frac{b'}{1 + 2q_2} \\
& + (1/2 - q_2)\sqrt{(1/2)^2 - \left(q_-^{(1)}\right)^2} \sin\left(p_1 - p_3 + b\left(q_-^{(1)}\right)\right) \frac{-b'}{1 - 2q_2} \\
\rightarrow & \frac{q_1 - q_1}{2\sqrt{(1/2)^2 - q_1^2}} \cos\left(p_1 + b(q_1)\right) \\
& + \sqrt{(1/2)^2 - q_1^2} \sin\left(p_1 + b(q_1)\right) \frac{b' - b'}{2} \quad \text{when} \quad C_1 = C_2 = 0 \\
= & 0, \quad\quad\quad\quad\quad\quad\quad\quad\quad\quad\quad\quad\quad\quad\quad\quad\quad\quad (7.71)
\end{aligned}
$$

as required. Using the same type of canonical transformations considered in the case of C_1, we can show that the Poisson bracket of C_2 with the remaining generators also vanishes. Thus we conclude that $\{C_2, M_j\} = \{C_2, N_k\} = 0$ when $C_1 = C_2 = 0$.

This shows that the generators satisfy all of the constraints imposed by subsystem independence.

References

1. Bohm, D.: Quantum Theory. Prentice Hall, New York (1951)
2. Einstein, E., Podolsky, B., Rosen, N.: Can quantum-mechanical description of physical reality be considered complete? Phys. Rev. **47**, 777–780 (1935)
3. Bell, J.S.: On the Einstein Podolsky Rosen paradox. Physics **1**, 195–200 (1964)
4. Tung, W.-K.: Group Theory in Physics. World Scientific, Singapore (1985)
5. Marsden, J.E., Ratiu, T.S.: Introduction to Mechanics and Symmetry. A Basic Exposition of Classical Mechanical Systems. Texts in Applied Mathematics, vol. 17. Springer, New York (1994)
6. Brunner, N., Cavalcanti, D., Pironio, S., Scarani, V., Wehner, S.: Bell nonlocality. Rev. Mod. Phys. **86**, 419–478 (2014)

Part III
Hybrid Quantum-Classical Systems

Chapter 8
Hybrid Quantum-Classical Ensembles

Abstract The problem of defining hybrid systems comprising quantum and classical components is highly nontrivial, and the approaches that have been proposed to solve this problem run into various types of fundamental difficulties. The formalism of configuration-space ensembles is able to overcome many of these difficulties, allowing for a general and consistent description of interactions between quantum and classical ensembles. Such hybrid ensembles have a number of desirable features; e.g., quantum-classical interactions do not blur the fundamental distinction between the quantum and classical components; configuration separability is satisfied; and non-relativistic systems are Galilean invariant whenever the interaction potential itself is Galilean invariant. After demonstrating general properties of hybrid ensembles, we consider their application to the description of measurement of a quantum system by a classical apparatus, including examples of position and spin measurement; the scattering of a classical particle from a quantum superposition; and the definitions of Gaussian and coherent ensembles for quantum-classical oscillators. Finally, we generalise quantum Wigner functions to hybrid ensembles.

8.1 Introduction

In Chap. 3 we considered interactions between two arbitrary ensembles on configuration space, and discussed separability, entanglement and measurement properties. In this chapter we specialise to joint 'quantum-classical' ensembles, comprising a quantum component and a classical component. Due to the physically distinct nature of the two components, we will also refer to these as *hybrid* ensembles.

There are a number of reasons to model the interaction of quantum and classical systems. Some of them are practical: for certain applications, it is convenient to assume that some components of a physical system—e.g., atomic nuclei, the electromagnetic field, the spacetime metric—can be modelled classically [1–4] while the rest of the system is modelled according to quantum mechanics. For such applications, it would be desirable to have an approach which is free of inconsistencies and allows for well defined approximation schemes.

© Springer International Publishing Switzerland 2016

M.J.W. Hall and M. Reginatto, *Ensembles on Configuration Space*,

Fundamental Theories of Physics 184, DOI 10.1007/978-3-319-34166-8_8

Other reasons stem from open problems in the foundations of physics. A good example can be found in measurement theory: in the standard Copenhagen interpretation of quantum mechanics, the measuring apparatus must be described in classical terms [5, 6] and this implies a coupling of some sort between a quantum system and the apparatus which is treated as a classical system. Gravity provides another example: since there is no full quantum theory yet, one would like to know to what extent the quantization of gravity is forced upon us by consistency arguments alone [7], and one way to do this is to investigate how far one can get with a hybrid system in which the gravitational field remains classical while matter is assumed to consist of quantum fields. Dyson has argued that it might be impossible in principle to observe the existence of individual gravitons, and this has lead him to the conjecture that "the gravitational field described by Einstein's theory of general relativity is a purely classical field without any quantum behaviour" [8]. His observations regarding the impossibility of detecting gravitons have been supported by detailed calculations [9, 10]. If Dyson's conjecture is true, hybrid models become unavoidable.

There are, of course, obstacles that have to be dealt with when modeling the interaction of quantum and classical systems. First of all, quantum mechanics and classical mechanics are usually formulated using very different mathematical structures. Thus, it becomes necessary to find a framework that can encompass both types of systems. The formalism of ensembles on configuration space provides such a common mathematical language. In addition to the mathematical aspects, there are difficult conceptual issues; i.e., the idealizations that often serve as building blocks of theories are different for classical and quantum systems. What aspects of quantum and classical mechanics should be preserved in the description of a hybrid system?

Many models have been proposed to describe interacting classical and quantum ensembles (a brief review of various approaches is presented in Sect. 9.1). They have typically run into various types of fundamental difficulties such as non-conservation of energy; absence of back-reaction of the quantum system on the classical system; nonlocality; negative probabilities; an inherent inability to describe all interactions of interest; and incorrect equations of motion for noninteracting independent systems. Thus, in the past it has not been clear as to whether a satisfactory description was even possible. The approach based on configuration space ensembles overcomes many difficulties arising in previous approaches [11, 12]. For example [13]:

1. Probability and energy are conserved.
2. There is back-reaction of the quantum system on the classical system.
3. The correct quantum and classical equations of motion are recovered in the limit of noninteracting independent systems.
4. The two "minimal requirements for a quantum-classical formulation" specified by Salcedo [14, 15] are satisfied: a Lie bracket may be defined on the set of observables, and this Lie bracket is equivalent to a Poisson bracket for classical observables and to $(i\hbar)^{-1}$ times the quantum commutator for quantum observables.

5. "Configuration separability" is satisfied: the quantum configuration statistics are invariant under any canonical transformation applied to the classical component, and vice versa (see also Sect. 3.3.3).
6. The "definite benchmark ... for an acceptable quantum-classical hybrid system" specified by Peres and Terno [16] is satisfied: the expectation values for the position and momentum observables of linearly-coupled quantum and classical oscillators obey the classical equations of motion.
7. Generalised Ehrenfest relations are satisfied.
8. Galilean invariance is satisfied for interaction potentials of the form $V(x, q, t) = V(|x - q|)$, where x and q are the configuration space coordinates of the classical and quantum components

To our knowledge, no other formulation satisfies all of these properties. In particular, requirement 4 is highly non-trivial [14, 15, 17], and therefore provides a critical test for any formulation of hybrid quantum-classical systems.

We briefly recall the basic elements of the formalism of ensembles on a joint configuration space in Sect. 8.2, and show how they can be applied to describing interactions between quantum and classical ensembles. After demonstrating general properties of hybrid ensembles in Sect. 8.3, we consider their application to the description of measurement of a quantum system by a classical apparatus in Sect. 8.4; the scattering of a classical particle from a quantum superposition in Sect. 8.5; and the definition of Gaussian and coherent ensembles for quantum-classical oscillators in Sect. 8.6. Finally, we propose a definition of hybrid Wigner functions, and give some applications thereof, in Sect. 8.7.

This results of this chapter provide a basis for the discussion of the consistency of the formalism and its comparison with other approaches to quantum-classical interactions in Chap. 9, and are generalised in Chap. 11 to describe the consistent coupling of quantum matter to classical spacetime.

8.2 Quantum-Classical Ensembles

8.2.1 Quantum and Classical Mechanics on Configuration Space

We have seen in previous chapters that quantum and classical systems are treated on an equal footing within the formalism of ensembles on configuration space, with differences being primarily due to the forms of their respective observables and ensemble Hamiltonians. Consider the example of particles. In either case an ensemble of physical systems is described by a probability density $P(x)$ on configuration space that evolves according to an action principle. Thus there is a canonically conjugate function $S(x)$ on configuration space, and an ensemble Hamiltonian $\mathscr{H}[P, S]$, such that

$$\frac{\partial P}{\partial t} = \frac{\delta \mathscr{H}}{\delta S}, \qquad \frac{\partial S}{\partial t} = -\frac{\delta \mathscr{H}}{\delta P}, \tag{8.1}$$

where $\delta/\delta f$ denotes the variational derivative with respect to function f (see Chap. 1).

For example, as described in Sect. 1.2, the classical and quantum ensemble Hamiltonians describing particles with mass m moving in a potential $V(x)$ are

$$\mathscr{H}_C[P, S] = \int dx\, P\left[\frac{|\nabla S|^2}{2m} + V(x)\right], \tag{8.2}$$

$$\mathscr{H}_Q[P, S] = \mathscr{H}_C[P, S] + \frac{\hbar^2}{4}\int dx\, P\frac{|\nabla \log P|^2}{2m}. \tag{8.3}$$

The equations of motion for $\mathscr{H}_C[P, S]$ follow as

$$\frac{\partial P}{\partial t} + \nabla\cdot\left(P\frac{\nabla S}{m}\right) = 0, \qquad \frac{\partial S}{\partial t} + \frac{|\nabla S|^2}{2m} + V = 0, \tag{8.4}$$

where the first equation is a continuity equation ensuring the conservation of probability, while the second equation is the classical Hamilton-Jacobi equation. In contrast, the equations of motion corresponding to $\mathscr{H}_Q[P, S]$ are equivalent to the quantum Schrödinger equation

$$i\hbar\frac{\partial \psi}{\partial t} = \frac{-\hbar^2}{2m}\nabla^2\psi + V\psi, \tag{8.5}$$

with

$$\psi := \sqrt{P}\, e^{iS/\hbar}. \tag{8.6}$$

Classical and quantum configuration ensembles are distinguished not only by the forms of their ensemble Hamiltonians, but by their corresponding sets of observables. As discussed in Sect. 2.3, classical observables have the form

$$C_f[P, S] = \int dx\, P(x)f(x, \nabla S) \tag{8.7}$$

where $f(x, p)$ is a function on the classical phase space, whereas quantum observables have the form

$$Q_{\hat{M}}[P, S] = \langle\psi|\hat{M}|\psi\rangle \tag{8.8}$$

where $\hat{M}$ is a Hermitian operator on the quantum Hilbert space and ψ is defined in Eq. (8.6) above. In both cases the numerical value of an observable is identified with its average value over the ensemble (see Sect. 2.2).

Note that the ensemble Hamiltonians in Eq. (8.2) and (8.3) are just particular instances of classical and quantum observables, with

$$\mathcal{H}_C[P, S] = C_H[P, S], \qquad \mathcal{H}_Q[P, S] = Q_{\hat{H}}[P, S], \qquad (8.9)$$

for the respective choices $H(x, p) = p{\cdot}p/(2m) + V(x)$ and $\hat{H} = -\hbar^2 \nabla^2/(2m) + V(x)$. Thus, in each case the numerical value of the ensemble Hamiltonian corresponds to the average ensemble energy (see also Sect. 1.4).

We will see in the next chapter, where we discuss the consistency of the formalism, that it is necessary to put some restrictions on the type of physical systems that may be modeled as classical 'particles'. These restrictions do not play a role for the material discussed in this chapter, but nevertheless it will be useful to include some remarks on this topic here. At a very fundamental level, classical particles are described by classical ensemble Hamiltonians and observables as discussed above. However, there are other requirements associated with "classicality." In particular, *when we talk about localised classical objects (for example, a measuring apparatus), we will always mean macroscopic objects.* One property of macroscopic objects is that they cannot be isolated from the environment: a macroscopic object cannot avoid scattering photons and other particles [18]. Thus the pointer of a measuring device will be continuously undergoing scattering processes. Another property of macroscopic objects is that they have very large numbers of degrees of freedom. When describing a classical measuring apparatus we will often discuss only *one* of its degrees of freedom, say the position of a pointer, but in reality there is an enormous number of "irrelevant" degrees of freedom, say 10^{23}. The degree of freedom that corresponds to the pointer is never completely isolated from all the other degrees of freedom of the apparatus. One must therefore keep in mind that classical objects, by virtue of being macroscopic, are *complex* objects which inevitably interact with the environment. For most applications this added complication can be neglected, but in some cases it will be crucial to take this into consideration.

8.2.2 Interacting Quantum and Classical Ensembles

Interactions between arbitrary ensembles have been previously discussed in Chap. 3. Here we wish to focus on interactions between classical and quantum ensembles in particular.

We will consider a classical ensemble with a continuous configuration space labelled by x, interacting with a quantum ensemble with an arbitrary configuration space labelled by q. The joint configuration space is therefore labelled by (q, x), and the hybrid joint ensemble is described by a pair of functions (P, S) on this configuration space, together with a suitable ensemble Hamiltonian $\mathcal{H}_{QC}[P, S]$.

As per Sect. 3.2, the quantum and classical components of a hybrid ensemble (P, S) are defined to be *independent* if and only if

$$P(q, x) = P_Q(q) P_C(x), \qquad S(q, x) = S_Q(q) + S_C(x), \qquad (8.10)$$

for two quantum and classical ensembles (P_Q, S_Q) and (P_C, S_C). Further, the hybrid ensemble is defined to be *noninteracting*, with respect to a given ensemble Hamiltonian $\mathcal{H}_{QC}$, if and only if all independent ensembles remain independent under evolution. Noninteraction is equivalent, via Eq. (3.4), to the joint ensemble Hamiltonian satisfying

$$\mathcal{H}_{QC}[P_Q P_C, S_Q + S_C] = \int dx\, P_X(x)\, \mathcal{H}_Q[P_Q, S_Q] + \int dq\, P_Q(q)\, \mathcal{H}_C[P_C, S_C] \tag{8.11}$$

for all quantum and classical ensembles (P_Q, S_Q) and (P_C, S_C), where $\mathcal{H}_Q$ and $\mathcal{H}_C$ are suitable quantum and classical ensemble Hamiltonians respectively. If Eq. (8.11) is *not* satisfied, $\mathcal{H}_{QC}$ will be said to describe *interacting* quantum and classical ensembles.

Example: Interacting Particles As a first example of interacting quantum and classical ensembles, let q denote the configuration space coordinate of a quantum particle of mass m, and x denote the configuration coordinate of a classical particle of mass M. A joint ensemble Hamiltonian corresponding to the interaction potential $V(q, x, t)$ is then given by

$$\mathcal{H}_{QC}[P, S] = \int dq\, dx\, P \left[\frac{|\nabla_q S|^2}{2m} + \frac{|\nabla_x S|^2}{2M} + V(q, x, t) + \frac{\hbar^2}{4} \frac{|\nabla_q \log P|^2}{2m} \right]. \tag{8.12}$$

For $V(q, x) \equiv 0$ there is no interaction between the quantum and classical parts of the composite ensemble, and Eq. (8.11) is satisfied with $\mathcal{H}_{QC}$ corresponding to the sum of $\mathcal{H}_Q$ and $\mathcal{H}_C$ in Eqs. (8.2) and (8.3). More generally, $\mathcal{H}_{QC}$ is seen to correspond to the sum of a quantum term, a classical term, and an interaction term

$$\mathcal{H}_I[P, S] = \int dq\, dx\, P\, V(q, x, t) = \langle V \rangle. \tag{8.13}$$

The Hamiltonian equations of motion corresponding to Eq. (8.12) follow via Eq. (8.1), and are given by

$$\frac{\partial P}{\partial t} + \nabla_q \cdot \left(P \frac{\nabla_q S}{m} \right) + \nabla_x \cdot \left(P \frac{\nabla_x S}{M} \right) = 0, \tag{8.14}$$

$$\frac{\partial S}{\partial t} + \frac{|\nabla_q S|^2}{2m} + \frac{|\nabla_x S|^2}{2M} + V - \frac{\hbar^2}{2m} \frac{\nabla_q^2 P^{1/2}}{P^{1/2}} = 0, \tag{8.15}$$

(see Appendix A.1 to this book regarding the evaluation of the required functional derivatives). These are coupled partial differential equations of first-order in time (of a type commonly encountered in hydrodynamics), and may be numerically integrated to solve for $P(q, x, t)$ and $S(q, x, t)$ providing that P and S are specified at some initial time t_0. Several analytic solutions are considered later in this chapter.

Example: Classical Particle Interacting with a Quantum Spin As an example of a *discrete* ensemble of quantum systems interacting with an ensemble of classical particles, consider the case where the z-component $\hat{\sigma}_z$ of an ensemble of spin-1/2 particles is linearly coupled to the momentum of an ensemble of one-dimensional classical particles. Modelling such an interaction has been attempted previously, but with a number of fundamental difficulties arising [11, 19].

We choose the configuration space of the quantum spin to be $\{\pm 1\}$, corresponding to the outcomes of a measurement of $\hat{\sigma}_z$, and the configuration space of the classical particle to be the real number line. From Eqs. (8.7) and (8.8) the quantum spin observable $\hat{\sigma}_z$ and the classical momentum observable p are represented by

$$Q_{\hat{\sigma}_z} = \sum_{s=\pm 1} P(s)\, s = \langle s \rangle, \qquad C_p = \int dx\, P(x)\, \frac{\partial S}{\partial x} = \langle p \rangle,$$

for quantum and classical ensembles respectively. This suggests, for a hybrid ensemble (P, S) on the joint configuration space $\{(s, x)\}$, representing a linear coupling between σ_z and p via an interaction ensemble Hamiltonian of the form

$$\mathcal{H}_I[P, S] = \kappa(t) \sum_{s=\pm 1} \int dx\, P(s, x)\, s\, \frac{\partial S(s, x)}{\partial x} =: \kappa(t)\,\langle sp \rangle, \qquad (8.16)$$

where $\kappa(t)$ quantifies a (possibly time-dependent) coupling strength. This ensemble Hamiltonian will be further discussed below in the context of measurement interactions. Note that the total ensemble Hamiltonian will be given by adding appropriate terms $\mathcal{H}_Q$ and $\mathcal{H}_C$ to $\mathcal{H}_I$, describing purely quantum and purely classical contributions to the evolution.

The form of $\mathcal{H}_I$ above may alternatively be motivated by writing down the interaction ensemble Hamiltonian corresponding to $\hat{M} = \kappa(t)\hat{\sigma}_z\hat{p}$ for a one-dimensional spin-1/2 *quantum* particle, via Eq. (8.8), and taking the classical limit $\hbar \to 0$ in the resulting expression. This method has the advantage of immediately implying that $\mathcal{H}_I$ preserves the normalisation and positivity of the probability distribution $P(s, x)$, as an immediate consequence of it being preserved under the fully quantum evolution, without the need to explicitly check the corresponding conditions in Sect. 1.4.

8.3 Some General Properties

In this section, we introduce certain useful concepts, such as conditional wavefunctions and density operators. We will also discuss some general desirable features of hybrid systems, such as configuration separability and invariance under Galilean transformations for non-relativistic systems.

Quantum vs Classical: an Invariant Distinction It might be thought that the interaction between the quantum and classical components of a hybrid ensemble could render the labels 'quantum' and 'classical' meaningless: would not the quantum component pick up 'classical' features, and vice versa, due to the interaction? Surprisingly perhaps, this is not the case.

As discussed in Sect. 3.3, it is straightforward to extend observables for individual ensembles to observables on a joint ensemble. In particular, using the definition of extended observables in Eqs. (3.12) and (3.13), the classical and quantum observables in Eqs. (8.7) and (8.8) extend to the respective forms

$$\tilde{C}_f[P, S] = \int dq dx \, P(q, x) f(x, \nabla_x S) \tag{8.17}$$

$$\tilde{Q}_{\hat{M}} = \langle \psi_{QC} | \hat{M} | \psi_{QC} \rangle, \tag{8.18}$$

where integration over q is replaced by summation over any discrete components and the *hybrid* wave function $\psi_{QC}(q, x)$ is defined by

$$\psi_{QC} := \sqrt{P} \, e^{iS/\hbar}. \tag{8.19}$$

It follows that the respective 'classical' and 'quantum' natures of these observables are fully preserved under evolution, as a consequence of the invariance of Poisson brackets under the extension of observables as per Eq. (3.16). In particular, this property implies, via Eqs. (2.22) and (2.28), that

$$\{\tilde{C}_f, \tilde{C}_g\} = \tilde{C}_{\{f,g\}}, \qquad \{\tilde{Q}_{\hat{M}}, \tilde{Q}_{\hat{N}}\} = \tilde{Q}_{[\hat{M},\hat{N}]/i\hbar}, \tag{8.20}$$

where $\{f, g\}$ denotes the usual Poisson bracket for phase space functions $f(x, p)$ and $g(x, p)$. Further, since Poisson brackets are automatically preserved under Hamiltonian evolution, these relations hold at all times. Thus:

> Interaction does not blur the fundamental distinction between the classical and quantum components of a hybrid ensemble.

In particular, the classical observables form a Lie algebra corresponding to the usual Poisson bracket on phase space, and the quantum observables form a Lie algebra corresponding to the usual quantum commutator on Hilbert space, with both algebras being independently preserved under evolution of the ensemble.

Configuration Separability As discussed in Sect. 3.3.3, all joint ensembles satisfy the property of configuration separability, i.e., the configuration statistics of either component are invariant under a canonical transformation of the other component. To translate this to hybrid ensembles, denote the quantum and classical configuration probability densities by

$$P_C(x) := \int dq \, P(q, x), \qquad P_Q(q) := \int dx \, P(q, x), \tag{8.21}$$

where integration over q is replaced by summation for discrete regions of the configuration space. Configuration separability is then the property:

A local unitary transformation on the quantum component of a hybrid ensemble does not affect the probability density $P_C(x)$ of the classical component. Conversely, a local canonical transformation on the classical component does not affect the probability density $P_Q(q)$ of the quantum component.

While this property follows as an immediate consequence of the general result in Eq. (3.21) for joint ensembles, it is of interest to derive it directly for a quantum-classical ensemble (P, S) [11]. Consider first a canonical transformation generated by an arbitrary quantum observable $\tilde{Q}_{\hat{M}}$. Any classical configuration observable is of the form $\tilde{C}_{g(x)}$ (e.g., $g(x) = x$), with $\tilde{C}_{g(x)} = \int dq\, dx\, Pg = \int dq\, dx\, \bar{\psi}_{QC}\, g\psi_{QC}$ via Eqs. (8.17) and (8.19). The general Poisson bracket formula in Eq. (2.27) then gives

$$\{\tilde{C}_{g(x)}, \tilde{Q}_{\hat{M}}\} = \frac{2}{\hbar} \operatorname{Im}\left\{\int dq\, dx\, \bar{\psi}_{QC}\, g(x)\hat{M}\psi_{QC}\right\} = 0, \qquad (8.22)$$

since $\hat{M}$ acts only on the quantum component of the hybrid wavefunction and so commutes with $g(x)$. Hence, the expectation value of $\tilde{C}_{g(x)}$ is unchanged by the transformation for any function $g(x)$, implying that $P_C(x)$ is invariant as claimed. Conversely, consider a canonical transformation of the classical component, generated by an arbitrary classical observable $\tilde{C}_f$. Any quantum configuration observable is of the form $Q_{g(\hat{q})}$, with $Q_{g(\hat{q})} = \int dq\, dx\, \bar{\psi}_{QC} g(\hat{q})\psi_{QC} = \int dq\, dx\, P\, g(q)$ via Eq. (8.18). The general Poisson bracket formula in Eq. (2.5) then gives

$$\{\tilde{Q}_{g(\hat{q})}, \tilde{C}_f\} = \int dq\, dx\left(\frac{\delta \tilde{Q}_{g(\hat{q})}}{\delta P}\frac{\delta \tilde{C}_f}{\delta S} - \frac{\delta \tilde{Q}_{g(\hat{q})}}{\delta S}\frac{\delta \tilde{C}_f}{\delta P}\right)$$
$$= -\int dq\, dx\, g(q)\nabla_x \cdot \left(P\frac{\partial f(x, \nabla_x S)}{\partial(\nabla_x S)}\right) = 0, \qquad (8.23)$$

using integration by parts with respect to x. Hence the expectation value of $\tilde{Q}_{g(\hat{q})}$ is unchanged by the transformation, implying that $P_Q(q)$ is invariant as claimed.

Reduced Density Operators and Phase Space Densities The concept of reduced or 'improper' mixtures was introduced for general joint ensembles in Sect. 4.2.4, and is of interest for hybrid ensembles in particular. First, for a hybrid quantum-classical ensemble (P, S), consider the corresponding reduced mixture of *quantum* configuration ensembles defined by

$$\mathscr{W}_Q := \{(P_x, S_x);\, w_x\}, \qquad (8.24)$$

with

$$P_x(q) := P(q, x)/P_C(x), \quad S_x(q) := S(q, x), \quad w_x := P_C(x), \qquad (8.25)$$

as per the definitions in Eqs. (4.14) and (4.15), where $P_C(x)$ is the marginal classical probability density in Eq. (8.21). The corresponding *conditional wave function* for the quantum component, for a given classical configuration x, follows via Eq. (8.6) as

$$\psi_x(q) := [P_x(q)]^{1/2} \, e^{iS_x(q)/\hbar}, \qquad (8.26)$$

and the statistics of the reduced mixture $\mathcal{M}_Q$ are therefore equivalent to the *conditional density operator*

$$\hat{\rho}_{Q|C} := \int dx \, P_C(x) \, |\psi_x\rangle\langle\psi_x| = \text{tr}_C[|\psi_{QC}\rangle\langle\psi_{QC}|]. \qquad (8.27)$$

Here the second equality formally follows as per the analogous result in Eq. (4.16) for quantum joint ensembles, where ψ_{QC} is the hybrid wave function defined in Eq. (8.19).

One may similarly consider the corresponding reduced mixture of *classical* configuration ensembles, defined by

$$\mathcal{W}_C := \{(P_q, S_q); w_q'\}, \qquad (8.28)$$

with

$$P_q(x) := P(q, x)/P_Q(q), \quad S_q(x) := S(q, x), \quad w_q' := P_Q(q), \qquad (8.29)$$

where $P_Q(q)$ is the marginal quantum probability density in Eq. (8.21). The statistics of this mixture are equivalent, via Eq. (4.5), to the classical phase space density

$$\rho_{C|Q}(x, p) := \int dq \, P_Q(q) \, P_q(x) \, \delta(p - \nabla_x S_q) = \int dq \, P(q, x) \, \delta(p - \nabla_x S). \quad (8.30)$$

It is important to note that $\psi_x(q)$ and $\hat{\rho}_{Q|C}$ do *not* satisfy the Schrödinger and quantum Liouville equations in general. Nor does $\rho_{C|Q}$ satisfy the classical Liouville equation. This is an expected property of 'improper' mixtures: they only contain partial information about a joint ensemble (see Sect. 4.2.4).

> A hybrid ensemble requires both $P(q, x)$ and $S(q, x)$ for its full description, and is not equivalently described by the quantum density operator $\hat{\rho}_{Q|C}$, nor by the classical phase space density $\rho_{C|Q}(x, p)$, in general.

Even so, it will be seen in this chapter and the next that reduced mixtures and the conditional density operator $\hat{\rho}_{Q|C}$ remain useful concepts in discussions of measurement and decoherence. They also make a natural appearance as marginals of hybrid Wigner functions (see Sect. 8.7).

Stationary Ensembles As discussed in Sect. 2.4, stationary ensembles are defined to be ensembles with time-independent dynamical properties. As shown there, this is equivalent to the conditions

$$\frac{\partial P}{\partial t} = 0, \qquad \frac{\partial S}{\partial t} = -E, \qquad (8.31)$$

for some constant E, equal to the average ensemble energy. Stationary ensembles generalise the notion of quantum stationary states, and are a crucial element in providing a unified approach to classical and quantum thermodynamics on configuration space (see Chap. 4).

As an example of stationary hybrid ensembles [11], consider the case of classical and quantum ensembles of particles, evolving under the joint ensemble Hamiltonian $\mathscr{H}_{QC}$ in Eq. (8.12), for a *translation-invariant* potential $V(q, x, t) \equiv V(q - x)$. We will minimise the classical contribution to the average energy via the ansatz $\nabla_x S = 0$. It follows via the equations of motion in Eqs. (8.14) and (8.15), and making a change of variables from q to $r := q - x$, that the corresponding stationary states satisfy

$$\nabla_r \cdot \left(P \frac{\nabla_r S}{m} \right) = 0, \qquad \frac{|\nabla_r S|^2}{2m} + V(r) - \frac{\hbar^2}{2m} \frac{\nabla_r^2 \sqrt{P}}{\sqrt{P}} = E, \qquad (8.32)$$

for some constant E. These are equivalent to the time-independent Schrödinger equation

$$\left[\frac{-\hbar^2}{2m} \nabla_r^2 + V(r) \right] \psi = E\psi, \qquad (8.33)$$

with $\psi := P^{1/2} e^{iS/\hbar}$.

The ansatz requires that S does not depend on x, and hence (for $V \neq 0$) it must be independent of r. It follows that if $\phi_n(r)$ denotes any *real-valued* energy eigenfunction corresponding to a solution of Eq. (8.33) (there is always a complete set of such eigenfunctions), with eigenvalue E_n, then there is a corresponding stationary ensemble described by

$$P(q, x, t) = P_C(x) [\phi_n(q - x)]^2, \qquad S(q, x, t) = -E_n t. \qquad (8.34)$$

Here $P_C(x)$ is an arbitrary probability density on the classical configuration space, and is equal to the marginal probability density in Eq. (8.21). The corresponding numerical value of the ensemble Hamiltonian $\mathscr{H}_{QC}$ may be calculated as $\mathscr{H}_{QC} = E_n$. Further, the conditional wavefunction follows from Eqs. (8.26) and (8.34) as the displaced energy eigenstate

$$\psi_x(q, t) = \phi_n(q - x)e^{-iE_n t/\hbar}. \tag{8.35}$$

Thus:

> Stationary hybrid ensembles for translation invariant interaction potentials have quantized energies, corresponding to eigenstates of quantum particles that are subject to the potential $V(q - x)$ with probability $P_C(x)$.

Centre-of-Mass and Relative Motion Stationary hybrid ensembles are seen to have simple relationships to purely quantum ensembles (as do the 'coherent' ensembles in Sect. 8.6). However, more generally, hybrid ensembles can behave very differently from both quantum and classical ensembles.

For example, again considering the ensemble Hamiltonian $\mathscr{H}_{QC}$ in Eq. (8.12), describing interacting classical and quantum ensembles of particles, one can make a change of coordinates for the joint configuration space from (q, x) to the centre-of-mass and relative coordinates:

$$\bar{\mathbf{x}} := \frac{mq + Mx}{m + M}, \qquad r := q - x. \tag{8.36}$$

Rewriting Eq. (8.12) with respect to these coordinates, for a rotationally and translationally invariant interaction potential $V(q, x, t) \equiv V(|q - x|)$, and defining the total mass M_T and relative mass μ by

$$M_T := m + M, \qquad \mu := \frac{mM}{m + M}, \tag{8.37}$$

yields [11]

$$\mathscr{H}_{QC} = \underbrace{\int d\bar{x} dr \, P \left[\frac{|\nabla_{\bar{x}} S|^2}{2M_T} + \frac{\hbar^2 m}{4(m + M)} \frac{|\nabla_{\bar{x}} \log P|^2}{2M_T} \right]}_{\text{(i) Quantum-like term: Free center-of-mass motion}}$$

$$+ \underbrace{\int d\bar{x} dr \, P \left[\frac{|\nabla_r S|^2}{2\mu} + \frac{\hbar^2 M}{4(m + M)} \frac{|\nabla_r \log P|^2}{2\mu} + V(|r|) \right]}_{\text{(ii) Quantum-like term: Relative-motion in a potential}}$$

$$- \underbrace{\frac{2\mu}{M_T} \int d\bar{x} dr \, P \left[\frac{\hbar^2 M}{4(m + M)} \frac{(\nabla_{\bar{x}} \log P \cdot \nabla_r \log P)}{2\mu} \right]}_{\text{(iii) Interaction term}}. \tag{8.38}$$

Comparison with the quantum ensemble Hamiltonian in Eq. (8.3) shows that, relative to the new coordinates, the hybrid ensemble Hamiltonian comprises (i) a quantum-like term corresponding to free centre-of-mass motion, but with a rescaled Planck constant

$$\hbar_{\overline{X}} := [m/(m+M)]^{1/2}\,\hbar; \tag{8.39}$$

(ii) a quantum-like term corresponding to relative motion in a central potential $V(|r|)$, but with a rescaled Planck constant

$$\hbar_R := [M/(m+M)]^{1/2}\,\hbar; \tag{8.40}$$

and (iii) an intrinsic interaction term that couples the gradients of P with respect to $\bar{x}$ and r. The presence of this last term implies, in stark contrast to classical-classical and quantum-quantum interactions, a surprising property:

> The centre-of-mass motion and the relative motion do not decouple for quantum-classical interactions.

Indeed, the only limits in which the intrinsic interaction term can be ignored are (a) in the limit of an infinite classical (quantum) mass, for which the ensemble Hamiltonian reduces to the sum of a classical (quantum) centre-of-mass term and a quantum (classical) relative motion term; and (b) in the limit $\hbar \to 0$, for which the ensemble Hamiltonian reduces to the sum of two classical terms.

Galilean-Invariant Interactions The non-decoupling of centre-of-mass and relative motion, as above, raises a question as to whether hybrid ensembles of nonrelativistic quantum and classical particles might violate fundamental physical symmetries, such as invariance under Galilean transformations. Here we show that this is not the case [11].

In particular, restricting to classical and quantum particles moving in three dimensions, consider the hybrid ensemble Hamiltonian $\mathcal{H}_{QC}$ in Eq. (8.12) for the case of a potential that transforms as a scalar under time and space translations and under rotations, i.e.,

$$V(q, x, t) \equiv V(|q - x|). \tag{8.41}$$

It is straightforward to show that the equations of motion in Eqs. (8.14) and (8.15) are then invariant under the general Galilean transformation

$$q \to q' = \mathsf{R}q - ut + a, \tag{8.42}$$

$$x \to x' = \mathsf{R}x - ut + a, \tag{8.43}$$

$$t \to t' = t + \tau, \tag{8.44}$$

for rotation matrix R, constant vectors u and a, and constant τ, provided that P and S transform as

$$P'(q', x', t') = P(q, x, t) \tag{8.45}$$

$$S'(q', x', t') = S(q, x, t) + \frac{1}{2}(m+M)|u|^2 t - u^T \mathsf{R}(mq + Mx) + c \tag{8.46}$$

for some constant c. Thus:

> Galilean invariance is satisfied for hybrid ensembles of particles whenever the interaction potential $V(q, x, t)$ itself is Galilean invariant.

We note that the forms of the generating observables for Galilean transformations have been discussed in Sect. 2.5.

Topological Constraint for Continuous Configuration Spaces In the case of quantum ensembles of particles, the topological constraint that $\oint_C dS/h$ must be an integer for all loops C in configuration space corresponds to single-valuedness of the wave function $\psi = P^{1/2}e^{iS/\hbar}$ [20, 21]. It is important to note that the same constraint may also be applied to purely classical ensembles, as was done for example by Schiller [22] when he formulated his theory of 'quasi-classical mechanics', which deals with a particular class of classical ensembles (those that can be associated with WKB solutions). Thus, constraints of this type are consistent with the equations of motion derived from either classical or quantum ensemble Hamiltonians, amounting only to a reduction of the space of physically allowed solutions, and are logically independent of the choice of the ensemble Hamiltonian. In the case that the quantum component has a continuous configuration space, it is natural to also impose the topological constraint that $\oint_{C_x} dS/h$ is an integer for all loops C_x in the joint configuration space that correspond to some fixed classical configuration x, corresponding to single-valuedness of the conditional wavefunction in Eq. (8.26).

8.4 Measurement of a Quantum System by a Classical Apparatus

In the standard Copenhagen interpretation of quantum mechanics, it has been repeatedly emphasized that any objective account of a physical experiment must be given in classical terms. Thus, for example, Bohr stated that (page 89 of Ref. [5])

> "…the point is that in each case we must be able to communicate to others what we have done and what we have learned, and that therefore the functioning of the measuring instruments must be described within the framework of classical physical ideas."

while Heisenberg wrote that (page 46 of Ref. [6])

> "…the concepts of classical physics form the language by which we describe the arrangements of our experiments and state the results."

Hence, if the Copenhagen interpretation is to be taken seriously, it follows that any dynamical description of the measurement process should be able to be formulated, at least approximately, in terms of an interaction between classical and quantum systems. It is of obvious interest to attempt this via the formalism of configuration ensembles on configuration space.

A generalisation of von Neumann's quantum measurement model, describing a measurement interaction between a given configuration ensemble and a 'pointer' ensemble, has been given in Sect. 3.4. Specialising to the case of a quantum ensemble interacting with a classical ensemble of one-dimensional pointers, the hybrid ensemble Hamiltonian describing the measurement interaction in Eq. (3.25) reduces to the form

$$\mathcal{H}_I = \kappa(t) \int dq dx \, P(q, x) f(q) \frac{\partial S(q, x)}{\partial x}, \tag{8.47}$$

where $\kappa(t)$ is a coupling constant and integration over q is replaced by summation for any discrete ranges. This interaction represents a linear coupling between the momentum of the pointer and the observable corresponding to the quantum operator $f(\hat{q})$ (see Sect. 3.4).

We set $f(q) = q$. If the measurement interaction acts over a sufficiently short timescale, such that other contributions to the total ensemble Hamiltonian can be neglected, the equations of motion reduce to

$$\frac{\partial P}{\partial t} = \frac{\delta \mathcal{H}_I}{\delta S} = -\kappa(t) q \frac{\partial P}{\partial x} \qquad \frac{\partial S}{\partial t} = -\frac{\delta \mathcal{H}_I}{\delta P} = -\kappa(t) q \frac{\partial P}{\partial x}, \tag{8.48}$$

which may immediately solved to give

$$P(q, x, T) = P(q, x - Kq, 0), \qquad S(q, x, T) = S(q, x - Kq, 0), \tag{8.49}$$

where T is the interaction time and $K := \int_0^T dt \, \kappa(t)$. Thus, as expected, the interaction correlates the classical pointer position with the quantum configuration: for a given value of q the pointer position is displaced by Kq.

Note that it is natural to assume that, prior to the measurement, the quantum and classical ensembles are independent as per Eq. (8.10), i.e., the initial joint ensemble is of the form

$$P(q, x, 0) = P_Q(q) P_C(x), \qquad S(q, x, 0) = S_Q(q) + S_C(x). \tag{8.50}$$

Thus, initially, the quantum ensemble is described by the wave function $\psi_0 = \sqrt{P_Q} \, e^{iS_Q/\hbar}$, and the pointer ensemble by the configuration space functions (P_C, S_C).

Example: Measurement of Position Consider first the case where the measurement interaction correlates an ensemble of classical pointers with the positions of an ensemble of one-dimensional quantum particles. Assuming the ensembles are initially independent as per Eq. (8.50), it follows that the distribution of pointer positions following the measurement is given by

$$P(x, T) = \int dq \, P(q, x, T) = \int dq \, P_Q(q) P_C(x - Kq). \tag{8.51}$$

Thus, the pointer distribution becomes convolved with the initial quantum distribution.

In the ideal case that the initial pointer position is sharply localised at $x = 0$ for each member of the ensemble, i.e., $P_C(x) = \delta(x)$, the final pointer distribution simplifies to

$$P(x, T) = K^{-1} P_Q(x/K) = K^{-1}|\psi_0(x/K)|^2. \tag{8.52}$$

Thus, the pointer is perfectly correlated with the initial quantum distribution, up to a scaling factor, in the limit of an ideal measurement.

The conditional quantum density operator in the ideal limit is easily determined via Eqs. (8.24)–(8.27). In particular, noting that $P(q, x, T) = P_Q(q)\delta(x - Kq) = K^{-1}P_Q(x/K)\delta(q - x/K)$, one finds

$$\hat{\rho}_{Q|C} = \int dq\, P_Q(q)\, |q\rangle\langle q|. \tag{8.53}$$

Thus, the reduced mixture of the quantum ensemble is 'decoherent' with respect to position, i.e., it is diagonal in the position basis.

Example: Measurement of Spin An ensemble of spin-1/2 particles may be described by a discrete configuration space, $\{1, -1\}$, corresponding to the eigenvalues of $\hat{\sigma}_z$, as discussed in Sect. 8.2.2 above. The measurement interaction Hamiltonian in Eq. (8.47) then reduces to the coupling between quantum spin and classical momentum in Eq. (8.16), identifying $q = s = \pm 1$.

We will assume that the pointer and spin ensembles are initially independent as per Eq. (8.50), i.e.,

$$P(\pm 1, x, 0) = w_\pm P_C(x), \qquad S(\pm 1, x, 0) = \gamma_\pm + S_C(x), \tag{8.54}$$

where $\psi_\pm = \sqrt{w_\pm}\, e^{i\gamma_\pm/\hbar}$ is the initial spin wave function in the $\hat{\sigma}_z$ basis. The pointer distribution after the measurement then follows from Eq. (8.49) as

$$P(x, T) = \sum_{s=\pm 1} P(s, x, T) = w_+ P_C(x - K) + w_- P_C(x + K). \tag{8.55}$$

Hence, the initial probability density $P_C(x)$ is displaced by K with probability w_+, and by $-K$ with probability w_-, where $w_\pm$ denotes the initial probability of spin up/down in the z-direction.

Further, if the initial pointer probability density $P_C(x)$ has a spread which is small with respect to K, then the final probability density will have two nonoverlapping peaks, with $P_C(x - K)P_C(x + K) = 0$. The conditional density operator after the measurement follows via Eqs. (8.26) and (8.27) as

$$\rho_{Q|C} = w_+ |+\rangle\langle+| + w_- |-\rangle\langle-|, \tag{8.56}$$

and hence 'decoheres' with respect to the $\hat{\sigma}_z$ basis.

Consistency with Copenhagen Interpretation The above examples are consistent with the elements of the Copenhagen interpretation discussed at the beginning of this section. In particular: (i) the measuring apparatus is described classically, as is required for the unambiguous communication and comparison of physical results; and (ii) information about quantum ensembles is obtained via an appropriate interaction with an ensemble of classical measuring apparatuses, which correlates the classical configuration with a corresponding quantum property.

It is also seen that there is a conditional decoherence of the quantum ensemble relative to the classical ensemble, which depends upon the nature of the quantum-classical interaction. This relevance of this property for describing measurements on quantum systems will be further discussed in Chap. 9, where the need to take into account the many degrees of freedom of 'real' measuring apparatuses will also be addressed.

Finally, note that a frequently criticised shortcoming of the Copenhagen interpretation is 'where' to place the quantum-classical cut. In the above approach this has a natural solution: the cut is placed at the point where an objective description of the measuring apparatus is required, corresponding to explicitly describing the apparatus by a classical ensemble. This is in direct contrast to the measurement problem that arises in approaches that attempt to describe the measuring apparatus as a quantum object.

8.5   Scattering of Classical Particles by Quantum Superpositions

Consider a gedanken experiment that involves the scattering of two non-relativistic particles, one a classical particle (the projectile) and the other one a quantum particle (the target) [13]. The particles are assumed to interact in some way, e.g., via gravitational attraction or a Coulombic interaction. Note that this scattering experiment may be interpreted as a type of measurement, in the sense that information about the state of the quantum particle can be inferred from the position of the classical particle after the interaction has taken place.

This gedanken experiment is most interesting when the quantum system is prepared in such a way that the initial amplitude for the quantum particle (i.e., as $t \to -\infty$, when the two particles are very far from each other so that the interaction term can be neglected), has two peaks of equal magnitude, A and B, that are well separated. Thus, if a position measurement of the type described in Sect. 8.4 was made well before the scattering interaction, then the quantum particle would have been 'found' at the location of peak A (with probability 1/2) or at the location of peak B (with probability 1/2). Consider now the case where the classical particle comes very close to peak A and remains at all times at a very large distance from peak B. What happens when the classical particle scatters from the quantum particle?

A "naive approach" is to use a picture in which the position of the classical particle is deterministically specified at all times, where this well-defined position is used to model the scattering interaction. In such a picture, there appear to be two mutually exclusive types of available models: (a) the classical particle is strongly scattered half the time, corresponding to a quantum particle at the location of peak A, and very weakly scattered the other half the time, corresponding a quantum particle at the location of peak B; or (b) the classical particle simultaneously "sees" half a quantum particle at the location of peak A and half a quantum particle at the location of peak B, and thus always undergoes a degree of scattering equal to about one half of the strong scattering in (a).

In option (a) the superposition state of the quantum particle is effectively replaced by a classical mixture, whereas in option (b) it is effectively replaced by a mean-field potential (of the type seen in semiclassical gravity [23]; see also Chap. 9). The two possibilities thus not only give conflicting predictions, but are somewhat *ad hoc*. Further, they give no insight into possible classical-classical or quantum-quantum limits of a quantum-classical scattering process.

The source of difficulties is, clearly, the use of a "naive approach". In contrast, the formalism of ensembles on configuration space is able to give a detailed (albeit statistical) account of quantum-classical scattering, with classical particles represented by ensembles rather than individual systems, where this account supports neither of the naive options (a) and (b).

To discuss the general qualitative aspects of configuration ensemble approach, we will consider gravitational scattering of a classical particle of mass M from a quantum particle of mass m. Hence, the hybrid ensemble Hamiltonian $\mathscr{H}_{QC}$ has the form Eq. 8.12, with interaction potential $V(q, x, t) \equiv G\frac{mM}{|q-x|}$, and the equations of motion for the joint ensemble follow immediately via Eqs. (8.14)–(8.15) as

$$\frac{\partial P}{\partial t} + \nabla_q \cdot \left(P\frac{\nabla_q S}{m} \right) + \nabla_x \cdot \left(P\frac{\nabla_x S}{M} \right) = 0, \qquad (8.57)$$

$$\frac{\partial S}{\partial t} + \frac{|\nabla_q S|^2}{2m} + \frac{|\nabla_x S|^2}{2M} + G\frac{mM}{|q - x|} - \frac{\hbar^2}{2m}\frac{\nabla_q^2 P^{1/2}}{P^{1/2}} = 0. \qquad (8.58)$$

We already know via Eq. (8.38) that the centre-of-mass and relative motions of the particles do not decouple under these equations, in contrast to classical-classical and quantum-quantum scattering. Qualitative features of these equations tell us further what to expect of the solutions:

1. When $|\frac{\hbar^2}{2m}\frac{\nabla_q^2 P^{1/2}}{P^{1/2}}| \ll |G\frac{mM}{|q-x|}|$, we can neglect the term proportional to $\frac{\hbar^2}{2m}\frac{\nabla_q^2 P^{1/2}}{P^{1/2}}$ in (8.58). This will be the case when the mass m of the quantum system is large enough for this inequality to be valid. But if this is the case, (8.58) reduces to the classical Hamilton-Jacobi equation and we end up with the equations of a joint classical-classical ensemble. This shows that the formalism has the correct classical limit.

2. If the interaction term $V(q, x) = G\frac{mM}{|q-x|}$ that appears in (8.58) is very small (say at $t \to -\infty$ when the two particles are very far from each other), and there is no initial correlation between the particles, then the non-linearity in the equations of the quantum particle will amount to only a small perturbation and the superposition principle will remain valid for the quantum ensemble to a very good approximation. This means that the formalism has the correct quantum limit. Notice, however, that the equations are non-linear when the interaction term is taken into consideration: the quantum superposition principle is thus expected to break down when the interaction between the classical and quantum particles is strong.

3. The ensembles are assumed to be independent prior to the interaction (i.e., at $t \to -\infty$), when the two particles are very far from each other, and thus

$$P^{(-\infty)}(q, x) = P_Q^{(-\infty)}(q) P_C^{(-\infty)}(x), \tag{8.59}$$

$$S^{(-\infty)}(q, x) = S_Q^{(-\infty)}(q) + S_C^{(-\infty)}(x), \tag{8.60}$$

analogously to Eq. (8.50). Hence, before the interaction, the hybrid ensemble consists of two independent classical and quantum components. However, after the interaction these components will no longer be independent, i.e.,

$$P^{(+\infty)}(q, x) \neq P_Q^{(+\infty)}(q) P_C^{(+\infty)}(x), \tag{8.61}$$

$$S^{(+\infty)}(q, x) \neq S_Q^{(+\infty)}(q) + S_C^{(+\infty)}(x), \tag{8.62}$$

The resulting correlations between the classical and quantum components means that a measurement of position on either component will nontrivially update the description of the *joint* ensemble. The scattering interaction thus generates an analogue of entanglement between the components (see also Sect. 3.2.3 for further discussion of entanglement for joint ensembles).

Case (a) of the "naive approach" corresponds to "either-or" outcomes that implicitly assume that there is no entanglement in quantum-classical systems, in contrast to the approach based on configuration ensembles. The latter approach is also fundamentally different from standard semiclassical gravity, corresponding to case (b) of the "naive approach": in particular, there is no 'mean-field' representation of a quantum gravitational field in the equations of motion (8.57) and (8.58). Thus, the predictions of the configuration ensemble approach differ substantially from the outcomes predicted using the "naive approach."

The comparison between mean-field approaches and hybrid interactions is further discussed in Chap. 9, and the configuration ensemble approach is extended to the gravitational coupling of quantum field ensembles with classical spacetime ensembles in Chap. 11.

8.6 Hybrid Oscillators and Gaussian Ensembles

The harmonic oscillator is ubiquitous in physical models, and hence it is natural to compare the hybrid version to its fully quantum and fully classical counterparts. Here we investigate the equations of motion for a quantum-classical ensemble of oscillators, and show that the definitions of 'Gaussian' and 'coherent' states are straightforward to generalise to this case.

In particular, consider a joint ensemble describing an n-dimensional quantum particle of mass m joined by a spring to an n-dimensional classical particle of mass M. The corresponding composite ensemble Hamiltonian then has the form of $\mathcal{H}_{QC}$ in Eq. (8.12), with interaction potential

$$V(q, x, t) = \frac{1}{2}k|q - x|^2. \tag{8.63}$$

It is convenient to define hybrid coordinates $\xi := (q, x)$, and block matrices

$$\mathsf{C} := \begin{pmatrix} k\mathsf{I} & -k\mathsf{I} \\ -k\mathsf{I} & k\mathsf{I} \end{pmatrix}, \quad \mathsf{U} := \begin{pmatrix} m^{-1}\mathsf{I} & 0 \\ 0 & M^{-1}\mathsf{I} \end{pmatrix}, \quad \mathsf{E} := \begin{pmatrix} \mathsf{I} & 0 \\ 0 & 0 \end{pmatrix} \tag{8.64}$$

(where I denotes the $n \times n$ identity matrix), allowing $\mathcal{H}_{QC}$ to be rewritten as

$$\mathcal{H}_{\text{osc}} = \int d\xi\, P\left[\frac{1}{2}(\nabla S)^T \mathsf{U} \nabla S + \frac{1}{2}\xi^T \mathsf{C}\xi + \frac{\hbar^2}{8}(\nabla \log P)^T \mathsf{E}\mathsf{U}\mathsf{E}(\nabla \log P)\right]. \tag{8.65}$$

Here ∇ denotes $\nabla_\xi \equiv (\nabla_q, \nabla_x)$. Note that the classical-classical oscillator corresponds to replacing E by 0, and that the quantum-quantum oscillator corresponds to replacing E by 1. This is useful for comparisons between the three cases. The Hamiltonian equations of motion in Eqs. (8.14) and (8.15) then have the form [11]

$$\frac{\partial P}{\partial t} + \nabla \cdot (P\mathsf{U}\nabla S) = 0, \tag{8.66}$$

$$\frac{\partial S}{\partial t} + \frac{1}{2}(\nabla S)^T \mathsf{U}\nabla S + \frac{1}{2}\xi^T \mathsf{C}\xi - \frac{\hbar^2}{2}\frac{\nabla \cdot (\mathsf{E}\mathsf{U}\mathsf{E}\nabla\sqrt{P})}{\sqrt{P}} = 0. \tag{8.67}$$

8.6.1 Gaussian Ensembles

In analogy with quantum oscillators, it is natural to consider hybrid ensembles having Gaussian probability densities, i.e., with

$$P(\xi, t) = \frac{\sqrt{\det \mathsf{K}}}{(2\pi)^{n/2}} e^{-\frac{1}{2}(\xi-\alpha)^T \mathsf{K}(\xi-\alpha)} \tag{8.68}$$

for some (possibly time-dependent) positive definite symmetric matrix K and vector $\boldsymbol{\alpha}$. It is straightforward to check that a Gaussian probability density is consistent with the equations of motion if and only if S has the quadratic form

$$S(\boldsymbol{\xi}, t) = \frac{1}{2}(\boldsymbol{\xi} - \boldsymbol{\alpha})^T \mathsf{L}(\boldsymbol{\xi} - \boldsymbol{\alpha}) + \boldsymbol{\beta} \cdot (\boldsymbol{\xi} - \boldsymbol{\alpha}) + s, \tag{8.69}$$

where L is a symmetric matrix, $\boldsymbol{\beta}$ is a vector, and s is a scalar (all possibly time-dependent). We will call ensembles with P and S of this form *Gaussian*.

The equations of motion for K, L, $\boldsymbol{\alpha}$, $\boldsymbol{\beta}$ and s may be found by substituting the above forms of P and S into Eqs. (8.66)–(8.67), and equating coefficients of the respective quadratic, linear and constant terms with respect to $\boldsymbol{\xi} - \boldsymbol{\alpha}$. After some straightforward algebra (requiring the formula $(d/dt) \det \mathsf{K} = \det \mathsf{K} \, \mathrm{tr}[\dot{\mathsf{K}}\mathsf{K}^{-1}]$), one obtains

$$\dot{\boldsymbol{\alpha}} = \mathsf{U}\boldsymbol{\beta}, \qquad \dot{\boldsymbol{\beta}} = -\mathsf{C}\boldsymbol{\alpha}, \tag{8.70}$$

$$\dot{\mathsf{K}} + \mathsf{KUL} + \mathsf{LUK} = 0, \qquad \dot{\mathsf{L}} + \mathsf{LUL} + \mathsf{C} = \frac{\hbar^2}{2}\mathsf{KEUEK}, \tag{8.71}$$

$$\dot{s} = \frac{1}{2}(\boldsymbol{\alpha} \cdot \dot{\boldsymbol{\beta}}) - \frac{\hbar^2}{4}\mathrm{tr}[\mathsf{EUEK}]. \tag{8.72}$$

Note that the first three equations are independent of the projection matrix E, and hence are also valid for classical-classical and quantum-quantum oscillators. The last equation is not of physical interest to solve for s in general, as the expectation value of any quantum or classical observable is independent of s via Eqs. (8.17) and (8.18).

The equations for $\dot{\boldsymbol{\alpha}}$ and $\dot{\boldsymbol{\beta}}$ are precisely those corresponding to a classical-classical oscillator with configuration $\boldsymbol{\alpha}$, momentum $\boldsymbol{\beta}$, and phase space Hamiltonian

$$H(\alpha, \beta) = \frac{1}{2}\boldsymbol{\beta}^T \mathsf{U}\boldsymbol{\beta} + \frac{1}{2}\boldsymbol{\alpha}^T \mathsf{C}\boldsymbol{\alpha}. \tag{8.73}$$

Thus, solving these equations for $\boldsymbol{\alpha}$ and $\boldsymbol{\beta}$ is equivalent to solving the classical equations of motion. It follows from Eq. (8.68) that:

> The average configuration of a Gaussian ensemble of hybrid oscillators is given by the trajectory in configuration space of a fully classical oscillator.

This link to classical motion is a special case of the generalised Ehrenfest theorem derived in Chap. 9. Using the forms of C and U in Eq. (8.64), the general solutions for $\boldsymbol{\alpha}$ and $\boldsymbol{\beta}$ are

$$\boldsymbol{\alpha} = (c, c) + \mu \cos(\omega_\mu t + \phi) \, (d/m, -d/M), \tag{8.74}$$

$$\boldsymbol{\beta} = -\mu\omega_\mu \sin(\omega_\mu t + \phi) \, (d, -d), \tag{8.75}$$

as may be checked by direct substitution, where c and d are arbitrary n-vectors, ϕ is an arbitrary constant, μ is the reduced mass defined in Eq. (8.37), and $\omega_\mu := \sqrt{k/\mu}$.

It is noteworthy that the frequency ω_μ associated with the average motion is determined by the reduced mass, even though the centre-of-mass and relative motions do not decouple as discussed in Sect. 8.3. The correlation between these degrees of freedom has been numerically investigated by Chua et al. [24]. Properties of Gaussian ensembles are further investigated in Sect. 8.7 below.

8.6.2 Coherent Ensembles

To define a hybrid analogue of quantum coherent states, we will be guided by the covariance and minimum uncertainty properties of such states [11]. First, quantum coherent states are linear in phase, corresponding to $L = 0$ in Eq. (8.69). The first equality in Eq. (8.71) then implies that the matrix K is constant, and the left hand side of the second equality reduces to C. Solving the reduced equality for K by writing it out in block-matrix form yields the general form

$$K = \frac{2}{\hbar}\sqrt{\frac{m}{k}}C + \begin{pmatrix} 0 & 0 \\ 0 & A \end{pmatrix}, \tag{8.76}$$

where A is any nonnegative symmetric $n \times n$ matrix. Substituting Eqs. (8.74) and (8.76) into Eq. (8.68) yields the corresponding solution

$$P(q, x, t) = P_A(x - x_t)(\sqrt{km}/\pi\hbar)^{n/2}e^{-\sqrt{km}|q-x-(q_t-x_t)|^2/\hbar} \tag{8.77}$$

for the probability density, where

$$q_t = c + (\mu/m)d\cos\left(\omega_\mu t + \phi\right), \qquad x_t = c - (\mu/M)d\cos\left(\omega_\mu t + \phi\right), \tag{8.78}$$

denote the quantum and classical components of the average configuration α, and $P_A(x)$ denotes the Gaussian probability density

$$P_A(x) := (2\pi)^{-n/2}(\det A)^{1/2}e^{-\frac{1}{2}x^T A x}. \tag{8.79}$$

Second, the uncertainty of the classical configuration x is clearly minimised in the limit that $P_A(x)$ approaches a delta-function (i.e., $A^{-1} \to 0$). This limit will be therefore be taken to define coherent ensembles. The corresponding probability density follows as

$$P(q, x, t) = \delta(x - x_t)(\sqrt{km}/\pi\hbar)^{n/2}e^{-\sqrt{km}|q-q_t|^2/\hbar}. \tag{8.80}$$

Substituting Eqs. (8.74)–(8.76) into Eqs. (8.69) and (8.72) further yields the linear form

$$S(q, x, t) = -\frac{1}{2}n\hbar\omega_m t + \frac{|d|^2}{4}\sqrt{k\mu}\sin 2(\omega_\mu t + \phi)$$
$$- d \cdot (q - x)\sqrt{k\mu}\sin(\omega_\mu t + \phi) \tag{8.81}$$

for S, up to an arbitary additive constant, where $\omega_m := \sqrt{k/m}$. Thus $P = P_Q P_C$ and $S = S_Q + S_C$ as per Eq. (8.10), implying:

> Coherent hybrid ensembles decompose into *independent* quantum and classical ensembles, with the position of the classical particle fully determined by the trajectory x_t, and the position of the quantum particle described by a Gaussian probability density P of width $\hbar/2\sqrt{km}$ centred on the trajectory q_t. The trajectories x_t and q_t correspond to the motion of two coupled *classical* oscillators.

Finally, note that the choice $d \equiv 0$ satisfies Eq. (8.31), and hence corresponds to a *stationary* ensemble. For this choice only the first term of $S(q, x, t)$ in Eq. (8.81) is nonzero, and the numerical value of the composite ensemble Hamiltonian follows from Eq. (8.31) as

$$\tilde{H}_{\text{osc}} = -\frac{\partial S}{\partial t} = \frac{1}{2}n\hbar\omega_m. \tag{8.82}$$

This may be recognised as the zero-point energy of an n-dimensional quantum harmonic oscillator of mass m, as expected from Eq. (8.33).

8.7 Hybrid Wigner Functions

Wigner functions play an important role in characterising quantum ensembles [25, 26]. Here we briefly show how they can be generalised to describe hybrid ensembles, and investigate the relation of hybrid Wigner functions to the reduced quantum and classical mixtures of a hybrid ensemble, and to uncertainty relations and covariance matrices for Gaussian ensembles.

8.7.1 Definition and Basic Properties

For simplicity we will consider a hybrid system corresponding to a one-dimensional quantum particle with position q, and a one-dimensional classical particle with position x. Now, the Wigner function of a quantum pure state $\psi_Q(q)$ is given by [25, 26]

$$W_Q(q, p | \psi_Q) := \frac{1}{\pi \hbar} \int dz \, e^{2ipz/\hbar} \bar{\psi}_Q(q + z) \, \psi_Q(q - z), \qquad (8.83)$$

and has many properties similar to a joint probability distribution for the quantum position and momentum observables $\hat{x}$ and $\hat{p}$. For example, the marginals of W_Q give the correct quantum probability densities of these observables for the state ψ_Q, and W_Q obeys the classical Liouville equation in the limit $\hbar \to 0$ (and exactly for the special case of quadratic Hamiltonians) [26].

The classical analogue of the Wigner function of a classical system is therefore the joint probability density on phase space. Since a classical configuration ensemble (P_C, S_C) associates a momentum $k = \partial_x S_C$ with position x (see Chap. 1), we therefore define the corresponding classical Wigner function by

$$W_C(x, k) := \delta(k - \partial_x S_C(x)) \, P_C(x). \qquad (8.84)$$

Note that this is precisely the phase space density of a 'pure' classical mixture, as per Eq. (4.7) of Chap. 4.

Guided by the above forms of the quantum and classical Wigner functions, we now define the *hybrid* Wigner function of a quantum-classical ensemble (P, S) by the joint phase space function

$$W_{QC}(q, x, p, k) := \frac{\delta(k - \partial_x S)}{\pi \hbar} \int dz \, e^{2ipz/\hbar} \bar{\psi}_{QC}(q + z, x) \, \psi_{QC}(q - z, x), \quad (8.85)$$

where ψ_{QC} is the hybrid wave function defined in Eq. (8.19). This definition leads to several nice properties, as will now be shown.

Independent Ensembles For a hybrid ensemble comprising independent quantum and classical ensembles as per Eq. (8.10), i.e., with $(P, S) \equiv (P_Q P_C, S_Q + S_C)$, the hybrid Wigner function simplifies to

$$\begin{aligned}
W_{QC}(q, x, p, k) &= \frac{\delta(k - \partial_x S_C)}{\pi \hbar} \int dz \, e^{2ipz/\hbar} P_C(x) \, \bar{\psi}_Q(q + z) \, \psi_Q(q - z) \\
&= W_Q(q, p) \, W_C(x, k). \qquad (8.86)
\end{aligned}$$

Thus it factorises into the purely quantum and purely classical Wigner functions defined in Eqs. (8.83) and (8.84).

Classical and Quantum Marginals It is natural to consider the marginal properties of the hybrid Wigner function. First, if we integrate over the quantum position and momentum degrees of freedom, q and p, we obtain

$$W_{C|Q}(x, k) := \int dq dp\, W_{QC}(q, x, p, k)$$

$$= \int dq \frac{\delta(k - \partial_x S)}{\pi \hbar} \int dz \left[\int dp\, e^{2ipz/\hbar} \right] \bar{\psi}_{QC}(q + z, x)\, \psi_{QC}(q - z, x)$$

$$= \int dq\, \delta(k - \partial_x S)\, |\psi_{QC}(q, x)|^2$$

$$= \int dq\, P(q, x)\, \delta(k - \partial_x S) = \rho_{C|Q}(x, k), \tag{8.87}$$

where $\rho_{C|Q}(x, k)$ is the conditional phase space density in Eq. (8.30), describing the reduced mixture of classical configuration ensembles associated with the hybrid ensemble. This phase space density correctly describes the statistics of all classical observables on the ensemble (see Sect. 4.2.2).

Similarly, integration over the classical position and momentum degrees of freedom yields, using Eq. (8.83),

$$W_{Q|C}(q, p) := \int dx dk\, W_{QC}(q, x, p, k)$$

$$= \int dx \frac{1}{\pi \hbar} \int dz\, e^{2ipz/\hbar} \bar{\psi}_{QC}(q + z, x)\, \psi_{QC}(q - z, x) \int dk \delta(k - \partial_x S)$$

$$= \int dx\, P_C(x) \frac{1}{\pi \hbar} \int dz\, e^{2ipz/\hbar} \bar{\psi}_x(q + z)\, \psi_x(q - z)$$

$$= \int dx\, P_C(x)\, W_Q(q, p|\psi_x) = W_{\hat{\rho}_{Q|C}}(q, p). \tag{8.88}$$

Here, $P_C(x) := \int dq P(q, x)$, $\psi_x(q)$ is the conditional wave function defined in Eq. (8.26), and $\hat{\rho}_{Q|C}$ is the conditional density operator in Eq. (8.27) with corresponding Wigner function $W_{\hat{\rho}_{Q|C}}(q, p)$. Note that $\hat{\rho}_{Q|C}$ is equivalent to the reduced mixture of quantum ensembles associated with the hybrid ensemble, and hence correctly describes the statistics of all quantum observables for the ensemble. Since the marginal $W_{Q|C}(q, p)$ is just the Wigner function of $\hat{\rho}_{Q|C}$, it therefore provides an equivalent representation of the quantum statistics [26].

> The classical and quantum marginals of the hybrid Wigner function fully describe the respective statistics of classical and quantum observables for a hybrid ensemble.

We note it is also straightforward to integrate the hybrid Wigner function over the quantum and classical momentum variables, p and k, to recover the joint configuration probability density:

$$W(q, x) := \int dp dk\, W_{QC}(q, x, p, k) = P(q, x). \tag{8.89}$$

Thus, the configuration marginal of the hybrid Wigner function fully describes the joint configuration statistics.

8.7.2   Hybrid Wigner Functions for Gaussian Ensembles

The forms of P and S for a Gaussian hybrid ensemble are given in Eqs. (8.68) and (8.69), with $\xi \equiv (q, x)$. We are interested in one-dimensional particles ($n = 1$), and hence, dropping explicit time dependence, these equations reduce to

$$P(\xi) = \frac{\sqrt{\det K}}{2\pi} \, e^{-\frac{1}{2}(\xi - \alpha)^T K(\xi - \alpha)}, \tag{8.90}$$

$$S(\xi) = \frac{1}{2}(\xi - \alpha)^T L(\xi - \alpha) + \beta \cdot (\xi - \alpha) + s, \tag{8.91}$$

where K and L are symmetric 2×2 matrices, α and β are 2-vectors, and s is a scalar.

To evaluate the corresponding hybrid Wigner function, we note first that, defining $\xi' := \xi - \alpha$ and $\gamma(z) := (z, 0)^T$, one has

$$
\begin{aligned}
P(q - z, x)\, P(q + z, x) &= \frac{\det K}{4\pi^2} e^{-\frac{1}{2}(\xi' - \gamma)^T K(\xi' - \gamma) - \frac{1}{2}(\xi' + \gamma)^T K(\xi' + \gamma)} \\
&= \frac{\det K}{4\pi^2} e^{-\xi'^T K \xi'} \, e^{-\gamma^T K \gamma} \\
&= e^{-K_{11} z^2} P(q, x)^2.
\end{aligned} \tag{8.92}
$$

One also finds that

$$
\begin{aligned}
S(q - z, x) - S(q + z, x) &= -2\gamma^T L \xi' - 2\beta \cdot \gamma \\
&= -2\gamma^T [L(\xi - \alpha) + \beta] \\
&= -2\gamma^T \nabla_\xi S \\
&= -2z \partial_q S.
\end{aligned} \tag{8.93}
$$

Substitution into Eq. (8.85) then yields

$$
\begin{aligned}
W_{QC}(q, x, p, k) &= \frac{\delta(k - \partial_x S)}{\pi \hbar} P(q, x) \int dz \, e^{2ipz/\hbar} \, e^{-2iz\partial_q S/\hbar} \, e^{-\frac{1}{2}K_{11} z^2} \\
&= P(q, x)\, \delta(k - \partial_x S) \frac{1}{\hbar\sqrt{2\pi K_{11}}} e^{-\frac{2}{\hbar^2 K_{11}}(p - \partial_q S)^2}.
\end{aligned} \tag{8.94}
$$

It follows that one also has the conditional momentum density

$$W_{QC}(p, k | q, x) = \delta(k - \partial_x S) \frac{1}{\hbar\sqrt{2\pi K_{11}}} e^{-\frac{2}{\hbar^2 K_{11}}(p - \partial_q S)^2}. \tag{8.95}$$

Noting that $\partial_x S$ and $\partial_q S$ are each linear functions of q and x, we thus have:

> The hybrid Wigner function of a Gaussian ensemble is a Gaussian probability density on a joint quantum-classical phase space, with the classical momentum density perfectly correlated with a linear function of q and x.

Further, for coherent ensembles (i.e., with $L = 0$), integrating the hybrid Wigner function over q, x and k, and noting that $K_{11} = (\Delta q)^2$ from Eq. (8.90), one finds the uncertainty relation

$$\Delta q\, \Delta p = \frac{\hbar}{2}. \tag{8.96}$$

for the quantum position and momentum parameters. Thus:

> The quantum component of a coherent ensemble saturates the Heisenberg uncertainty relation.

8.7.3 Covariance Matrix

Finally, the hybrid Wigner function can be used to determine corresponding mean values and covariances of q, x, p and k, and is fully determined by these quantities in the Gaussian case. It is of interest to determine the connection between these quantities and observables of the hybrid ensemble.

The mean values associated with the hybrid Wigner function parameters are simply given by the corresponding quantum and classical averages, as follows immediately from the marginal distributions in Eqs. (8.87)–(8.89). Thus,

$$\langle q \rangle_W = \tilde{Q}_{\hat{q}}, \qquad \langle x \rangle_W = \tilde{C}_x, \qquad \langle p \rangle_W = \tilde{Q}_{\hat{p}}, \qquad \langle k \rangle_W = \tilde{C}_k, \tag{8.97}$$

where $\langle \cdot \rangle_W$ denotes the average evaluated with respect to $W_{QC}(q, x, p, k)$. For Gaussian ensembles these averages are just the components of the vectors α and β in Eqs. (8.90) and (8.91).

Evaluation of the covariances of the hybrid Wigner function parameters is more involved. We first define the hybrid covariance matrix C by

$$C_{jk} := \langle v_j v_k \rangle_{W_{QC}} - \langle v_j \rangle_{W_{QC}} \langle v_k \rangle_{W_{QC}}, \tag{8.98}$$

where v denotes the 4-vector (q, x, p, k). Letting $\mathrm{Cov}_\rho(a, b) := \langle ab^T \rangle_\rho - \langle a \rangle_\rho \langle b^T \rangle_\rho$ denote the $n \times n$ covariance matrix of n-vectors a and b relative to density ρ, and defining $\pi := (p, k)$, it follows that C has the block matrix form

$$C = \mathrm{Cov}_W(v, v) = \left(\frac{X \mid Y}{Y^T \mid Z}\right),\tag{8.99}$$

with

$$X = \mathrm{Cov}_W(\xi, \xi),\qquad Y = \mathrm{Cov}_W(\xi, \pi),\qquad Z = \mathrm{Cov}_W(\pi, \pi),\tag{8.100}$$

where we recall that $\xi \equiv (q, x)$.

The components of X follow immediately via Eq. (8.89) as $\mathrm{Cov}_P(\xi_m, \xi_n)$, i.e., they can be calculated directly from the configuration probability density P. The diagonal components of Z are also straightforward to evaluate for any hybrid ensemble (P, S), using the classical and quantum marginal densities in Eqs. (8.87) and (8.88):

$$Z_{11} = \mathrm{Var}_{W_{Q|C}} p = \tilde{Q}_{\hat{p}^2} - (\tilde{Q}_{\hat{p}})^2 = \mathrm{Var}_P(\partial_q S) + \frac{\hbar^2}{4}\int dq dx\,\frac{(\partial_q P)^2}{P},\tag{8.101}$$

$$Z_{22} = \mathrm{Var}_W k = \mathrm{Var}_P(\partial_x S).\tag{8.102}$$

The diagonal components of Y can similarly be evaluated via the classical and quantum marginal densities, and using the property that moments of a quantum Wigner function correspond to symmetrised moments of the corresponding operators [26], yielding

$$Y_{11} = \mathrm{Cov}_{W_{Q|C}}(q, p) = \tilde{Q}_{(\hat{q}\hat{p}+\hat{p}\hat{q})/2} - \tilde{Q}_{\hat{q}}\tilde{Q}_{\hat{p}} = \mathrm{Cov}_P(q, \partial_q S),\tag{8.103}$$

$$Y_{22} = \mathrm{Cov}_{W_{C|Q}}(x, k) = \mathrm{Cov}_P(q, \partial_x S).\tag{8.104}$$

Further, the off-diagonal components of Y may be evaluated using the marginal densities

$$W_{QK}(q, k) := \int dx dp\, W_{QC}(q, x, p, k) = \int dx\, P(q, x)\,\delta(k - \partial_x S),\tag{8.105}$$

$$W_{XP}(x, p) := \int dq dk\, W_{QC}(q, x, p, k) = P_C(x)\int dq\, W_Q(x, p|\psi_x),\tag{8.106}$$

with the conditional wavefunction $\psi_x(q)$ as defined in Eq. (8.26), to give

$$Y_{12} = \mathrm{Cov}_{W_{QK}}(q, k) = \mathrm{Cov}_P(q, \partial_x S),\tag{8.107}$$

$$Y_{12} = \mathrm{Cov}_{W_{XP}}(x, p) = \mathrm{Cov}_P(q, \partial_q S),\tag{8.108}$$

where the last equality follows using

$$\langle xp \rangle_{W_{XP}} = \int dx\, x\, P_C(x) \int dq\, p\, W_Q(q, p|\psi_x)$$

$$= \int dx\, x\, P_C(x) \langle \psi_x|\hat{p}|\psi_x \rangle$$

$$= \langle \psi_{QC}|\hat{x}\hat{p}|\psi_{QC} \rangle = \langle x\, \partial_q S \rangle_P.$$

Finally, the off-diagonal components of $\mathbf{Z}$ may be calculated using

$$\langle pk \rangle_W = \int dq dx dp\, p\, (\partial_x S)\, \frac{1}{\pi \hbar} \int dz\, e^{2ipz/\hbar} \bar{\psi}_{QC}(q+z, x)\, \psi_{QC}(q-z, x)$$

$$= \int dq dx\, (\partial_x S)\, P_C(x) \int dp\, p\, W_Q(q, p|\psi_x)$$

$$= \int dq dx\, (\partial_x S)\, P_C(x)\, (\partial_q S)\, |\langle x|\psi_x \rangle|^2$$

$$= \langle (\partial_x S)(\partial_q S) \rangle_P \tag{8.109}$$

where the second last equality follows using the property $\int dp\, p\, W_Q(q, p|\psi_Q) = (\partial_q S_Q)\, P_Q(x)$ for any wave function $\psi_Q = P_Q^{1/2} e^{iS_Q/\hbar}$ [27], to give

$$Z_{12} = Z_{21} = \mathrm{Cov}_W(p, k) = \mathrm{Cov}_P(\partial_q S, \partial_x S). \tag{8.110}$$

Substitution of the above expressions into Eq. (8.99) gives the general expression

$$C = \left(\begin{array}{c|c} \mathrm{Cov}_P(\xi, \xi) & \mathrm{Cov}_P(\xi, \nabla_\xi S) \\ \hline \mathrm{Cov}_P(\nabla_\xi S, \xi) & \mathrm{Cov}_P(\nabla_\xi S, \nabla_\xi S) + \frac{\hbar^2}{4} \left(\begin{array}{cc} \langle (\partial_q P)^2/P^2 \rangle_P & 0 \\ 0 & 0 \end{array} \right) \end{array} \right) \tag{8.111}$$

for the hybrid Wigner covariance matrix, in terms of P and S. It is of interest to note that each of the individual covariances is the same form that would be obtained using the exact uncertainty principle in Sect. 5.4, with the quantum contributions determined by adding a nonclassical momentum fluctuation to a classical momentum (see Chap. 5).

For the particular case of a hybrid Gaussian ensemble, as per Eqs. (8.90) and (8.91), it follows that

$$C = \left(\begin{array}{c|c} K^{-1} & K^{-1}L \\ \hline LK^{-1} & LK^{-1}L + \frac{\hbar^2}{4} \left(\begin{array}{cc} K_{11} & 0 \\ 0 & 0 \end{array} \right) \end{array} \right). \tag{8.112}$$

The same result can obtained, of course, via substitution of the explicit form in Eq. (8.94) for the hybrid Wigner function into Eq. (8.99). It would be of interest to consider the steering of a quantum ensemble by a classical ensemble for hybrid Gaussian ensembles, in analogy to the case of quantum Gaussian states considered by Wiseman et al. [28].

References

1. Boucher, W., Traschen, J.: Semiclassical physics and quantum fluctuations. Phys. Rev. D **37**, 3522–3532 (1988)
2. Makri, N.: Time-dependent quantum methods for large systems. Ann. Rev. Phys. Chem. **50**, 167–191 (1999)
3. Hawking, S.W.: Particle creation by black holes. Commun. Math. Phys. **43**, 199–220 (1975)
4. Kiefer, C.: Quantum Gravity. Oxford University Press, Oxford (2012)
5. Bohr, N.: Atomic Physics and Human Knowledge. Wiley, New York (1958)
6. Heisenberg, W.: Physics and Philosphy. Allen and Unwin, London (1958). Chaps. 3, 8
7. Albers, M., Kiefer, C., Reginatto, M.: Measurement analysis and quantum gravity. Phys. Rev. D **78**, 064051 (2008)
8. Dyson, F.: The world on a string. N. Y. Rev. Books **51**(8), 16–19 (2004)
9. Rothman, T., Boughn, S.: Can gravitons be detected? Found. Phys. **36**, 1801–1825 (2006)
10. Boughn, S., Rothman, T.: Aspects of graviton detection: graviton emission and absorption by atomic hydrogen. Class. Quantum Grav. **23**, 5839–5852 (2006)
11. Hall, M.J.W., Reginatto, M.: Interacting classical and quantum ensembles. Phys. Rev. A **72**, 062109 (2005)
12. Hall, M.J.W.: Consistent classical and quantum mixed dynamics. Phys. Rev. A **78**, 042104 (2008)
13. Reginatto, M., Hall, M.J.W.: Quantum-classical interactions and measurement: a consistent description using statistical ensembles on configuration space. J. Phys.: Conf. Ser. **174**, 012038 (2009)
14. Caro, J., Salcedo, L.L.: Impediments to mixing classical and quantum dynamics. Phys. Rev. A **60**, 842–852 (1999)
15. Salcedo, L.L.: Comment on "a quantum-classical bracket that satisfies the Jacobi identity" [J. Chem. Phys. 124, 201104 (2006)]. J. Chem. Phys. **126**, 057101 (2007)
16. Peres, A., Terno, D.R.: Hybrid classical-quantum dynamics. Phys. Rev. A **63**, 022101 (2001)
17. Agostini, F., Caprara, S., Ciccotti, G.: Do we have a consistent non-adiabatic quantum-classical mechanics? Europhys. Lett. **78**, 30001 (2007)
18. Joos, E., Zeh, H.D., Kiefer, C., Giulini, D.J.W., Kupsch, J., Stamatescu, I.-O.: Decoherence and the appearance of a classical world in quantum theory, 2nd edn. Springer, New York (2003)
19. Diosi, L., Gisin, N., Strunz, W.T.: Quantum approach to coupling classical and quantum dynamics. Phys. Rev. A **61**, 022108 (2000)
20. Takabayasi, T.: On the formulation of quantum mechanics associated with classical pictures. Prog. Theor. Phys. **8**, 143–182 (1952)
21. Takabayasi, T.: Remarks on the formulation of quantum mechanics with classical pictures and on relations between linear scalar fields and hydrodynamical fields. Prog. Theor. Phys. **9**, 187–222 (1953)
22. Schiller, R.: Quasi-classical theory of the nonspinning electron. Phys. Rev. **125**, 1100–1108 (1962)
23. Page, D.N., Geilker, C.D.: Indirect evidence for quantum gravity. Phys. Rev. Lett. **47**, 979–982 (1981)
24. Chua, A.J.K., Hall, M.J.W., Savage, C.M.: Interacting classical and quantum particles. Phys. Rev. A **85**, 022110 (2011)
25. Wigner, E.: On the quantum correction for thermodynamic equilibrium. Phys. Rev. **40**, 749–759 (1932)
26. Hillery, M., O'Connell, R.F., Scully, M.O., Wigner, E.P.: Distribution functions in physics: fundamentals. Phys. Rep. **106**, 121–167 (1984)
27. Hall, M.J.W.: Exact uncertainty relations. Phys. Rev. A **64**, 052103 (2001)
28. Wiseman, H.M., Jones, S.J., Doherty, A.C.: Steering, entanglement, nonlocality, and the Einstein-Podolsky-Rosen paradox. Phys. Rev. Lett. **98**, 140402 (2007)

Chapter 9
Consistency of Hybrid Quantum-Classical Ensembles

Abstract The formalism of ensembles on configuration space allows for a general description of interactions between quantum and classical ensembles. In this chapter, we consider such hybrid ensembles and focus on consistency requirements for models of quantum-classical interactions. We show how the configuration ensemble approach is able to satisfy desirable properties such as a Lie algebra of observables and Ehrenfest relations, while evading no-go theorems based in part on such properties. We then discuss issues concerning locality. It is found that, in principle, noninteracting ensembles of quantum and classical particles can be associated with nonlocal energy flows and nonlocal signaling. However, it is shown that such effects can be suppressed by a requirement of 'classicality', that localised classical systems have a very large number of degrees of freedom. Measurement aspects are also discussed and again 'classicality' plays an important role, this time ensuring an effective and irreversible decoherence. Finally, comparisons are briefly made with elements of the mean-field approach to quantum-classical interactions.

9.1 Introduction

In Chap. 8 we showed how the general formalism of ensembles on configuration space can be applied to successfully describe quantum-classical interactions. We demonstrated that this description has a number of nice properties, and considered examples such as measurement of a quantum spin by a classical pointer, scattering of a classical particle from a quantum superposition, and hybrid harmonic oscillators. In this chapter we will address the question of the *consistency* of the description.

Finding a physically consistent approach to modelling interactions between quantum and classical systems is a highly nontrivial task. Many proposals have been made, but all have concomitant difficulties of some sort, which we now briefly discuss.

First, in the *mean-field* approach, phase space coordinates of a classical system appear as parameters in a quantum Hamiltonian operator. This operator directly specifies the evolution of the quantum system in the usual way, while its average over the quantum degrees of freedom specifies a classical Hamiltonian for the classical parameters [1, 2]. However, while computationally useful as a semiclassical approx-

© Springer International Publishing Switzerland 2016

M.J.W. Hall and M. Reginatto, *Ensembles on Configuration Space*,
Fundamental Theories of Physics 184, DOI 10.1007/978-3-319-34166-8_9

imation to a fully quantum model, the classical system evolves deterministically. Thus, the mean-field approach cannot couple any quantum uncertainties into the classical parameters, where such a coupling is required, for example, if measurement and scattering interactions are to lead to a multiplicity of possible outcomes [1]. Nevertheless, Elze has shown that the mean-field approach does satisfy several basic consistency criteria [3], and hence it will be compared in more detail to the configuration ensemble approach in Sect. 9.6 below.

Second, the *phase space* approach relies on modelling the classical system by a set of mutually commuting 'phase space' observables on some Hilbert space [4], and allowing a unitary interaction with the quantum system. The most sophisticated model of this type is by Sudarshan and co-workers [5–8], in which the interaction Hamiltonian depends on non-observable operators on the classical Hilbert space. However, while this model has many interesting properties, the classical observables remain 'classical' only for a limited class of interactions, which does not include the standard Stern–Gerlach measurement interaction [7]. Peres and Terno have further shown that this approach does not reproduce the correct classical limit for quantum-classical oscillators, and indeed may result in a runaway increase of the classical oscillator amplitude [9, 10]. Diosi et al. have proposed a variation on the phase space approach, in which the classical phase space parameters are mapped to a set of coherent states rather than to a set of orthonormal states [11]. However, this variation doe not yield the classical equations of motion in the limit of no interaction, and intrinsically imposes quantum uncertainty relations upon the classical system.

Third, the *trajectory* approach is based on the deBroglie–Bohm formulation of quantum mechanics, in which quantum systems are described by an ensemble of trajectories acted on by a 'quantum potential' [12–14]. Interaction with a classical system is incorporated by modifying the equations of motion for the Bohmian trajectories in various ways [15–17]. This approach incorporates backreaction on the classical system, and has been found useful for semiclassical calculations in quantum chemistry. However, like the mean-field approach the classical motion is deterministic, and moreover does not respect energy conservation [18, 19].

Finally, various counterexamples and no-go theorems show that other proposed types of quantum-classical interaction lead to at least one of the following problems: negative probabilities; the absence of any backreaction on the classical system from the interaction; or to the loss of the correspondence principle in the classical limit [1, 9, 11, 20–23].

In contrast to the above-mentioned approaches, the approach in Chap. 8, based on ensembles on configuration space, avoids these problems and satisfies many further desirable properties. Some of these have been listed in Sect. 8.1 and/or discussed in Sect. 8.3. Here we will focus on consistency requirements for models of quantum-classical interactions, such as the existence of a dynamical bracket, a form of Ehrenfest's theorem, statistical completeness, and locality. We find that the last of these does restrict the application of our approach in some scenarios. In particular, as was previewed in Sect. 8.2.1, we must restrict the notion of 'classicality' for localised classical systems with controllable interactions, such as particles and measurement apparatuses (although not for classical fields such as gravity), to systems with a large number of interacting degrees of freedom. Violations of locality are then suppressed by essentially the same mechanism that ensures measurement irreversibility.

In Sects. 9.2 and 9.3 we show how the configuration ensemble approach is able to satisfy desirable properties such as a Lie algebra of observables and Ehrenfest relations, while evading no-go theorems based in part on such properties. In Sect. 9.4 we discuss locality aspects of the approach. It is found that, in principle, noninteracting ensembles of quantum and classical particles can be associated with nonlocal energy flows and nonlocal signaling. However, it is shown that such effects can be suppressed by a requirement of 'classicality.' Measurement aspects are discussed in Sect. 9.5, and again 'classicality' plays an important role—this time ensuring an effective and irreversible decoherence. Finally, comparisons are briefly made with elements of the mean-field approach in Sect. 9.6.

9.2 Dynamical Bracket Considerations

9.2.1 Two Minimal Conditions

We begin our analysis of the consistency of our approach to hybrid quantum-classical interactions by considering two minimal conditions proposed by Salcedo for the set of observables of any quantum-classical model [24]:

(i) A dynamical bracket may be defined on the set of observables, under which the observables form a Lie algebra, and
(ii) The dynamical bracket reduces to the classical Poisson bracket for any two classical observables, and to $(i\hbar)^{-1}$ times the quantum commutator for any two quantum observables.

These conditions have been previously justified on physical grounds by Caro and Salcedo [21] and are highly nontrivial. None of the alternative proposals discussed in Sect. 9.1 satisfy both conditions [21, 24, 25], with one exception: the mean-field approach [3] (see also Sect. 9.6 below). For example, in the phase space approach the classical observables commute under the dynamical bracket by construction. As a further example, the dynamical brackets proposed by Aleksandrov [26] and Prezhdo [27] fail conditions (i) and (ii) respectively. These minimal conditions are, therefore, a critical test for any hybrid theory.

Fortunately, this test is easily passed by the configuration ensemble approach [23]:

The description of hybrid systems via the formalism of ensembles on configuration space satisfies the minimal conditions (i) and (ii) for dynamical brackets, as a straightforward consequence of the guaranteed existence of a Poisson bracket.

Checking Condition (i) Recall first that in the configuration space formalism all physical systems, including hybrid quantum-classical systems, are described by a probability density P on the configuration space of the system that evolves according to an action principle (see Chap. 1). This implies the existence of a canonically conjugate function S on the configuration space, and Hamiltonian equations of motion of the form

$$\frac{\partial P}{\partial t} = \frac{\delta \mathcal{H}}{\delta S}, \qquad \frac{\partial S}{\partial t} = -\frac{\delta \mathcal{H}}{\delta P}, \tag{9.1}$$

where $\mathcal{H}[P, S]$ is the ensemble Hamiltonian. These equations may alternatively be written as $\partial P/\partial t = \{P, \mathcal{H}\}$ and $\partial S/\partial t = \{S, \mathcal{H}\}$, where the Poisson bracket $\{,\}$ is defined by (see Sect. 2.2)

$$\{A, B\} := \int d\xi \left[\frac{\delta A}{\delta P} \frac{\delta B}{\delta S} - \frac{\delta B}{\delta P} \frac{\delta A}{\delta S} \right] \tag{9.2}$$

for any two functions $A[P, S]$ and $B[P, S]$. Here ξ parameterises the configuration space of the system, integration is replaced by summation over any discrete components of ξ, and the functional derivatives are replaced by a partial derivative over such discrete components. Thus, the Poisson bracket plays the role of a dynamical bracket in the configuration ensemble formalism, with the rate of change of a (possibly time-dependent) observable $A[P, S, t]$ following via Eqs. (9.1) and (9.2) as

$$\frac{dA}{dt} = \frac{\delta A}{\delta P} \frac{\partial P}{\partial t} + \frac{\delta A}{\delta S} \frac{\partial S}{\partial t} + \frac{\partial A}{\partial t} = \{A, \mathcal{H}\} + \frac{\partial A}{\partial t} \tag{9.3}$$

(see also Sect. 2.2).

Further, as discussed in Sect. 2.2, observables are represented by a set of functionals of P and S which satisfy

$$A[P, S + c] = A[P, S], \qquad \frac{\delta A}{\delta S} = 0 \text{ if } P(\xi) = 0, \qquad A[\lambda P, S] = \lambda A[P, S]. \tag{9.4}$$

The first two requirements ensure conservation and normalisation of probability under canonical transformations generated by $A[P, S]$, and the third requirement implies that the numerical value of $A[P, S]$ may be interpreted as the average of the observable over the ensemble.

The Poisson bracket is well known to be a Lie bracket [28]. In particular, it satisfies the defining properties of linearity, antisymmetry, and the Jacobi identity. Further, as shown in Sect. 2.2, the requirements in Eq. (9.4) are preserved by the Poisson bracket. Hence, one can consistently assume, without loss of generality, that the set of observables form a closed set with respect to the Poisson bracket, implying that the minimal condition (i) above is trivially satisfied.

Checking Condition (ii) To determine whether the second minimal condition (ii) is also satisfied, it is necessary to first recall the definitions of 'classical' and 'quantum' observables for hybrid ensembles. Such ensembles have a joint configuration space parameterised by $\xi = (q, x)$, where q labels a complete basis set of the Hilbert space of the quantum system, and x labels the position of the classical system. For any function $f(x, p)$ on the phase space of the classical system, the corresponding classical observable for the hybrid ensemble is defined as per Eq. (8.17) of Chap. 8:

$$\tilde{C}_f[P, S] := \int dq dx\, P(q, x) f(x, \nabla_x S), \qquad (9.5)$$

where integration over q is replaced by summation over any discrete components. Further, for any Hermitian operator $\hat{M}$ on the Hilbert space of the quantum system, the corresponding quantum observable for the hybrid ensemble is defined as per Eqs. (8.18) and (8.19):

$$\tilde{Q}_{\hat{M}}[P, S] := \langle \psi_{QC} | \hat{M} | \psi_{QC} \rangle, \qquad (9.6)$$

where the hybrid wave function $\psi_{QC}(q, x)$ is defined by

$$\psi_{QC} := \sqrt{P}\, e^{iS/\hbar}. \qquad (9.7)$$

Note that $\tilde{C}_f$ and $\tilde{Q}_{\hat{M}}$ have the form of classical and quantum averages, respectively.

It is easily checked that the classical and quantum observables satisfy the general requirements in Eq. (9.4). Further, one has immediately from Eq. (8.20) (see also Sect. 2.3) that

$$\{\tilde{C}_f, \tilde{C}_g\} = \tilde{C}_{\{f,g\}}, \qquad \{\tilde{Q}_{\hat{M}}, \tilde{Q}_{\hat{N}}\} = \tilde{Q}_{[\hat{M},\hat{N}]/i\hbar}, \qquad (9.8)$$

where $\{f, g\}$ denotes the usual Poisson bracket for phase space functions $f(x, p)$ and $g(x, p)$, and $[,]$ is the usual quantum commutator. Hence minimal condition (ii) above is also satisfied.

9.2.2 Evading No-Go Theorems: An Algebraic Loophole

It is of interest to remark on how the configuration ensemble approach is able to avoid 'no-go' theorems in the literature for the existence of a dynamical bracket for hybrid systems [20–22]. Such theorems require that the dynamical bracket satisfy further conditions in addition to the minimal conditions (i) and (ii) above.

In particular, these theorems all require that the further condition holds:

(iii) The set of observables can be extended to form a product algebra, such that the product $A * B$ satisfies

$$\tilde{C}_f * \tilde{C}_g = C_{fg}, \qquad \tilde{Q}_{\hat{M}} * \tilde{Q}_{\hat{N}} = \tilde{Q}_{\hat{M}\hat{N}}, \qquad (9.9)$$

for classical and quantum observables respectively, and the Leibniz rule $\{A, B * C\} = \{A, B\} * C + B * \{A, C\}$.

However, unlike conditions (i) and (ii), this third condition has no compelling physical support. For example, the product of two non-commuting Hermitian operators is not a Hermitian operator, and hence $\tilde{Q}_{\hat{M}\hat{N}}$ is not an observable. Thus, the existence of such a product algebra clearly goes beyond the domain of observable quantities, and so cannot be justified on physical grounds. It follows that such 'no-go' theorems have no import for the configuration ensemble approach, in which no such product of observables is defined or required for physical predictions.

The above point emphasises a more general advantage of the configuration ensemble approach, as a good starting point for axiomatising physical theories. Once the basic concept of a probability density on a configuration space, evolving according to an action principle, has been accepted, then a dynamical Lie algebra arises for free as per Eq. (9.2). No additional algebraic structure, such as orthomodular posets or C^*-algebras, needs to be postulated. This advantage has already been exploited in Chaps. 5–7, where three different axiomatic approaches to quantum mechanics have been given within the framework of configuration ensembles.

9.3   Generalised Ehrenfest Relations and the Classical Limit

Oscillator Benchmark It is well known that the average position and momentum of linearly-coupled quantum oscillators obey the classical equations of motion. This motivated Peres and Terno to propose that this property should generalise to coupled quantum-classical oscillators, as a [9]

> definite benchmark ... for an acceptable quantum-classical hybrid formalism.

They went on to show that this benchmark fails in the 'phase space' approach to quantum-classical interactions [9, 10] (see also Sect. 9.1).

In contrast, the configuration ensemble approach to hybrid interactions not only satisfies the oscillator benchmark, but further satisfies a much stronger property: a natural generalisation of the quantum Ehrenfest relations.

Ehrenfest Relations In particular, consider a hybrid quantum-classical ensemble corresponding to a quantum particle of mass m interacting with a classical particle of mass M via a potential $V(q, x)$, where q and x denote the position configurations of the quantum and classical particles respectively. For simplicity, it will be assumed that both q and x are one-dimensional. The hybrid ensemble is therefore described by a probability density $P(q, x)$, a canonically conjugate field $S(q, x)$, and an ensemble Hamiltonian as per Eq. (8.12):

$$\mathscr{H}_{QC}[P, S] := \int dq \, dx \, P \left[\frac{(\partial_q S)^2}{2m} + \frac{(\partial_x S)^2}{2M} + V(q, x, t) + \frac{\hbar^2}{8m} \frac{(\partial_q P)^2}{P^2} \right], \quad (9.10)$$

where ∂_q and ∂_x denote the partial derivatives with respect to q and x respectively.

Now, the expectation values of the classical and quantum position observables follow from Eqs. (9.5–9.7) as

$$\langle x \rangle = C_x = \int dq\,dx\,P\,x, \qquad \langle q \rangle = Q_{\hat{q}} = \int dq\,dx\,P\,q. \qquad (9.11)$$

Similarly, distinguishing the classical and quantum momentum observables by the labels k and p, the expectation values of these observables follow as

$$\langle k \rangle = C_p = \int dq\,dx\,P\,\partial_x S, \qquad \langle p \rangle = Q_{\hat{p}} = \int dq\,dx\,P\,\partial_q S. \qquad (9.12)$$

The evolution of these expectation values may be calculated via Eqs. (9.1) and (9.10), and, as shown below, one finds [23]:

Hybrid Ehrenfest relations: The position and momentum observables of interacting quantum and classical particles satisfy

$$\frac{d}{dt}\langle x \rangle = M^{-1}\langle k \rangle, \qquad \frac{d}{dt}\langle k \rangle = -\langle \partial_x V \rangle, \qquad (9.13)$$

$$\frac{d}{dt}\langle q \rangle = m^{-1}\langle p \rangle, \qquad \frac{d}{dt}\langle p \rangle = -\langle \partial_q V \rangle. \qquad (9.14)$$

These relations imply that the centroid of a narrow initial probability density $P(q, x)$ will evolve classically for short timescales. Hence, in this sense, hybrid systems have a well defined classical limit.

The above relations are a clear generalisation of the standard Ehrenfest relations for quantum systems [29]. For the particular case of linearly-coupled classical and quantum oscillators, with

$$V(q, x) = \frac{1}{2}m\omega^2 q^2 + \frac{1}{2}M\Omega^2 x^2 + Kqx, \qquad (9.15)$$

Equations (9.13) and (9.14) simplify to the closed set of equations

$$\frac{d}{dt}\langle x \rangle = M^{-1}\langle k \rangle, \qquad \frac{d}{dt}\langle k \rangle = -M\Omega^2\langle x \rangle - K\langle q \rangle, \qquad (9.16)$$

$$\frac{d}{dt}\langle q \rangle = m^{-1}\langle p \rangle, \qquad \frac{d}{dt}\langle p \rangle = -m\omega^2\langle q \rangle - K\langle x \rangle. \qquad (9.17)$$

Thus, the average position and momentum values obey precisely the same equations of motion as those of two fully classical oscillators, as required by the oscillator benchmark proposed by Peres and Terno for the acceptability of any description of hybrid interactions [9].

Deriving the Relations To demonstrate the generalised Ehrenfest relations, note from Eqs. (9.3), (9.10) and (9.11) that

$$\frac{d}{dt}\langle x\rangle = \{\tilde{C}_x, \mathscr{H}_{QC}\} = -\int dq\,dx\,x\left[m^{-1}\partial_q(P\partial_q S) + M^{-1}\partial_x(P\partial_x S)\right], \quad (9.18)$$

where functional derivatives are evaluated as shown in Appendix A.1 of this book. Applying integration by parts with respect to q and x, to the first and second terms respectively, yields the first relation in Eq. (9.13). The first relation in Eq. (9.14) is derived in a similar manner.

To obtain the second relation in Eq. (9.13), note first that

$$\begin{aligned}
\frac{\delta\mathscr{H}_{QC}}{\delta P} &= \frac{(\partial_q S)^2}{2m} + \frac{(\partial_x S)^2}{2M} + V + \frac{\hbar^2}{2m}\frac{\partial P^{1/2}}{\partial P}\frac{\delta}{\delta P^{1/2}}\int dq\,dx\,(\partial_q P^{1/2})^2 \\
&= \frac{(\partial_q S)^2}{2m} + \frac{(\partial_x S)^2}{2M} + V - \frac{\hbar^2}{2m}\frac{\partial_q^2 P^{1/2}}{P^{1/2}}.
\end{aligned} \quad (9.19)$$

Equations (9.3), (9.10) and (9.12) then yield

$$\begin{aligned}
\frac{d}{dt}\langle k\rangle = &-\int dq\,dx\,(\partial_x S)\left[\partial_q\left(P\frac{\partial_q S}{m}\right) + \partial_x\left(\frac{P\partial_x S}{M}\right)\right] \\
&+ \int dq\,dx\,(\partial_x P)\left[\frac{(\partial_q S)^2}{2m} + \frac{(\partial_x S)^2}{2M} + V\right] \\
&- \frac{\hbar^2}{2m}\int dq\,dx\,(\partial_x P)\frac{\partial_q^2 P^{1/2}}{P^{1/2}}.
\end{aligned} \quad (9.20)$$

Using integration by parts in the first and second lines leads to cancellation of all terms involving S, with only a term $-\int dq dx\,P\partial_x V$ remaining. Further, the term in the third line simplifies via integration by parts to

$$\begin{aligned}
-\frac{\hbar^2}{m}\int dqdx\,(\partial_x P^{1/2})(\partial_q^2 P^{1/2}) &= \frac{\hbar^2}{m}\int dqdx\,(\partial_q\partial_x P^{1/2})\,(\partial_q P^{1/2}) \\
&= \frac{\hbar^2}{2m}\int dqdx\,\partial_x\left[(\partial_q P^{1/2})^2\right] = 0. \quad (9.21)
\end{aligned}$$

Hence the second relation in Eq. (9.13) immediately follows. The second relation in Eq. (9.14) is obtained by similar reasoning, with ∂_q replaced by ∂_x as appropriate.

9.4 Locality Considerations

9.4.1 Configuration and Momentum Separability

It is reasonable to expect, for a hybrid quantum-classical ensemble, that measurement of the classical configuration cannot detect whether or not a local transformation has been applied to the quantum component, and vice versa. As discussed in Sect. 8.3, this property does indeed hold:

> **Configuration separability**: The configuration statistics of the classical component of a hybrid ensemble are invariant under any local unitary transformation carried out on the quantum component. Conversely, the configuration statistics of the quantum component are invariant under any local canonical transformation carried out on the classical component.

Configuration separability may be equivalently be expressed in terms of observables:

$$\{\tilde{C}_{g(x)}, \tilde{Q}_{\hat{M}}\} = 0, \qquad \{\tilde{Q}_{g(\hat{q})}, \tilde{C}_f\} = 0, \tag{9.22}$$

as per Eqs. (8.22) and (8.23). Here g is any function of the classical and quantum configurations, $\hat{M}$ is an arbitrary Hermitian operator generating a unitary transformation on the quantum Hilbert space, and $f(q, p)$ is an arbitrary function generating a canonical transformation on the classical phase space. Thus, the statistics of x are invariant under the unitary transformation generated by $\hat{M}$, while the statistics of q are invariant under the phase space transformation generated by f.

Here we will show that a similar but weaker property holds for the momentum statistics of interacting quantum and classical particles. In particular, let x_m and k_m label the components of the classical position and momentum, and q_m and p_m label the components of the quantum position and momentum. Then one has:

> **Momentum separability**: The average classical momentum of a hybrid ensemble is invariant under local unitary transformations of the quantum component, and the average quantum momentum is invariant under local canonical transformations of the classical component, i.e.,
>
> $$\{\tilde{C}_{k_m}, \tilde{Q}_{\hat{M}}\} = 0, \qquad \{\tilde{C}_f, \tilde{Q}_{\hat{p}_m}\} = 0. \tag{9.23}$$
>
> for arbitrary Hermitian operators $\hat{M}$ and phase space functions $f(x, k)$.

The combination of configuration and momentum separability immediately implies the dynamical bracket relations

$$\{\tilde{C}_{x_m}, \tilde{Q}_{\hat{q}_n}\} = 0, \qquad \{\tilde{C}_{x_m}, \tilde{Q}_{\hat{p}_n}\} = 0, \tag{9.24}$$

$$\{\tilde{C}_{p_m}, \tilde{Q}_{\hat{q}_n}\} = 0, \qquad \{\tilde{C}_{p_m}, \tilde{Q}_{\hat{p}_n}\} = 0, \tag{9.25}$$

for the position and momentum observables of the hybrid ensemble. These relations were assumed by Salcedo in the proof of a no-go theorem for the existence of hybrid classical and quantum dynamics [20]. However, this theorem is inapplicable to the configuration ensemble approach because it requires a further nonphysical assumption, that observables form a product algebra, which does not hold in this approach (see also Sect. 9.2.2).

To prove the first relation in Eq. (9.23), note first that making a change of representation of the ensemble, from the pair (P, S) to the pair $(\psi_{QC}, \bar{\psi}_{QC})$, one has

$$\frac{\delta A}{\delta P} = \frac{\partial \psi_{QC}}{\partial P} \frac{\delta A}{\delta \psi_{QC}} + \frac{\partial \bar{\psi}_{QC}}{\partial P} \frac{\delta A}{\delta \bar{\psi}_{QC}} = \frac{1}{\bar{\psi}_{QC} \psi_{QC}} \operatorname{Re} \left\{ \psi_{QC} \frac{\delta A}{\delta \psi_{QC}} \right\}, \tag{9.26}$$

$$\frac{\delta A}{\delta S} = \frac{\partial \psi_{QC}}{\partial S} \frac{\delta A}{\delta \psi_{QC}} + \frac{\partial \bar{\psi}_{QC}}{\partial S} \frac{\delta A}{\delta \bar{\psi}_{QC}} = -\frac{2}{\hbar} \operatorname{Im} \left\{ \psi_{QC} \frac{\delta A}{\delta \psi_{QC}} \right\}, \tag{9.27}$$

and hence the Poisson bracket defined in Eq. (9.2) can be reexpressed in terms of the hybrid wave function and its conjugate as

$$\{A, B\} = \frac{2}{\hbar} \operatorname{Im} \left\{ \int dq dx \, \frac{\delta A}{\delta \psi_{QC}} \frac{\delta B}{\delta \bar{\psi}_{QC}} \right\}, \tag{9.28}$$

generalising the analogous result in Sect. 2.3.3 for quantum ensembles. Noting that the average classical momentum can be written in the 'quantum' form

$$\tilde{C}_{k_m} = \int dq dx \, P \, \partial_{x_m} S = \langle \psi_{QC} | \hat{k}_m | \psi_{QC} \rangle, \tag{9.29}$$

where ψ_{QC} is the hybrid wave function defined in Eq. (9.7) and $\hat{k}_m := (\hbar/i)(\partial/\partial x_m)$, it follows that

$$\{\tilde{C}_{k_m}, \tilde{Q}_{\hat{M}}\} = \frac{2}{\hbar} \operatorname{Im} \left\{ \int dq dx \, \overline{(\hat{k}_m \psi_{QC})} \hat{M} \psi_{QC} \right\} = \frac{1}{\hbar} \langle \psi_{QC} | [\hat{k}_m, \hat{M}] | \psi_{QC} \rangle = 0,$$

as desired.

Finally, to prove the second relation in Eq. (9.23), note that the average quantum momentum can be rewritten in the 'classical' form

$$\tilde{Q}_{\hat{p}_m} = \langle \psi_{QC} | \hat{p}_m | \psi_{QC} \rangle = \int dq dx\, P\, \partial_{q_m} S. \tag{9.30}$$

The definition of the Poisson bracket in Eq. (9.2), with the variational derivatives evaluated as per Eq. (A.6) in Appendix A.1 of this book, then yields

$$
\begin{aligned}
\{\tilde{C}_f, \tilde{Q}_{\hat{p}_m}\} &= \int dq dx \left[\frac{\delta \tilde{C}_f}{\delta P} \frac{\delta \tilde{Q}_{\hat{p}_m}}{\delta S} - \frac{\delta \tilde{Q}_{\hat{p}_m}}{\delta P} \frac{\delta \tilde{C}_f}{\delta S} \right] \\
&= \int dq dx \left[-f(x, \nabla_x S)\,(\partial_{q_m} P) + (\partial_{q_m} S)\, \nabla_x \cdot \left(P \frac{\partial f}{\partial \nabla_x S} \right) \right] \\
&= \int dq dx\, P \left[(\partial_{q_m} f(x, \nabla_x S)) - (\partial_{q_m} \nabla_x S) \cdot \frac{\partial f}{\partial \nabla_x S} \right] \\
&= 0
\end{aligned}
$$

as required.

9.4.2 Strong Separability

It is natural to ask, in light of the configuration and momentum separability properties in Eqs. (9.22) and (9.23), whether the *strong separability* property

$$\{\tilde{C}_f, \tilde{Q}_{\hat{M}}\} = 0 \quad ? \tag{9.31}$$

also holds for hybrid ensembles. That is, does the dynamical bracket between arbitrary classical and quantum observables vanish? It turns out that the answer to this question is in the negative. This leads to a significant constraint on what can be modelled as a 'classical system', to avoid the undesirable possibility of nonlocal signaling between noninteracting quantum and classical ensembles.

It is important to note that strong separability *does* hold in an important case. In particular, it is valid whenever the quantum and classical components are *independent*, i.e., when

$$P(q, x) = P_Q(q)\, P_C(x), \qquad S(q, x) = S_Q(q) + S_C(x). \tag{9.32}$$

Since independent ensembles remain independent when they are noninteracting, i.e., when the ensemble Hamiltonian splits into distinct quantum and classical contributions $\mathcal{H}_{QC} = \mathcal{H}_Q + \mathcal{H}_C$ (see Sect. 3.2), one has:

Strong separability is satisfied at all times by the observables of noninteracting independent ensembles.

To prove the above statement, note first that the strong separability property for independent ensembles has been demonstrated for arbitrary joint ensembles in Sect. 3.3.3, and hence holds for hybrid ensembles in particular. An explicit proof for the hybrid case may also be given as follows [23]. From Eqs. (9.5) and (9.7) one can rewrite classical observables in terms of the hybrid wave function as

$$\tilde{C}_f = \int dq\, dx\, \bar{\psi}_{QC}\psi_{QC} f(x, k), \qquad k := \frac{\hbar}{2i}\left(\frac{\nabla_x \psi_{QC}}{\psi_{QC}} - \frac{\nabla_x \bar{\psi}_{QC}}{\bar{\psi}_{QC}}\right), \qquad (9.33)$$

and hence, dropping the QC subscript for convenience, one has

$$\frac{\delta \tilde{C}_f}{\delta \psi} = \bar{\psi} f + \bar{\psi}\psi\,(\nabla_k f) \cdot \frac{\partial k}{\partial \psi} - \nabla_x \cdot \left(\bar{\psi}\psi\,(\nabla_k f) \cdot \frac{\partial k}{\partial(\nabla_x \psi)}\right)$$

$$= \bar{\psi} f - \frac{\hbar}{2i}\frac{\bar{\psi}}{\psi}\,(\nabla_k f \cdot \nabla_x \psi) - \frac{\hbar}{2i}\nabla_x \cdot \left(\bar{\psi}\nabla_k f\right). \qquad (9.34)$$

Moreover, from Eq. (9.6) one has $\delta \tilde{Q}_{\hat{M}}/\delta \bar{\psi} = \hat{M}\psi$. Hence, since the independence condition Eq. (9.32) is equivalent to the factorisation $\psi(q, x) = \psi_Q(q)\psi_C(x)$ of the hybrid wavefunction, it follows that

$$\int dq\, dx\, \frac{\delta \tilde{C}_f}{\delta \psi}\frac{\delta \tilde{Q}_{\hat{M}}}{\delta \bar{\psi}} = \int dq\, \bar{\psi}_Q \hat{M}\psi_Q \left\{\int dx\, \bar{\psi}_C \psi_C f\right.$$

$$\left. - \frac{\hbar}{2i}\int dx\, \left[\bar{\psi}_C(\nabla_k f \cdot \nabla_x \psi_C) + \psi_C \nabla_x \cdot (\bar{\psi}_C \nabla_k f)\right]\right\}$$

$$= \int dq\, \bar{\psi}_Q \hat{M}\psi_Q \int dx\, \bar{\psi}_C \psi_C f(x, \nabla_x S_C) \qquad (9.35)$$

for independent ensembles, where integration by parts has been used to obtain the final result. This expression is clearly real, implying immediately from the general Poisson bracket formula in Eq. (9.28) that

$$\{\tilde{C}_f, \tilde{Q}_{\hat{M}}\} = \frac{2}{\hbar}\,\text{Im}\left\{\int dq\, dx\, \frac{\delta \tilde{C}_f}{\delta \psi}\frac{\delta \tilde{Q}_{\hat{M}}}{\delta \bar{\psi}}\right\} = 0 \qquad (9.36)$$

for independent ensembles.

The failure of strong separability to hold more generally is demonstrated by the following counterexample [23]. In particular, for classical and quantum particles having masses M and m respectively, consider the 'free' ensemble Hamiltonians

$$\mathcal{H}_C^0 = \int dq dx \, P \, \frac{|\nabla_x S|^2}{2M}, \quad \mathcal{H}_Q^0 = \int dq dx \, P \left[\frac{|\nabla_q S|^2}{2m} + \frac{\hbar^2}{8m} \frac{|\nabla_q P|^2}{P^2} \right], \quad (9.37)$$

corresponding to classical and quantum observables $\tilde{C}_f$ and $\tilde{Q}_{\hat{M}}$, respectively, with $f(x, p) = p \cdot p/(2M)$ and $\hat{M} = \hat{p} \cdot \hat{p}/(2m)$. Evaluating their Poisson bracket via Eq. (9.2) yields

$$\{\mathcal{H}_C^0, \mathcal{H}_Q^0\} = \frac{\hbar^2}{2mM} \int dq \, dx \, P \, (\nabla_x S) \cdot \nabla_x (P^{-1/2} \nabla_q^2 P^{1/2}) \neq 0, \quad (9.38)$$

i.e., it does not vanish identically. Such violations of strong separability have been referred to as 'ghost interactions' by Salcedo [30]. Note, however, that this Poisson bracket *does* vanish whenever the quantum and classical configurations are uncorrelated, i.e., whenever $P(q, x) = P_Q(q) P_C(x)$.

9.4.3 Implications of Strong Separability Violation

The violation of strong separability for some observables places a strong constraint on what types of quantum-classical interactions can be consistently described in the configuration ensemble formalism. In particular, it will be seen that while this violation is harmless for intrinsically interacting systems, it leads to nonlocal energy flows and to the possibility of action at a distance between noninteracting ensembles of quantum and classical particles.

No Implications for Intrinsically Interacting Systems First, it should be noted that in the particular case where the hybrid ensemble describes quantum matter coupled to classical spacetime [1, 31] (see also Chap. 11), a violation of strong separability in Eq. (9.31) is irrelevant to locality issues, as there is no sense in which interaction between the systems can be 'switched off'—matter bends space and space curves matter, and so a change in one component is fully expected to drive a change in the other component. This corresponds to the direct *multiplicative* coupling of the metric tensor to the fields in the corresponding ensemble Hamiltonian (see Chap. 11).

Similar remarks applies to any hybrid ensemble for which the ensemble Hamiltonian can never be reduced to a simple sum of quantum and classical terms. As discussed in Sect. 3.2.2 this is equivalent to a direct coupling between the components at all times, i.e., to an intrinsic inseparability.

Nonlocal Energy Flow Between Noninteracting Particles In contrast, to see how action at a distance can arise for systems that can be uncoupled, consider a hybrid ensemble with two well-separated noninteracting components, and an ensemble Hamiltonian of the form

$$\mathcal{H}[P, S] = \mathcal{H}_C + \mathcal{H}_Q \quad (9.39)$$

It follows from Eq. (9.3) that

$$\frac{d\mathcal{H}_C}{dt} = \{\mathcal{H}_C, \mathcal{H}_C + \mathcal{H}_Q\} + \frac{\partial \mathcal{H}_C}{\partial t} = \{\mathcal{H}_C, \mathcal{H}_Q\} + \frac{\partial \mathcal{H}_C}{\partial t}. \tag{9.40}$$

Hence, if the Poisson bracket $\{\mathcal{H}_C, \mathcal{H}_Q\}$ does not vanish, then energy can flow from one ensemble to the other, despite their spatial separation. Using a subscript to denote this nonlocal contribution to the average energy flow, one has

$$\left(\frac{d\mathcal{H}_C}{dt}\right)_{\text{nonlocal}} = \{\mathcal{H}_C, \mathcal{H}_Q\} = -\left(\frac{d\mathcal{H}_Q}{dt}\right)_{\text{nonlocal}}. \tag{9.41}$$

The case of two noninteracting quantum and classical particles evolving under local (possibly time-dependent) potentials, with

$$\mathcal{H}_C = \mathcal{H}_C^0 + \int dq dx\, P\, V_C(x, t), \qquad \mathcal{H}_Q = \mathcal{H}_Q^0 + \int dq dx\, P\, V_Q(q, t), \quad (9.42)$$

provides a simple and instructive example, where $\mathcal{H}_Q^0$ and $\mathcal{H}_C^0$ are the 'free' ensemble Hamiltonians in Eq. (9.37). In this case the nonlocal contribution to the rate of change of the average classical energy follows via Eqs. (9.38), (9.41) and configuration separability as

$$\left(\frac{d\mathcal{H}_C}{dt}\right)_{\text{nonlocal}} = \{\mathcal{H}_C^0, \mathcal{H}_Q^0\} = \frac{\hbar^2}{2mM} \int dq\, dx\, P\, (\nabla_x S) \cdot \nabla_x (P^{-1/2} \nabla_q^2 P^{1/2}). \tag{9.43}$$

Note that the flow is second order in $\hbar$, and inversely proportional to the classical mass M, and hence is expected to be very small relative to typical classical energies. Note also that the average flow is independent of the local potentials V_Q and V_C (it may similarly be shown to be independent of electromagnetic potentials).

We remark that the nonlocal energy flow can also be rewritten in the suggestive form

$$\left(\frac{d\mathcal{H}_C}{dt}\right)_{\text{nonlocal}} = \int dq dx\, P\, F_{C|Q} \cdot v_C, \tag{9.44}$$

where the 'classical velocity' and 'quantum-induced classical force' are defined by

$$v_C := \frac{\nabla_x S}{M}, \qquad F_{C|Q} := \frac{\hbar^2}{2m} \nabla_x \left(\frac{\nabla_q^2 P^{1/2}}{P^{1/2}}\right), \tag{9.45}$$

respectively. These are analogous to similar expressions in Bohmian mechanics for two quantum particles, where v_C corresponds to the velocity of the C-particle, and $F_{C|Q}$ to the force on the C-particle due to the quantum potential of the Q-particle [12, 14]. Thus the energy flow in Eq. (9.44) may be interpreted in terms of work done on the classical ensemble by such a force.

We will see in Sect. 9.4.4 that by restricting the definition of 'classical' systems to have many degrees of freedom leads to the effective vanishing of such nonlocal energy flows.

Nonlocal Signaling between Noninteracting Particles The nonlocal energy flow to the classical particle in Eq. (9.43) is independent of the local quantum potential V_Q (it may similarly be shown to be independent of any electromagnetic potential acting on the quantum particle). Hence, an attempt to send a signal from a quantum ensemble of particles to a classical ensemble of particles, by changing V_Q over a short time interval $[0, \delta t]$, will not lead to any change in the average classical energy to first order in δt. Similarly, there will be no change to second order: one has

$$
\begin{aligned}
\frac{d^2 \mathcal{H}_C}{dt^2} &= \{\{\mathcal{H}_C^0, \mathcal{H}_Q^0\}, \mathcal{H}_C + \mathcal{H}_Q\}\} + \frac{\partial^2 \mathcal{H}_C}{\partial t^2} \\
&= \{\{\mathcal{H}_C^0, \mathcal{H}_Q^0\}, \mathcal{H}_C + \mathcal{H}_Q^0\}\} + \left\{\{\mathcal{H}_C^0, \mathcal{H}_Q^0\}, \tilde{C}_{V_Q}\right\} + \frac{\partial^2 \mathcal{H}_C}{\partial t^2} \\
&= \{\{\mathcal{H}_C^0, \mathcal{H}_Q^0\}, \mathcal{H}_C + \mathcal{H}_Q^0\}\} + \frac{\partial^2 \mathcal{H}_C}{\partial t^2},
\end{aligned}
\tag{9.46}
$$

which is again independent of V_Q. Here the last line follows using Eq. (9.43) and via integration by parts to evaluate

$$
\left\{\{\mathcal{H}_C^0, \mathcal{H}_Q^0\}, \tilde{C}_{V_Q}\right\} = \frac{\hbar^2}{2mM} \int dq\, dx\, \nabla_x \cdot \left[P\nabla_x(P^{-1/2}\nabla_q^2 P^{1/2})\right] V_Q(q, t) = 0.
\tag{9.47}
$$

However, after some calculation, the *third* time derivative is found to have the form

$$
\frac{d^3 \mathcal{H}_C}{dt^3} = \left\{\{\mathcal{H}_Q^0, \tilde{C}_{V_Q}\}, \{\mathcal{H}_C^0, \mathcal{H}_Q^0\}\right\} + \dots,
\tag{9.48}
$$

where the dots indicate contributions independent of V_Q. The first term on the right does not vanish in general. Hence, signaling from the quantum particle to the classical particle is possible, in principle, via controlling V_Q to affect the average classical energy.

It has been suggested that such signaling via the violation of strong separability could be avoided via the physically reasonable assumption that *the only observables accessible to direct measurement are classical configuration observables* [23]. However, while such observables—such as the pointer position of a classical measuring apparatus—automatically satisfy the configuration separability property $\{\tilde{C}_{g(x)}, \tilde{Q}_{\hat{M}}\} = 0$ as per Eq. 9.22, it was later discovered that this assumption is *not* sufficient to avoid nonlocal signaling [32].

In particular, for noninteracting quantum and classical particles, i.e., with an ensemble Hamiltonian as per Eqs. (9.39) and (9.42), consider the time evolution of a classical observable $\tilde{C}_{g(x)}$, where g is any function of the classical configuration x. From Eq. (9.3) and configuration separability we have

$$\frac{d\tilde{C}_{g(x)}}{dt} = \{\tilde{C}_{g(x)}, \mathcal{H}_C + \mathcal{H}_Q\} = \{\tilde{C}_{g(x)}, \mathcal{H}_C\} = \{\tilde{C}_{g(x)}, \mathcal{H}_C^0\}, \qquad (9.49)$$

with the last equality following using $\{\tilde{C}_{g(x)}, \tilde{C}_{V_C}\} = \tilde{C}_{\{g(x), V_C\}} = 0$ via Eq. (9.8). Hence there is no effect of the quantum component of the ensemble Hamiltonian on the first time derivative of the classical configuration observable. Further, the second time derivative follows, using configuration separability with respect to $\tilde{C}_{g(x)}$ and $\tilde{C}_{V_Q}$ and the Jacobi identity, as

$$\frac{d^2\tilde{C}_{g(x)}}{dt^2} = \{\{\tilde{C}_{g(x)}, \mathcal{H}_C^0\}, \mathcal{H}_C + \mathcal{H}_Q^0 + \tilde{C}_{V_Q}\}$$
$$= \{\{\tilde{C}_{g(x)}, \mathcal{H}_C^0\}, \mathcal{H}_C\} + \{\tilde{C}_{g(x)}, \{\mathcal{H}_C^0, \mathcal{H}_Q^0\}\}. \qquad (9.50)$$

While the last term is seen to depend on the nonlocal energy flow in Eq. (9.43), there is again no dependence on the local potential V_Q, and hence no signaling effects arise to second order. One similarly finds that no such effects arise to third order. However, after a lengthy calculation one obtains

$$\frac{d^4\tilde{C}_{g(x)}}{dt^4} = \left\{\left\{\left\{\tilde{C}_{g(x)}, \{\mathcal{H}_C^0, \mathcal{H}_Q^0\}\right\}, \mathcal{H}_Q^0\right\}, \tilde{C}_{V_Q}\right\} + \cdots$$
$$= \frac{\hbar^2}{2m^2M} \int dq dx\, P\,(\nabla_q V_Q) \cdot \left(\nabla_q \frac{\delta A[P]}{\delta P}\right) + \cdots, \qquad (9.51)$$

where the dots indicate terms with no dependence on V_Q, and we define $A[P] := \int dq dx\, P\,(\nabla_x g) \cdot \nabla_x(P^{-1/2}\nabla_q^2 P^{1/2})$.

Thus, the fourth time derivative of the classical configuration observable has an explicit dependence on the local quantum potential V_Q. This allows nonlocal signalling, in principle, by one observer making a choice between two potentials V_Q and V_Q' in the vicinity of the quantum ensemble, and a second observer estimating the fourth time derivative of a classical configuration observable for the classical ensemble (e.g., by measuring the observable at four closely separated times on four respective subensembles). This has been numerically confirmed for the case of a Gaussian hybrid ensemble (see Sect. 8.6.1), with the evolution of the classical observable $\tilde{C}_{x^2} = \langle x^2 \rangle$ found to depend on the value of the spring constant k of a one-dimensional oscillator potential $V_Q = \frac{1}{2}kq^2$ acting on the quantum particle [32].

We conclude that:

> The violation of strong separability has no implications for intrinsically inter-
> acting hybrid systems, such as classical gravitational fields coupled to quan-
> tum matter fields. However, it implies that the application of the configura-
> tion ensemble approach to hybrid systems must be restricted in some way, to
> avoid observable nonlocal energy transfer and signaling between noninteract-
> ing quantum and classical particles.

A suitable restriction is discussed below.

9.4.4 Suppression of Strong Separability Violation via "Classicality"

We first note that if nature provides us with systems that behave as classical particles,
they are likely to correspond to macroscopic objects (e.g., measuring devices, as in
the Copenhagen interpretation of quantum mechanics), i.e., systems with interactions between many internal and/or external degrees of freedom. It turns out that
effects such as nonlocal energy flows and non-local signalling are suppressed for
such objects, making them unobservable. It will be seen in Sect. 9.5 that the same
property also guarantees effective decoherence in measurements of quantum systems
by classical apparatuses.

We therefore make the assumption:

> **Classicality**: Localised classical objects are always macroscopic; i.e., with a
> large number of internal and external degrees of freedom.

Thus "classicality" implies, in addition to having classical equations of motion,
that a classical object (for example, a measuring device) is macroscopic. This a very
reasonable assumption, supported by observation. It is known from experiment that
microscopic objects are described by quantum theory. Therefore, a classical object
cannot be microscopic.

How are macroscopic objects different from microscopic objects? One important
property of macroscopic objects (for example, classical measuring devices) is that
they cannot be isolated from the environment: a macroscopic object cannot avoid
scattering photons and other particles [33]. Thus, for example, the pointer of a mea-

suring device will be continuously undergoing scattering processes, implying there are a large number of external degrees of freedom. Another important property of macroscopic objects is that they have a very large number of internal degrees of freedom. When describing a classical measuring apparatus, we often discuss only *one* of its degrees of freedom, say the position of a pointer, but in reality there is an enormous number of "irrelevant" degrees of freedom, say 10^{23}, which must be treated statistically if they are modeled at all (usually they are simply neglected, as we have also done previously).

The important feature of classicality for our purposes is that it implies most classical observables are effectively unobservable: it is impossible to make a measurement of detailed functions of the underlying degrees of freedom. As pointed out by Peres in a similar context, even the constants of the motion of a large classical system are effectively unmeasurable [34], which is only exacerbated if information is being carried away via interactions with external degrees of freedom such as photons. The interactions involved are simply uncontrollable (and irreversible) for all practical purposes.

In particular, for a noninteracting quantum system interacting with a macroscopic classical system, there will be an enormous number of relevant degrees of freedom x. Hence, while changing the local quantum potential V_Q will lead to a nonlocal change in properties of the configuration x as discussed in Sect. 9.4.3, this correlation involves so many degrees of freedom (involving photons heading off to infinity) that it becomes impossible to observe—the correlation is too spread out among all the degrees of freedom. Conversely, one cannot in practice control the local classical potential V_C to induce an observable change in the quantum system, since there are so many diverse classical degrees of freedom correlated in a complex way.

It is interesting to consider the effect of classicality from the perspective of the forms of nonlocal energy flow in Eqs. (9.43) and (9.44), which generalise in the simplest case to a sum over the classical degrees of freedom $(x^{(1)}, \ldots x^{(N)})$:

$$\left(\frac{d\mathscr{H}_C}{dt}\right)_{\text{nonlocal}} = \sum_n \frac{\hbar^2}{2mM_n} \int dq dx^{(1)} \ldots dx^{(N)} \, P \, (\nabla_{x^{(n)}} S) \cdot \nabla_{x^{(n)}} (P^{-1/2} \nabla_q^2 P^{1/2})$$

$$= \sum_n \int dq dx^{(1)} \ldots dx^{(N)} \, P \, F_{C|Q}^{(n)} \cdot v_C^{(n)} \tag{9.52}$$

The 'quantum-induced forces' $F_{C|Q}^{(n)}$ scale as $\hbar^2/M_m$, so that the corresponding 'accelerations' scale as $(\hbar/M_m)^2$, and are not expected to have any simple correlation with the corresponding 'velocities' $v_C^{(n)}$. Hence the nonlocal energy flow will be made up of N effectively random contributions, thus making a contribution scaling as $\hbar^2/\sqrt{N}$ relative to the local classical energy (since the latter scales as N). For $N \sim 10^{23}$ this is simply unobservable in practice. We conclude more generally that:

Violations of strong separability are effectively unobservable under the assumption of classicality.

We will see in the following section that classicality also plays a important role in measurements.

9.5 Measurement Considerations

Examples of measurements of quantum configuration observables, such as position and spin, via interaction with an ensemble of classical pointers, have been described in Sect. 8.4, including the conditional decoherence of the quantum component relative to the classical component. Here it is noted there is a simple model for describing the indirect measurement of *any* quantum observable via interaction with a classical pointer. The mechanism by which the macroscopic nature of such a pointer, i.e., its classicality, leads from conditional decoherence (an 'improper mixture') to effective decoherence (a 'proper mixture'), is also discussed.

9.5.1 General Measurement Model

The measurement interaction in Sect. 8.4 of the previous chapter has the simple generalisation [31]

$$\mathscr{H}_I := \kappa(t) \int dq\, dx\, \bar{\psi}_{QC}(q, x) \left(\frac{\hbar}{i} \frac{\partial}{\partial x}\right) \hat{M} \psi_{QC}(q, x). \tag{9.53}$$

Here $\kappa(t)$ is a (possibly time-dependent) coupling constant, x denotes the position of a one-dimensional classical pointer (with integration over q replaced by summation over any discrete values), $\hat{M}$ denotes the quantum observable being measured, and ψ_{QC} is the hybrid wavefunction defined in Eq. (9.7). It may be checked that this reduces to the form of the interaction ensemble Hamiltonian in Eq. (8.47) in the special case $\hat{M} = f(\hat{q})$.

Assuming that the measurement takes place over a sufficiently short time period, $[0, T]$, such that $\mathscr{H}_0$ can be ignored during the measurement, it is straightforward to check that the equations of motion in Eq. (9.1) during the interaction are equivalent to the hybrid Schrödinger equation

$$i\hbar \frac{\partial \psi_{QC}}{\partial t} = \kappa(t) \left(\frac{\hbar}{i} \frac{\partial}{\partial x}\right) \hat{M} \psi_{QC}, \tag{9.54}$$

analogous to the standard von Neumann measurement model for a quantum pointer. For an initially independent ensemble at time $t = 0$ as per Eq. (9.32); i.e., $\psi_{QC}(q, x, 0) = \psi_Q(q)\psi_C(x)$, this equation may be trivially integrated to give

$$\psi_{QC}(q, x, T) = \sum_n \langle q|\hat{\Pi}_n|\psi_Q\rangle \, \psi_C(x - K\lambda_n) \qquad (9.55)$$

at the end of the measurement interaction, where $K = \int_0^T dt\, \kappa(t)$ and $\hat{M}$ has the spectral decomposition

$$\hat{M} = \sum_n \lambda_n \, \hat{\Pi}_n. \qquad (9.56)$$

Thus, λ_n denotes an eigenvalue of $\hat{M}$, and Π_n denotes the projection onto the corresponding eigenspace. The pointer probability distribution after measurement follows as

$$
\begin{aligned}
P(x, T) &= \int dq\, |\psi_{QC}(q, x, T)|^2 \\
&= \sum_{m,n} \langle \psi_Q|\hat{\Pi}_m \left(\int dq\, |q\rangle\langle q| \right) \hat{\Pi}_n|\psi_Q\rangle \, \bar{\psi}_C(x - K\lambda_m)\, \psi_C(x - K\lambda_n) \\
&= \sum_n p(\lambda_n|\psi_Q)\, P_C(x - K\lambda_n), \qquad (9.57)
\end{aligned}
$$

using $\hat{\Pi}_m\hat{\Pi}_n = \delta_{mn}\hat{\Pi}_n$, where $p(\lambda_n|\psi_Q) := \langle \psi_Q|\hat{\Pi}_n|\psi_Q\rangle$ is the usual quantum probability associated with eigenvalue λ_n for the initial quantum state $|\psi_Q\rangle$.

Hence, the measurement displaces the initial pointer probability density $P_C(x)$ by an amount $K\lambda_n$ with probability $p(\lambda_n|\psi_Q)$, thus correlating the position of the pointer with the eigenvalues of M. In particular, choosing a sufficiently narrow initial distribution $P_C(x)$ (e.g., a delta-function), the displaced probability densities will be nonoverlapping, corresponding to a 'good' measurement: each eigenvalue λ_n will be perfectly correlated with the measured pointer position. This generalises Eq. (8.55), for the special case of a quantum spin measurement:

> The measurement of any quantum observable may be modelled via interaction with a strictly classical measuring apparatus, followed by a direct measurement of the classical configuration.

The existence of such a model may be regarded as supporting the Copenhagen interpretation, in which measurement pointers and the like are regarded as truly 'classical' objects (see also Sect. 8.4).

9.5.2 Ineffective Decoherence: Improper Mixtures

In the above measurement model, consider again a 'good' measurement; i.e., with interaction strength sufficiently large to give nonoverlapping final pointer distributions

$$P_C(x - K\lambda_m) P_C(x - K\lambda_n) = 0 \quad \text{for} \quad m \neq n. \tag{9.58}$$

The reduced or 'improper' mixture of quantum ensembles corresponding to the hybrid wave function at time T then follows from Eqs. (8.27) and (9.57) as

$$\hat{\rho}_{Q|C} := \int dx \, \text{tr}_C[|\psi_{QC}\rangle\langle\psi_{QC}|] = \sum_n \hat{\Pi}_n |\psi_Q\rangle\langle\psi_Q|\hat{\Pi}_n = \sum_n w_n |\psi_n\rangle\langle\psi_n|, \tag{9.59}$$

with $w_n := p(\lambda_n|\psi_Q)$ and $|\psi_n\rangle := \hat{\Pi}_n|\psi_Q\rangle/\sqrt{w_n}$. Note this generalises Eqs. (8.53) and (8.56) for position and spin measurements.

As discussed in Sects. 4.2.4 and 8.3, reduced mixtures are not a complete representation of the hybrid ensemble. In other words, the hybrid ensemble is not equivalent to preparing one of the mutually orthogonal states $|\psi_n\rangle$ of the quantum ensemble with probability w_n, essentially because the hybrid ensemble is described by a 'pure' state $|\psi_{QC}\rangle$ which need not evolve equivalently to such a mixture. Indeed, the hybrid wave function could in principle evolve back into a 'pure' initial state.

Hence, while the conditional density operator is diagonal or decoherent with respect to $\hat{M}$, this is not sufficient to assume that the hybrid ensemble has decohered or 'collapsed', i.e., that the measurement has prepared a proper mixture of hybrid ensembles. The decoherence is ineffective without a further mechanism of some sort. This is an analogue of the well-known measurement problem in quantum mechanics, but with an important advantage: because the configuration ensemble approach *only* deals with ensembles, not with individual members, there is no requirement to 'explain' individual outcomes.

We will now show how the notion of 'classicality' discussed in Sect. 9.4.4 leads to *effective* decoherence. The discussion is analogous to that for the quantum measurement problem.

9.5.3 Macroscopic Measuring Devices Revisited

Measuring devices are, by construction, robust and deterministic (i.e., non-chaotic), otherwise they would not be able to provide reliable information about the quantity that is being measured. They have a limited, well defined range of operation, and are usually operated in the linear regime and follow well defined equations of motion. For example, think of an old-fashioned Geiger–Müller counter with a pointer as readout device. There is typically amplification associated with measuring devices (e.g., the cascade of charged particles in the Geiger–Müller tube, which creates an electrical current that can then be measured), and thus irreversibility is built into the device, but it turns out that this aspect is not crucial to our discussion.

For simplicity, we consider a measurement of the spin of a quantum particle as carried out by a classical measuring device, and assume that the device is macroscopic and that it has all the desirable properties that we listed in the previous paragraph. Thus the example is similar to the example of spin measurement that was considered previously in Sect. 8.4. We will consider the degree of freedom of the pointer and neglect the remaining "irrelevant" degrees of freedom of the measuring apparatus. However, we now formulate the problem in two dimensions instead of one dimension; that is, the pointer will have coordinates $r = (x, y)$. Furthermore, we will assume that the pointer is a macroscopic object, say a small sphere attached by a thin rod to the rest of the apparatus, and that it interacts with its environment. The environment will consist of photons that scatter from the pointer.

As a first step, we neglect the coupling of the pointer with the environment, which will be added later. More precisely, we model the case in which the z-component $\hat{\sigma}_z$ of a spin-1/2 particle is linearly coupled to the momentum in the y-direction of the classical particle. The ensemble Hamiltonian is thus given by

$$\mathcal{H}_{\text{spin}}[P, S] = \sum_{\alpha=\pm 1} \int dx\, dy\, P(\alpha, x, y) \frac{|\nabla S(\alpha, x, y)|^2}{2M}$$

$$+ \kappa(t) \sum_{\alpha=\pm 1} \int dx\, dy\, \alpha P(\alpha, x, y) \frac{\partial S(\alpha, x, y)}{\partial y}. \qquad (9.60)$$

We assume as we did before that the measurement interaction occurs over a sufficiently short time period $[0, T]$ that the first term of $\mathcal{H}_{\text{spin}}$ can be ignored. The equations of motion during the interaction are then

$$\frac{\partial P(\alpha, x, y, t)}{\partial t} = -\alpha\, \kappa(t) \frac{\partial P(\alpha, x, y)}{\partial y},$$

$$\frac{\partial S(\alpha, x, y, t)}{\partial t} = -\alpha\, \kappa(t) \frac{\partial S(\alpha, x, y)}{\partial y}, \qquad (9.61)$$

which can be integrated to give

$$P(\alpha, x, y, T) = P(\alpha, x, y - \alpha K, 0), \quad S(\alpha, x, y, T) = S(\alpha, x, y - \alpha K, 0), \quad (9.62)$$

where $K := \int_0^T dt\, \kappa(t)$. Thus, as expected, the interaction directly correlates the pointer position y with the spin α in the z-direction.

We consider a scenario in which the apparatus is being used to *search* for the particle with spin, such that the measuring device moves slowly in the x-direction with constant velocity v_x (this assumption is not an essential one, but it helps when it comes to visualizing the experiment). For simplicity, we define coordinates in which the particle with spin is at the origin and we assume that the interaction takes place at time $t = 0$.

We assume furthermore that the classical and quantum ensembles are initially *independent*. Before the interaction takes place (i.e., for $t < 0$), the pointer is described by P and S given by

$$P_C(x, y, t) = \frac{1}{2\pi\sigma^2} \exp\left\{-\frac{1}{2}\frac{(x - v_x t)^2 + y^2}{\sigma^2}\right\}, \qquad S_C(x, y, t) = Mxv_x. \quad (9.63)$$

It is straightforward to check that S and P solve both the Hamilton–Jacobi and the continuity equations. The initial probability of spin up/down in the z-direction before the interaction takes place is set equal to w_+ with $w_\pm \geq 0$ and $w_+ + w_- = 1$.

After the interaction takes place, the marginal probability for the classical particle will be given by the mixture

$$P(x, y, T) = w_+ P_C(x - v_x t, y - K, 0) + w_- P_C(x - v_x t, y + K, 0), \quad (9.64)$$

where we have taken the limit where $T \rightarrow 0$ (while keeping K constant). Hence, the initial probability density $P_C(x, y, t)$ of the pointer is displaced in the y-direction by K with probability w_+, and by $-K$ with probability w_-. Since S is independent of y, it will not change: $S_C(x, y, t) = Mxv_x$ before *and* after the interaction.

We assume that the apparatus is designed so that the measurement gives a well defined result, which translates into the requirement that the initial pointer probability density $P_C(x)$ has a spread which is small with respect to K, so that the probabilities associated with each of the elements of the mixture do not overlap.

The state of the classical pointer immediately before and immediately after the interaction is described by the probability densities shown schematically in Fig. 9.1. Notice that the position of the classical pointer, which was previously known to within a region of radius $\sim\sigma$ before the interaction, is described after the interaction by a bimodal distribution in which the two peaks are displaced a distance $2K$ from each other and have relative magnitudes of $w_\pm$. The measuring device will continue moving slowly in the x-direction with constant velocity v_x.

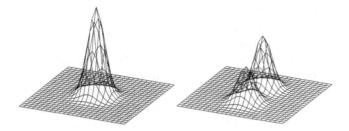

Fig. 9.1 Marginal probability density for the classical ensemble. *Left* Before the interaction, the classical state is localized. *Right* After the interaction, the classical ensemble is represented by a *mixture*. The initial probability density of the pointer is displaced in the positive y-direction with probability w_+, and in the negative y-direction with probability w_- (here we plot the mixture for the case $w_+ = 0.6$, $w_- = 0.4$)

9.5.4 Effective Decoherence and Regaining of Strong Separability

The next step is to take into consideration the effect of the environment. We assume that the macroscopic pointer moves through an environment that is filled with photons and that these particles are scattered by the pointer. In reality, there will be other quantum particles which contribute to the environment, but for the purpose of this section it will be sufficient to consider only photons. What may be the source of these photons? To read the position of a pointer, you need to shine light on it, and this is certainly one source. As the experiment is being carried out, the experimenter will be looking at the pointer to see if it moves, or perhaps a video camera will be directed to it to keep a record. But there will also be photons emitted by macroscopic objects that are close to the pointer, or photons emitted by other parts of the measuring device itself. These will be always present regardless of whether an experimentalist is there or not to look at the pointer.

A lower bound for the photons emitted by a nearby macroscopic object can be estimated by calculating the emission of photons due to blackbody radiation. The photon radiant emittance of a blackbody at a temperature of $300\,\mathrm{K}$ can be estimated from the Stefan Boltzamnn law and it is enormous, of the order of $10^{22}\ \mathrm{s}^{-1}\ \mathrm{m}^{-2}$. Thus one expects the pointer to be constantly bombarded by photons. To get a crude estimate, assume that this is roughly the flux of photons at the position of the pointer, and assume that the cross sectional area of the pointer is about $1\ \mathrm{mm}^2$. Then, photons will scatter from the pointer at a rate of $\sim 10^{13}\ \mathrm{s}^{-1}$. This should be compared to the time scale at which one can extract information from a measuring device: the fastest electronics that are available today have a time resolution of about ten picoseconds, that is, 10^{-11} s, but the limiting factor is typically the resolving time of the measuring device, which can be substantial (for a Geiger–Müller counter, for example, it is of the order of 10^{-4} s). This means that even under the best of circumstances a large number of photons will scatter from the pointer while a measurement takes place.

At the same time, it is important to keep in mind that the equations that describe the motion of the pointer are *not* affected by the photon environment: the average transfer of momentum will be null because the "cloud" of photons is isotropic, and the difference in momenta between the macroscopic pointer and a single photon is so large that one can neglect the recoil of the pointer as it scatters the photons.

We can study the effect of the coupling between the pointer and the environment using an approach that is similar to the analysis that explains the appearance of straight tracks in a Wilson cloud chamber when an atom undergoes radioactive decay with the emission of an α-particle. In the late 1920s, Gamow [35] and, independently, Gurney and Condon [36, 37] showed that one could explain alpha radioactivity using wave mechanics. However, in this theory the alpha particle is described as a spherical matter wave that emerges from the atomic nucleus, and this seemed at odds with the alpha particle tracks that are observed in a Wilson cloud chamber, which are always straight lines. The resolution was provided by Mott [38, 39], who calculated the scattering of alpha particles by the atoms in the Wilson cloud chamber quantum mechanically using Born's collision theory. He used first order perturbation theory to show that the probability of finding the alpha particle is concentrated on a sharp cone behind the first ionized atom, in the direction of the incoming matter wave. He then applied second order perturbation theory and considered the case in which a second atom becomes ionized after interacting with the alpha particle. Mott showed that both atoms had to lie inside the cone. The same will hold for all the other atoms that become ionized. The end result is that the probability of finding the alpha particle is concentrated along a straight line. In this way, an initial spherically symmetric probability distribution (the one corresponding to the spherical matter wave that emerges from the atomic nucleus) becomes, after interaction with the environment of the Wilson cloud chamber, a probability distribution that is concentrated on a straight line.

The technical details of our analysis are of course very different from the ones in Mott's argument, because we are considering a different, very idealized physical situation. In fact, our argument is rather simple. We suppose that a photon scatters from the macroscopic pointer at some time $t > 0$; i.e., after the interaction with the particle with spin. Since the probability of finding the pointer is given by Eq. (9.64), this can only happen *either* in the vicinity of the coordinates $(x = v_x t, y = +K)$ *or* in the vicinity of the coordinates $(x = v_x t, y = -K)$. After the photon is scattered, one needs to update the probability of finding the pointer via Bayes theorem. The net effect will be to select only *one* of the elements of the mixture. Let us assume for simplicity that the scattered photon provides a measurement that is sufficient to distinguish between the two elements of the mixture, but that the resolution is so poor that it can not do better than that. Then, in the first case, the probability of finding the pointer will have to be updated according to

$$P(x, y, t) \rightarrow P_C(x - v_x t, y - K, 0), \tag{9.65}$$

while in the second case, it will have to be updated according to

$$P(x, y, t) \rightarrow P_C(x - v_x t, y + K, 0). \tag{9.66}$$

Notice that it is not necessary to update $S_C(x, y, t)$ because we have assumed that we can neglect the recoil of the pointer as it scatters photons. Thus the measuring device will continue moving in the x-direction with velocity v_x.

The classical particle is now more localized than it was before: there is a "collapse" of the probability that results from Bayesian updating and only *one* of the elements of the mixture survives. This means that all further scattering events in which photons scatter from the pointer can only take place in the vicinity of the line $y = +K$ (in the first case, when Eq. (9.65) holds) or in the vicinity of the line $y = -K$ (in the second case, when Eq. (9.66) holds). Thus *the update via Bayes theorem leads to the prediction that the pointer will be seen either at $y = +K$ or at $y = +K$ even though it originally was in a mixed state that included both cases.* At the same time, only one of the two possibilities in the quantum sector survives and there is a corresponding "collapse" given by $(w_+, w_-) \to (1, 0)$ (in the first case when Eq. (9.65) holds) or $(w_+, w_-) \to (0, 1)$ (in the second case when Eq. (9.66) holds).

We see then that the improper mixture effectively collapses to a *proper* mixture via interactions with the environment. Therefore the classical and quantum components become *independent*. Thus strong separability is satisfied after the pointer interacts with the environment, at which point effects such as nonlocal energy flow and signaling via the quantum potential becomes *impossible* (see Sects. 9.4.2–9.4.4). This happens in an extremely short time scale because, as we mentioned above, the pointer will scatter photons at an enormously high rate. If one were to consider more realistic types of measurements and environments, the technical difficulties would become severe. But it seems clear that one would reach essentially the same type of conclusion. Thus, in the measurement context, the results in Sect. 9.4.4 are further strengthened:

> Under the 'classicality' assumption that the measurement apparatus is macroscopic, the hybrid ensemble effectively decoheres into a proper mixture of noninteracting independent ensembles. Strong separability is therefore satisfied for each element of the mixture, ruling out any possibility of nonlocal energy flow and nonlocal signaling after the measurement.

9.6 Comparisons with Mean-Field Approach

We conclude this chapter by very briefly comparing the configuration ensemble approach to quantum-classical interactions with the mean-field approach mentioned in Sect. 9.1. The idea of the latter approach is very simple [1, 2], and several equivalent reformulations and extensions of the approach have recently been given [3, 40, 41].

In particular, in the mean-field approach the classical system is described at any time by a single point (x, p) in a $2n$-dimensional classical phase space, and the quantum system by a wave function $\psi_Q(q) \equiv \langle q | \psi_Q \rangle$ on a quantum Hilbert space. Evolution of the hybrid system is described via a Hamiltonian operator $\hat{H}(x, p)$, parameterised by the classical coordinates, with

$$\dot{x} = \nabla_p \langle \psi_Q | \hat{H}(x, p) | \psi_Q \rangle, \quad \dot{p} = -\nabla_x \langle \psi_Q | \hat{H}(x, p) | \psi_Q \rangle, \quad i\hbar \frac{\partial}{\partial t} | \psi_Q \rangle = \hat{H}(x, p) | \psi_Q \rangle. \tag{9.67}$$

Thus, the classical particle sees a classical Hamiltonian corresponding to the averaged quantum Hamiltonian. Significantly, the classical particle follows a deterministic trajectory in phase space, rather than being described by an ensemble on configuration space.

Observables To compare the two approaches, it is necessary to define observables for the mean-field approach [3, 41]. We will do this in a way that facilitates the comparison. In particular, define the quantum configuration ensemble (P_Q, S_Q) via the polar decomposition $\psi_Q = P_Q^{1/2} e^{iS_Q/\hbar}$ of the wave function. Observables are then functionals $A[P_Q, S_Q, x, p]$ with a corresponding Poisson bracket

$$\{A, B\}_{MF} := \int dq \left[\frac{\delta A}{\delta P_Q} \frac{\delta B}{\delta S_Q} - \frac{\delta B}{\delta P_Q} \frac{\delta A}{\delta S_Q} \right] + \nabla_x A \cdot \nabla_p B - \nabla_p A \cdot \nabla_x B. \quad (9.68)$$

The above evolution equations are then equivalent to

$$\dot{x} = \{x, \mathscr{H}\}_{MF}, \quad \dot{p} = \{p, \mathscr{H}\}_{MF}, \quad \frac{\partial P_Q}{\partial t} = \{P_Q, \mathscr{H}\}_{MF}, \quad \frac{\partial S_Q}{\partial t} = \{S_Q, \mathscr{H}\}_{MF}, \quad (9.69)$$

with Hamiltonian

$$\mathscr{H}[P_Q, S_Q, x, p] := \langle \psi_Q | \hat{H}(x, p) | \psi_Q \rangle. \quad (9.70)$$

It is natural to define the *classical* and *quantum* observables by

$$\tilde{C}_f^{MF}[P_Q, S_Q, x, p] := \int dq\, P_Q\, f(x, p), \qquad \tilde{Q}_{\hat{M}}^{MF}[P_Q, S_Q, x, p] := \langle \psi_Q | \hat{M} | \psi_Q \rangle, \quad (9.71)$$

for arbitrary phase space functions $f(x, p)$ and Hermitian operators $\hat{M}$. It is then easy to show that [3]

$$\{\tilde{C}_f^{MF}, \tilde{C}_g^{MF}\} = \tilde{C}_{\{f,g\}}^{MF}, \qquad \{\tilde{Q}_{\hat{M}}^{MF}, \tilde{Q}_{\hat{N}}^{MF}\} = \tilde{Q}_{[\hat{M},\hat{N}]/i\hbar}^{MF}, \quad (9.72)$$

analogously to Eq. (9.8) for classical and quantum observables in the configuration ensemble formalism.

Dynamical Bracket Versus Homogeneity and Weak Values If the set of mean-field observables is closed under the Poisson bracket, then from Eq. (9.72) the two minimal conditions imposed on the dynamical bracket in Sect. 9.2.1 will be satisfied. However, the mean field approach must also include observables of the form of $\mathscr{H}$ above, so as to allow interactions between the classical system and the quantum ensemble; i.e., it must also include observables of the form [3]

$$A[P_Q, S_Q, x, p] = \left\{ \langle \psi_Q | \hat{M}(x, p) | \psi_Q \rangle, \langle \psi_Q | \hat{N}(x, p) | \psi_Q \rangle \right\}_{MF}$$
$$= (i\hbar)^{-1} \langle \psi_Q | [\hat{M}(x, p), \hat{N}(x, p)] | \psi_Q \rangle$$
$$+ (\nabla_x \langle \psi_Q | \hat{M}(x, p) | \psi_Q \rangle) \cdot (\nabla_p \langle \psi_Q | \hat{N}(x, p) | \psi_Q \rangle)$$
$$- (\nabla_p \langle \psi_Q | \hat{M}(x, p) | \psi_Q \rangle) \cdot (\nabla_x \langle \psi_Q | \hat{N}(x, p) | \psi_Q \rangle). \quad (9.73)$$

The last two terms imply that

$$A[\lambda P_Q, S_Q, x, p] \neq \lambda A[P_Q, S_Q, x, p] \tag{9.74}$$

in general, in contrast to the configuration ensemble scaling property in Eq. (9.4). Violation of the homogeneity property raises significant issues: it implies that observables do not have a general interpretation as expectation values (see Sect. 1.4.3); that they scale nonlinearly under a 'collapse' of the wave function from ψ_Q to $\lambda\psi_Q$; and also that weak values cannot be defined in general (see Sect. 2.4.2).

Ehrenfest relations It is straightforward to check that the analogue of the hybrid Ehrenfest relations in Eq. (9.13) also hold for the mean-field theory [3, 41]. Thus both approaches imply a well defined classical limit, and that the oscillator benchmark for hybrid theories is satisfied (see Sect. 9.3).

Strong separability It follows directly from Eq. (9.73) that [3]

$$\{\tilde{C}_f^{MF}, \tilde{Q}_{\hat{M}}^{MF}\}_{MF} = (i\hbar)^{-1}\langle[\psi_Q|f(x, p)\hat{1}, \hat{M}]|\psi_Q\rangle = 0. \tag{9.75}$$

Hence strong separability is guaranteed in the mean-field approach, in contrast to the configuration-ensemble approach, thus automatically evading issues of nonlocal energy flows and nonlocal signaling (see Sect. 9.4).

Measurement As remarked in the introduction to this chapter (see also Sect. 8.5), the mean-field approach is unable to correlate a classical pointer with the eigenvalues of a quantum observable, as the equations of motion imply that the pointer observable will always evolve deterministically to the same final value, for a given initial quantum wave function ψ_Q. Thus, unlike the configuration ensemble approach, the mean-field approach cannot couple quantum fluctuations into correlated classical observables, limiting its applicability as a fundamental theory.

References

1. Boucher, W., Traschen, J.: Semiclassical physics and quantum fluctuations. Phys. Rev. D **37**, 3522–3532 (1988)
2. Makri, N.: Time-dependent quantum methods for large systems. Ann. Rev. Phys. Chem. **50**, 167–191 (1999)
3. Elze, H.-T.: Linear dynamics of quantum-classical hybrids Phys. Rev. A **85**, 052109 (2012)
4. Koopman, B.O.: Hamiltonian systems and transformations in Hilbert space. Proc. Natl. Acad. Sci. U.S.A. **17**, 315–318 (1931)
5. Sudarshan, E.C.G.: Interaction between classical and quantum systems and the measurement of quantum observables. Pramana **6**, 117–126 (1976)
6. Sherry, T.N., Sudarshan, E.C.G.: Interaction between classical and quantum systems: a new approach to quantum measurement. I. Phys. Rev. D **18**, 4580–4589 (1978)
7. Sherry, T.N., Sudarshan, E.C.G.: Interaction between classical and quantum systems: a new approach to quantum measurement. II. Theoretical considerations. Phys. Rev. D **20**, 857–868 (1979)

8. Gautam, S.R., Sherry, T.N., Sudarshan, E.C.G.: Interaction between classical and quantum systems: a new approach to quantum measurement. III. Illustration. Phys. Rev. D **20**, 3081–3094 (1979)

9. Peres, A., Terno, D.R.: Hybrid classical-quantum dynamics. Phys. Rev. A **63**, 022101 (2001)

10. Terno, D.R.: Inconsistency of quantum-classical dynamics, and what it implies. Found. Phys. **36**, 102–111 (2006)

11. Diosi, L., Gisin, N., Strunz, W.T.: Quantum approach to coupling classical and quantum dynamics. Phys. Rev. A **61**, 022108 (2000)

12. Bohm, D.: A suggested interpretation of the quantum theory in terms of "Hidden" variables. II. Phys. Rev. **85**, 180–193 (1952)

13. Bell, J.S.: Speakable and Unspeakable in Quantum Mechanics. Cambridge University Press, Cambridge (1987)

14. Holland, P.R.: The Quantum Theory of Motion. Cambridge University Press, Cambridge (1993)

15. Gindensperger, E., Meier, C., Beswick, J.A.: Mixing quantum and classical dynamics using Bohmian trajectories. J. Chem. Phys. **113**, 9369–9372 (2000)

16. Prezhdo, O.V., Brooksby, C.: Quantum backreaction through the Bohmian particle. Phys. Rev. Lett. **86**, 3215–3219 (2001)

17. Burghardt, I., Parlant, G.: On the dynamics of coupled Bohmian and phase-space variables: a new hybrid quantum-classical approach. J Chem Phys. **120**, 3055–3058 (2004)

18. Salcedo, L.L.: Comment on "Quantum Backreaction through the Bohmian Particle". Phys. Rev. Lett. **90**, 118901 (2003)

19. Prezhdo, O., Brooksby, C.: Comment on "Quantum Backreaction through the Bohmian Particle." Prezhdo and Brooksby Reply. Phys. Rev. Lett. **90**, 118902 (2003)

20. Salcedo, L.L.: Absence of classical and quantum mixing. Phys. Rev. A **54**, 3657–3660 (1996)

21. Caro, J., Salcedo, L.L.: Impediments to mixing classical and quantum dynamics. Phys. Rev. A **60**, 842–852 (1999)

22. Sahoo, D.: Mixing quantum and classical mechanics and uniqueness of Planck's constant. J. Phys. A **37**, 997–1010 (2004)

23. Hall, M.J.W.: Consistent classical and quantum mixed dynamics. Phys. Rev. A **78**, 042104 (2008)

24. Salcedo, L.L.: Comment on "A quantum-classical bracket that satisfies the Jacobi identity" [J. Chem. Phys. 124, 201104 (2006)]. J. Chem. Phys. **126**, 057101 (2007)

25. Agostini, F., Caprara, S., Ciccotti, G.: Do we have a consistent non-adiabatic quantum-classical mechanics? Europhys. Lett. **78**, 30001 (2007)

26. Aleksandrov, I.V.: The statistical dynamics of a system consisting of a classical and a quantum subsystem. Z. Naturforsch. A **36**, 902–908 (1981)

27. Prezhdo, O.V.: A quantum-classical bracket that satisfies the Jacobi identity. J. Chem. Phys. **124**, 201104 (2006)

28. Goldstein, H.: Classical Mechanics. Addison-Wesley, New York (1950)

29. Merzbacher, E.: Quantum Mechanics, 3rd edn. Wiley, New York (1998)

30. Salcedo, L.L.: Statistical consistency of quantum-classical hybrids. Phys. Rev. A **85**, 022127 (2012)

31. Hall, M.J.W., Reginatto, M.: Interacting classical and quantum ensembles. Phys. Rev. A **72**, 062109 (2005)

32. Hall, M.J.W., Reginatto, M., Savage, C.M.: Nonlocal signaling in the configuration space model of quantum-classical interactions. Phys. Rev. A **86**, 054101 (2012)

33. Joos, E., Zeh, H.D., Kiefer, C., Giulini, D.J.W., Kupsch, J., Stamatescu, I.-O.: Decoherence and the Appearance of a Classical World in Quantum Theory, 2nd edn. Springer, New York (2003)

34. Peres, A.: Can we undo quantum measurements? Phys. Rev. D **22**, 879–883 (1980)

35. Gamow, G.: Zur Quantentheorie des Atomkernes. Z. Physik **51**, 204–212 (1928)

36. Gurney, R.W., Condon, E.U.: Wave mechanics and radioactive disintegration. Nature **122**, 439 (1928)

37. Gurney, R.W., Condon, E.U.: Quantum mechanics and radioactive disintegration. Phys. Rev **33**, 127–140 (1929)
38. Mott, N.F.: The wave mechanics of α-ray tracks. Proc. Roy. Soc. A **126**, 79–84 (1929)
39. Heisenberg, W.: The Physical Principle of the Quantum Theory. Dover, New York (1949)
40. Zhang, Q., Wu, B.: General approach to quantum-classical hybrid systems and geometric forces. Phys. Rev. Lett. **97**, 190401 (2006)
41. Alonso, J.L., Castro, A., Clemente-Gallardo, J., Cuchí, J.C., Echenique, P., Falceto, F.: Statistics and Nosé formalism for Ehrenfest dynamics. J. Phys. A **44**, 395004 (2011)

Part IV
Classical Gravitational Fields and Their Interaction with Quantum Fields

Chapter 10
Ensembles of Classical Gravitational Fields

Abstract We define ensembles on configuration space for classical gravitational fields that obey the Einstein equations. Our starting point is the Hamilton–Jacobi formulation of general relativity. After a brief review of the Einstein–Hamilton–Jacobi equation in the metric representation, we introduce the additional mathematical structure that is needed to formulate the theory of configuration space ensembles; i.e., a measure over the space of metrics and a probability functional. Then we define an appropriate ensemble Hamiltonian for the gravitational field, show that it leads to the correct equations, and recover the Einstein equations in the usual formulation. In addition, we show that the formalism of ensembles on configuration space provides a novel approach to solving the reconstruction problem; i.e., the derivation of the full set of Einstein equations from a Hamilton–Jacobi formulation of gravity. Having derived the equations for the general case, we move on to the simpler case of spherical symmetric spacetimes and derive the corresponding equations for midisuperspace models of spherically symmetric gravity. We consider the example of classical ensembles of black holes in this midisuperspace approximation.

10.1 Introduction

This chapter and the one that follows are devoted to gravity. We already introduced ensembles on configuration space for general relativistic gravitational fields in Chap. 5. There we considered the problem of quantization via an exact uncertainty principle and showed that this quantization procedure leads in the case of gravity to the Wheeler–DeWitt equation with a specific operator ordering. Now we look at other applications of the formalism. In this chapter we develop the theory of classical ensembles of gravitational fields and in Chap. 11 we consider hybrid systems where quantum matter fields and classical gravitational fields interact.

We will see that classical ensembles of gravitational fields are interesting in their own right. However, there is an additional reason for developing the theory of classical ensembles: namely, as preparation for the physical systems that we discuss in the next chapter, in which quantum matter fields couple to classical gravitational fields. The main motivation for studying such hybrid systems is the lack of a full theory of

© Springer International Publishing Switzerland 2016

M.J.W. Hall and M. Reginatto, *Ensembles on Configuration Space*,
Fundamental Theories of Physics 184, DOI 10.1007/978-3-319-34166-8_10

quantum gravity. As we discuss in more detail in the introduction to the next chapter, one would like to see to what extent a hybrid system may provide a consistent, satisfactory description of matter and gravitation. There is also the possibility that the gravitational field may not be quantum in nature, in which case hybrid models become unavoidable.

The most direct way of introducing ensembles of classical gravitational fields is via the Hamilton–Jacobi formulation of general relativity. This is the route that we follow in this chapter. We first write down the Einstein–Hamilton–Jacobi equation in the metric representation and afterwards introduce some additional mathematical structure, a measure over the space of metrics and a probability functional. Then we define an appropriate ensemble Hamiltonian for the gravitational field, show that it leads to the correct equations, and recover the Einstein equations in the usual formulation. Having derived the equations for the general case, we move on to the simpler case of spherical symmetric spacetimes and derive the corresponding equations for midisuperspace models of spherically symmetric gravity. Finally, in the last section of this chapter, we consider the example of classical ensembles of black holes.

10.2 Einstein–Hamilton–Jacobi Equation and Ensembles for Classical Gravitational Fields

General relativity provides a geometrical formulation of gravity. In the standard approach, the geometry of four-dimensional space-time is determined by the Einstein equations. An alternative, perhaps less familiar way of deriving this geometry is via the Einstein–Hamilton–Jacobi equation [1, 2], first proposed by Peres [3]. As the Einstein–Hamilton–Jacobi equation provides the most direct way of introducing ensembles of classical gravitational fields on configuration space (see Sect. 5.4.4), we first review this formulation.

10.2.1 Einstein–Hamilton–Jacobi Equation

The Hamilton–Jacobi formulation of general relativity is based on a 3+1 decomposition of space-time and results in equations for a functional $S[h_{ij}]$, where h_{ij} is the metric of the space-like hypersurface. To define this hypersurface, consider a four-dimensional space-time with metric $g_{\mu\nu}$,

$$ds^2 = g_{\mu\nu}dx^\mu dx^\nu \qquad \mu, \nu = 0, 1, 2, 3 \qquad x^0 \equiv t. \qquad (10.1)$$

The space-time metric $g_{\mu\nu}$ can be written in the form

$$g_{\mu\nu} = \begin{bmatrix} g_{00} & g_{0j} \\ g_{i0} & g_{ij} \end{bmatrix} = \begin{bmatrix} -N^2 + N_j N^j & N_j \\ N_i & h_{ij} \end{bmatrix} \qquad (10.2)$$

where N and N_j are known as the shift function and lapse vector. The line element on any hypersurfaces defined by a constant x^0 is given by $ds^2|_{dx^0=0} = h_{ij}dx^i dx^j$, thus the metric induced on the space-like hypersurface is given by h_{ij}, as required.

A Hamilton–Jacobi formulation for the gravitational field can be defined in terms of the functional equations [1, 2]

$$H = \kappa G_{ijkl}\frac{\delta S}{\delta h_{ij}}\frac{\delta S}{\delta h_{kl}} - \frac{1}{\kappa}\sqrt{h}\,(R - 2\lambda) = 0, \tag{10.3}$$

$$H_i = D_j\left(\frac{\delta S}{\delta h_{kj}}\right) = 0, \tag{10.4}$$

where $\kappa = 16\pi$ (in units where both the speed of light c and the gravitational constant G are equal to one, $c = G = 1$), $G_{ijkl} = (2h)^{-1/2}\left(h_{ik}h_{jl} + h_{il}h_{jk} - h_{ij}h_{kl}\right)$ is the DeWitt supermetric [1], h is the determinant of h_{ij}, R is the curvature scalar, λ is the cosmological constant, and D_j is the spatial covariant derivative.

The momentum constraints, Eq. (10.4), are equivalent to requiring invariance of the Hamilton–Jacobi functional S under spatial coordinate transformations [4]. To show this, consider an infinitesimal change of coordinates $x^k \to x^k + \varepsilon^k(x)$ and the corresponding transformation of the metric, $h_{kl} \to h_{kl} - (D_k\varepsilon_l + D_l\varepsilon_k)$. The variation of S can be expressed as

$$\delta_\varepsilon S = \int d^3x\,\frac{\delta S}{\delta h_{kl}}\delta h_{kl} = \int d^3x\left[D_k\left(\frac{\delta S}{\delta h_{kl}}\right)\varepsilon_l + D_l\left(\frac{\delta S}{\delta h_{kl}}\right)\varepsilon_k\right], \tag{10.5}$$

where in the second equality we used $\delta h_{kl} = -(D_k\varepsilon_l + D_l\varepsilon_k)$ and carried out an integration by parts. This shows that $\delta_\varepsilon S = 0$ requires

$$D_k\left(\frac{\delta S}{\delta h_{kl}}\right) = 0, \tag{10.6}$$

which is Eq. (10.4). Therefore, instead of basing the Hamilton–Jacobi theory of the gravitational field on Eqs. (10.3) and (10.4), we can give an equivalent formulation in which we keep Eq. (10.3), ignore Eq. (10.4), and require that S be invariant under the gauge group of spatial coordinate transformations. We will follow this approach here. One can show that S is also required to satisfy the condition [5]

$$\frac{\partial S}{\partial t} = 0 \tag{10.7}$$

and that there are no further constraints.

The Einstein–Hamilton–Jacobi equation, Eq. (10.3), is actually an infinite number of equations, one per three-dimensional point. As pointed out by Giulini [6], it is possible to introduce an alternative viewpoint in which Eq. (10.3) is regarded as an equation to be integrated with respect to a "test function" in which case we are dealing with *one* equation for *each* choice of lapse function N,

$$\int d^3x \, NH = 0;$$
(10.8)

i.e., for each choice of foliation [6, 7]. Such an alternative viewpoint is extremely useful when searching for solutions: although it may be impossible to find the general solution (which requires solving the Einstein–Hamilton–Jacobi equation for all choices of lapse functions), it may be possible to find particular solutions for specific choices; for example, the choice $S \sim \int d^3x \, \sqrt{h}$ is a solution that describes de Sitter spacetime in a flat foliation [7].

We have seen that a solution of the Einstein–Hamilton–Jacobi equation $H = 0$ is a functional $S[h_{kl}]$ of the metric h_{kl} on a three dimensional space-like hypersurface. How do you get to a four-dimensional spacetime and show invariance under spacetime coordinate transformations? This problem, which is known as the *reconstruction problem*, is discussed in Appendix 1 of this chapter. One can show that h_{kl} satisfies the rate equation

$$\frac{\partial h_{ij}}{\partial t} = NG_{ijkl}\frac{\delta S}{\delta h_{kl}} + D_iN_j + D_jN_i,$$
(10.9)

where the lapse function N and the shift function N_j are arbitrary functions of the coordinates. Different choices of N and N_j correspond to different choices of gauge (i.e., different choices of coordinates for the metric of the four-dimensional space-time). In Appendix 1 of this chapter, we derive this equation in the standard way by embedding the hypersurface in a four-dimensional spacetime [8]. Gerlach has shown that it can also be derived from the Einstein–Hamilton–Jacobi formalism [4]. Below, we will show that Eq. (10.9) follows from the formalism of ensembles on configuration space in a natural way.

10.2.2 Measure and Probability

To define classical ensembles for gravitational fields, it is necessary to introduce some additional mathematical structure: a measure Dh over the space of metrics h_{kl} and a probability functional $P[h_{kl}]$. We follow the exposition of Appendix B of Ref. [9].

A standard way of defining the measure [10, 11] is to introduce an invariant norm for metric fluctuations that depends on a parameter ω,

$$\|\delta h\|^2 = \int d^nx \, [h(x)]^{\omega/2} \, G^{ijkl} \, [h(x); \omega] \, \delta h_{ij}\delta h_{kl}$$
(10.10)

where n is the number of dimensions (in our case, $n = 3$) and

$$G^{ijkl} = \frac{1}{2} [h(x)]^{(1-\omega)/2} \left[h^{ik} h^{jl} + h^{il} h^{jk} + \lambda h^{ij} h^{kl} \right] \tag{10.11}$$

is a generalization of the inverse of the DeWitt supermetric (in [10] the particular case $\omega = 0$ is considered). This norm induces a local measure for the functional integration given by

$$\int d\mu \, [h] = \int \prod_x [\det G(h(x))]^{1/2} \prod_{i \geq j} dh_{ij}(x). \tag{10.12}$$

For the expression given in Eq. (10.11),

$$\det G(h(x)) \propto \left(1 + \frac{1}{2} \lambda n \right) [h(x)]^\sigma , \tag{10.13}$$

where $\sigma = (n+1)[(1-\omega)n - 4]/4$ (one needs to impose the condition $\lambda \neq -2/n$, otherwise the measure vanishes). Therefore, up to an irrelevant multiplicative constant, the measure takes the form

$$\int d\mu \, [h] = \int \prod_x \left[\sqrt{h(x)} \right]^\sigma \prod_{i \geq j} dh_{ij}(x). \tag{10.14}$$

Without loss of generality, one may set Dh equal to $d\mu \, [h]$ (i.e., one may set $\omega = 0$, which in turn implies $\sigma = 0$), since a term of the form $\left[\sqrt{h(x)} \right]^\sigma$ may be absorbed into the definition of $P[h_{kl}]$. This choice leads to the DeWitt measure for pure gravity in a four-dimensional space-time.

The probability functional $P[h_{kl}]$ has to satisfy certain conditions. It is natural to require that $\int Dh\, P$ be invariant under the gauge group of spatial coordinate transformations. Since the family of measures defined by Eq. (10.14) is invariant under spatial coordinate transformations [11, 12], the invariance of $\int Dh\, P$ leads to a condition on P that is similar to the one required of S. To show this, consider again an infinitesimal change of coordinates $x'^k = x^k + \varepsilon^k(x)$ and the corresponding transformation of the metric, $h_{kl} \to h_{kl} - (D_k \varepsilon_l + D_l \varepsilon_k)$. The variation of $\int Dh\, P$ can be expressed as

$$\delta_\varepsilon \int Dh\, P = \int Dh \int d^3x \left[D_k \left(\frac{\delta P}{\delta h_{kl}} \right) \varepsilon_l + D_l \left(\frac{\delta P}{\delta h_{kl}} \right) \varepsilon_k \right]. \tag{10.15}$$

Therefore, $\delta_\varepsilon \int Dh P = 0$ requires

$$D_k \left(\frac{\delta P}{\delta h_{kl}} \right) = 0, \tag{10.16}$$

i.e., the gauge invariance of P. In addition to Eq. (10.35), it will be assumed that

$$\frac{\partial P}{\partial t} = 0 \tag{10.17}$$

also holds, analogous to the condition for S given by Eq. (10.7).

Finally, it should be pointed out that one must factor out the infinite diffeomorphism gauge group volume out of the measure, to calculate finite averages using the measure Dh and the probability functional P. This can be achieved by fixing a particular gauge when carrying out the calculation of averages. This issue will not be discussed further here because it does not affect the derivation of the equations of motion which we discuss in the next section.

10.2.3 Classical Ensemble Hamiltonian for the Gravitational Field

An appropriate ensemble Hamiltonian for the gravitational field is given by [13]

$$\mathscr{H} = \int d^3x \, N \int Dh \, PH, \tag{10.18}$$

where H is given by Eq. (10.3) and N is the lapse function that appears in Eq. (10.2), with corresponding equations of motion

$$\frac{\partial P}{\partial t} = \frac{\Delta \mathscr{H}}{\Delta S}, \quad \frac{\partial S}{\partial t} = -\frac{\Delta \mathscr{H}}{\Delta P}, \tag{10.19}$$

where $\Delta/\Delta F$ denotes the variational derivative with respect to the functional F (see Appendix A of this book). With $\frac{\partial S}{\partial t} = \frac{\partial P}{\partial t} = 0$, the equations of motion take the form

$$\int d^3x \, N \, [H] = 0, \tag{10.20}$$

and

$$\int d^3x \, N \left[\frac{\delta}{\delta h_{ij}} \left(NPG_{ijkl} \frac{\delta S}{\delta h_{kl}} \right) \right] = 0. \tag{10.21}$$

Equation (10.20) is the Einstein–Hamilton–Jacobi equation and Eq. (10.21) a continuity equation. If we assume that N is an arbitrary function of the coordinates, the terms in square brackets must vanish and Eq. (10.20) reduces to Eq. (10.3), the usual form of the Einstein–Hamilton–Jacobi equation.

10.2.4 Rate Equation for the Metric Field

We consider solutions of Eq. (10.21) that are valid when N is an arbitrary function of the coordinates. Keeping in mind that $\frac{\partial P}{\partial t} = 0$, the most general rate equation for the metric h_{ij} consistent with the interpretation of Eq. (10.21) as a continuity equation is of the form

$$\delta h_{ij} = \left(\alpha G_{ijkl} \frac{\delta S}{\delta h_{kl}} + \delta_\varepsilon h_{ij} \right) \delta t \tag{10.22}$$

for some arbitrary function α. We have include the term $\delta_\varepsilon h_{ij} = - \left(D_i \varepsilon_j + D_j \varepsilon_i \right)$ which allows for gauge transformations of h_{kl}, which is permitted because the gauge transformations are assumed to leave $\int DhP$ invariant, as discussed before. In other words, the most general infinitesimal change δh_{ij} of h_{ij} will be a combination of motion along the "velocity field" $G_{ijkl} \frac{\delta S}{\delta h_{kl}}$ and a gauge transformation. This leads to the rate equation for h_{kl} in its standard form,

$$\frac{\partial h_{ij}}{\partial t} = N G_{ijkl} \frac{\delta S}{\delta h_{kl}} + D_i N_j + D_j N_i, \tag{10.23}$$

where we have written N and N_j in place of α and $-\varepsilon_j$ to agree with the usual notation. Equation (10.23) is identical to the equation derived from the ADM canonical formalism provided N is identified with the lapse function and N_k with the shift vector [8].

This shows that there is a *natural, well defined intrinsic concept of time* in the formalism. It appears as a consequence of the continuity equation. We will see in the next chapter that this concept of time is still valid for hybrid systems that describe quantum fields interacting with a classical gravitational field.

It is remarkable that the rate equation for the metric, Eq. (10.23), can be shown to be a direct consequence of applying the theory of ensembles on configuration space to classical general relativity. The derivation presented here is related to the one carried out by Gerlach for pure gravity using a Hamilton–Jacobi formulation [4]. Gerlach's derivation, however, is much more involved than ours. In part, this is due to the fact that we have at our disposal mathematical structure which goes beyond the Einstein–Hamilton–Jacobi equation, namely the additional concepts, introduced in the previous sections, which are needed for the description of ensembles of gravitational fields.

The equations that determine an ensemble of gravitational fields are
- The Einstein–Hamilton–Jacobi equation.
- The continuity equation.
- The rate equation for the metric field.

There is a well defined, intrinsic concept of time which derives from the continuity equation.

10.3 Spherically Symmetric Gravity

Finding solutions of the Einstein–Hamilton–Jacobi equation can be very difficult. Therefore, when dealing with specific problems, it is convenient to consider whether there are any symmetries that can simplify the formulation of the problem. This leads in general to midisuperspace models or, in the case in which the symmetries are used to get rid of all field degrees of freedom, to minisuperspace models. Ensembles on configuration space can be defined for both midisuperspace and minisuperspace models. In this section, we consider the case of spherical symmetry and the corresponding equations for the midisuperspace model.

In the case of spherical symmetry, the line element may be written in the form

$$g_{\mu\nu}dx^{\mu}dx^{\nu} = -N^2dt^2 + \Lambda^2(dr + N_r dt)^2 + R^2 d\Omega^2. \tag{10.24}$$

The lapse function N and the shift function N_r are functions of the radial coordinate r and the time coordinate t. The configuration space for the gravitational field consists of two fields, R and Λ. Under transformations of r, R behaves as a scalar and Λ as a scalar density. Spherically symmetric gravity is discussed in detail in a number of papers, mostly in reference to the canonical quantization of black hole spacetimes. For discussions using the metric representation, see for example [14–16]. For discussions of the Einstein–Hamilton–Jacobi equation in the context of the WKB approximation of quantized spherically symmetric gravity, see for example [17, 18].

We set $c = G = 1$ as before. The Einstein–Hamilton–Jacobi equation for the case of vacuum gravity takes the form $H_{\Lambda R} = 0$ with

$$H_{\Lambda R} = -\frac{1}{R}\frac{\delta S}{\delta R}\frac{\delta S}{\delta \Lambda} + \frac{\Lambda}{2R^2}\left(\frac{\delta S}{\delta \Lambda}\right)^2 + V, \tag{10.25}$$

where

$$V = \frac{RR''}{\Lambda} - \frac{RR'\Lambda'}{\Lambda^2} + \frac{R'^2}{2\Lambda} - \frac{\Lambda}{2}. \tag{10.26}$$

It follows from Eq. (10.25) that there is an inverse supermetric in the space of fields R, Λ given by

$$G^{ab}[R, \Lambda] = \begin{pmatrix} 0 & -\frac{1}{2R} \\ -\frac{1}{2R} & \frac{\Lambda}{2R} \end{pmatrix}, \tag{10.27}$$

where a, b range over $\{r, t\}$, and therefore that the determinant of G_{ab} satisfies $\det|G| \sim R^2$. The momentum constraint takes the form

$$\frac{\delta S}{\delta R}R' - \Lambda\left(\frac{\delta S}{\delta \Lambda}\right)' = 0 \tag{10.28}$$

where primes indicate derivatives with respect to r. We will assume that S is invariant under diffeomorphisms and thus that it automatically solves the momentum constraint.

An appropriate ensemble Hamiltonian for spherically symmetric gravity is given by

$$\mathcal{H} = \int dr\, \dot{N} \int d\mu\, [R, \Lambda]\, P H_{AR}. \tag{10.29}$$

If we apply Eq. (10.12) with $\sqrt{\det |G|} \sim R$, we get the measure $d\mu\, [R, \Lambda] = R\, DR\, D\Lambda$. However, as we pointed out before, any function of the fields that appears as a multiplicative factor in the measure may be absorbed into the definition of P. We make this choice and define the measure by

$$d\mu\, [R, \Lambda] = DR\, D\Lambda, \tag{10.30}$$

with the understanding that the multiplicative factor R has been absorbed into the definition of P.

With $\frac{\partial S}{\partial t} = \frac{\partial P}{\partial t} = 0$, and assuming N is an arbitrary function of the coordinate r, the equations of motion derived from the ensemble Hamiltonian of Eq. (10.29) are Eq. (10.25), the Einstein–Hamilton–Jacobi equation,

$$H_{AR} = -\frac{1}{R}\frac{\delta S}{\delta R}\frac{\delta S}{\delta \Lambda} + \frac{\Lambda}{2R^2}\left(\frac{\delta S}{\delta \Lambda}\right)^2 + V = 0, \tag{10.31}$$

and the continuity equation

$$\frac{\delta}{\delta R}\left(P\frac{1}{R}\frac{\delta S}{\delta \Lambda}\right) + \frac{\delta}{\delta \Lambda}\left(P\frac{1}{R}\frac{\delta S}{\delta R} - P\frac{\Lambda}{R^2}\frac{\delta S}{\delta \Lambda}\right) = 0. \tag{10.32}$$

In the case of spherical gravity, it is also possible to derive rate equations for the fields from the continuity equation. The analysis is similar to that we presented in the previous section when we considered the general case. The interpretation of Eq. (10.32) as a continuity equation is consistent with the rate equations

$$\dot{R} = -N\frac{1}{R}\frac{\delta S}{\delta \Lambda} + \delta_\varepsilon R,$$
$$\dot{\Lambda} = -N\left(P\frac{1}{R}\frac{\delta S}{\delta R} + P\frac{\Lambda}{R^2}\frac{\delta S}{\delta \Lambda}\right) + \delta_\varepsilon \Lambda, \tag{10.33}$$

where N is an arbitrary function and $\delta_\varepsilon R$ and $\delta_\varepsilon \Lambda$ are infinitesimal transformations of the fields which leave the integral of the probability $\int DR\, D\Lambda\, P$ invariant.

To derive explicit forms for $\delta_\varepsilon R$ and $\delta_\varepsilon \Lambda$, we will require that P be invariant under coordinate transformations. Then P satisfies a constraint equivalent to the momentum constraint satisfied by S,

$$\frac{\delta P}{\delta R} R' - \Lambda \left(\frac{\delta P}{\delta \Lambda}\right)' = 0. \qquad (10.34)$$

If we now multiply this expression by an arbitrary function N_r, integrate with respect to the fields and coordinates, and do an integration by parts, we get

$$\int dr \int DR \, D\Lambda \left[\frac{\delta P}{\delta R} N_r R' + \frac{\delta P}{\delta \Lambda} (N_r \Lambda)'\right] = 0, \qquad (10.35)$$

while if we do a gauge transformation that leaves $\int DR \, D\Lambda P$ invariant, we must have

$$\delta_\varepsilon \int dr \int DR \, D\Lambda P = \int dr \int DR \, D\Lambda \left[\frac{\delta P}{\delta R}\delta_\varepsilon R + \frac{\delta P}{\delta \Lambda}\delta_\varepsilon \Lambda\right] = 0 \qquad (10.36)$$

Therefore, we can identify

$$\delta_\varepsilon R = N_r R', \qquad (10.37)$$
$$\delta_\varepsilon \Lambda = (\Lambda N_r)', \qquad (10.38)$$

which leads to

$$\dot{R} = -\frac{N}{R}\frac{\delta S}{\delta \Lambda} + N_r R', \qquad (10.39)$$
$$\dot{\Lambda} = -\frac{N}{R}\frac{\delta S}{\delta R} + \frac{\Lambda}{R^2}\frac{\delta S}{\delta \Lambda} + (\Lambda N_r)'. \qquad (10.40)$$

These are the correct expressions for the rate equations for spherical gravity as derived using the Hamiltonian formalism [15]. However, here they have been derived directly from the formalism of ensembles in configuration space.

Ensemble of gravitational fields may be defined for the case of simplified midisuperspace models like spherically symmetric gravity.

10.4   Ensembles of Black Holes

As an example of configuration space ensembles in spherical gravity, we consider ensembles of black holes. We will restrict to non-rotating, uncharged black holes, which are characterized by only one parameter, the mass of the black hole.

The Hamilton–Jacobi formulation for black holes has been discussed in detail in the context of the WKB approximation of quantized spherically symmetric gravity, for example in Refs. [17, 18]. One remarkable feature is the existence of a general solution of the Einstein–Hamilton–Jacobi equation that can be written in closed form.

To derive this general solution, it will be convenient to simplify the constraint equation $H_{AR} = 0$ of spherical gravity so that we end up with an equivalent equation that is independent of $\frac{\delta S}{\delta R}$. To do this, we use the momentum constraint and write $\frac{\delta S}{\delta R}$ as

$$\frac{\delta S}{\delta R} = \frac{\Lambda}{R'} \left(\frac{\delta S}{\delta \Lambda} \right)'. \tag{10.41}$$

If we make this substitution in H_{AR} and multiply H_{AR} by R'/Λ, we get the equation

$$\frac{R'}{\Lambda} H_{AR} = \left[-\frac{1}{2R} \left(\frac{\delta S}{\delta \Lambda} \right)^2 + \frac{RR'^2}{2\Lambda^2} - \frac{R}{2} \right]' = 0, \tag{10.42}$$

or, equivalently,

$$\frac{1}{R} \left(\frac{\delta S}{\delta \Lambda} \right)^2 - \frac{RR'^2}{\Lambda^2} + R = 2m, \tag{10.43}$$

where m is an integration constant. We now solve for $\frac{\delta S}{\delta \Lambda}$ and define

$$Q[R, \Lambda] := \frac{\delta S}{\delta \Lambda} = R \sqrt{\frac{R'^2}{\Lambda^2} + \frac{2m}{R} - 1}. \tag{10.44}$$

Equation (10.43) is a functional differential equations that is solved by [18]

$$S = \int dr \left\{ \Lambda Q - RR' \cosh^{-1} \frac{R'}{\Lambda \sqrt{1 - \frac{2m}{R}}} \right\}. \tag{10.45}$$

It is straightforward to check that S satisfies both the Hamiltonian and momentum constraints. In this way, we arrive at a general solution which is valid for all choices of gauge. The corresponding momenta are given by

$$\frac{\delta S}{\delta \Lambda} = Q, \tag{10.46}$$

$$\frac{\delta S}{\delta R} = \frac{1}{Q} \left[\Lambda (m - R) + R \left(\frac{RR'}{\Lambda} \right)' \right]. \tag{10.47}$$

To describe an ensemble of black holes, it is necessary to define a probability functional P. This P has to be a solution of the continuity equation, but it is exceedingly difficult to find a P that solves the continuity equation with this choice of S. Therefore, instead of the general solution, Eq. (10.45), we will consider a *particular solution* that is valid for a particular slicing condition (i.e., choice of N and N_r).

We will assume Gaussian coordinate conditions, which amounts to setting $N = 1$ and $N_r = 0$. These conditions correspond to a particular gauge choice, made possible by the invariance of general relativity under arbitrary coordinate transformations. As shown in Appendix 2 of this chapter, the choice of Gaussian coordinate conditions leads to a particular form for the metric of a black hole, known as the Lemaître metric, and the rate equations in these coordinates leads to a very simple relation between R and Λ,

$$\Lambda = R'. \tag{10.48}$$

One can check that this relation implies that the potential term defined in Eq. (10.26) satisfies $V = 0$, thus only the kinetic energy term of the Hamiltonian constraint remains. This leads to a great simplification of the constraints, which now take the form

$$-\frac{1}{R}\frac{\delta S}{\delta R}\frac{\delta S}{\delta \Lambda} + \frac{\Lambda}{2R^2}\left(\frac{\delta S}{\delta \Lambda}\right)^2 = 0, \qquad \frac{\delta S}{\delta R}R' - \Lambda\left(\frac{\delta S}{\delta \Lambda}\right)' = 0. \tag{10.49}$$

It is straightforward to check that both of these equations are solved by the choice

$$S = \int dr \sqrt{R}\Lambda. \tag{10.50}$$

This is the functional S which solves the constraint equations for the special case of Gaussian coordinate conditions. Notice that now the mass m of the black hole does not appear explicitly in the expression for S, Eq. (10.50).

If we use this expression for S in the continuity equation, we get

$$\frac{\delta}{\delta R}\left(P\frac{1}{\sqrt{R}}\right) - \frac{\delta}{\delta \Lambda}\left(P\frac{\Lambda}{2R\sqrt{R}}\right) = 0. \tag{10.51}$$

The equation is singular so we replace it by the equivalent but better behaved equation

$$\int dr\, \Phi\left\{\frac{\delta}{\delta R}\left(P\frac{1}{\sqrt{R}}\right) - \frac{\delta}{\delta \Lambda}\left(P\frac{\Lambda}{2R\sqrt{R}}\right)\right\} = 0. \tag{10.52}$$

where $\Phi(r)$ is a test function. Keeping in mind that Φ is arbitrary and P must be invariant under coordinate transformations, the general solution is given by

$$P \sim R\,\Pi[\mu]. \tag{10.53}$$

where Π is an arbitrary function of the variable μ which is defined by

$$\mu = \frac{1}{V}\int dr\sqrt{R}\Lambda. \tag{10.54}$$

for some constant V. One can check that P is a scalar, as required, because both R and μ are scalars.

As we mentioned in the previous section, when we defined the measure according to Eq. (10.30) we absorbed a multiplicative factor R into the definition of P. If we now undo this, we end up with $\Pi[\mu]$ as the natural probability functional. As shown in Appendix 2 of this chapter, $R\Lambda^2 = 2m$. We can make the numerical value of μ finite by setting $V = \int dr$, in which case the numerical value of μ is given by $\mu = \sqrt{2m}$.

Thus the probability functional $\Pi[\mu]$ *depends solely on the mass of the black hole*, that is, on the one parameter that determines the space-time completely. This is a very satisfying result: the ensemble is composed of black holes of different masses, with probabilities determined by $\Pi[\mu]$.

Appendix 1: The Reconstruction Problem

Given a solution of the Einstein–Hamilton–Jacobi equation; i.e., $\{h_{kl}, N, N_i, S\}$ defined on a three-dimensional space-like hypersurface, how do you go to a four-dimensional spacetime? This problem is known as the *reconstruction problem*. We will look into this now, paying particular attention to the issue of general covariance. We follow closely the presentation of Wald [8].

Assume that the hypersurface can be embedded into a four-dimensional space-time with metric $g_{\mu\nu}$,

$$ds^2 = g_{\mu\nu}dx^\mu dx^\nu \qquad \mu, \nu = 0, 1, 2, 3 \qquad x^0 \equiv t. \qquad (10.55)$$

The space-time metric $g_{\mu\nu}$ can be written in the form

$$g_{\mu\nu} = \begin{bmatrix} g_{00} & g_{0j} \\ g_{i0} & g_{ij} \end{bmatrix} = \begin{bmatrix} -M^2 + M_j M^j & M_j \\ M_i & h_{ij} \end{bmatrix} \qquad (10.56)$$

where the M, M_j are arbitrary functions (later they will be related to the shift function and lapse vector). Then, the line element on any hypersurfaces defined by constant t is given by $ds^2|_{dt=0} = h_{ij}dx^i dx^j$, which means that the metric induced on the hypersurface is indeed given by the spatial metric h_{ij}.

The extrinsic curvature K_{ij} of the hypersurface is defined by [8]

$$K_{ij} = \frac{1}{2M} \left(\frac{\partial h_{ij}}{\partial t} - D_i M_j - D_j M_i \right) \qquad (10.57)$$

or

$$\frac{\partial h_{ij}}{\partial t} = 2MK_{ij} + D_i M_j + D_j M_i, \qquad (10.58)$$

while the rate equation is of the form

$$\frac{\partial h_{ij}}{\partial t} = NG_{ijkl}\frac{\delta S}{\delta h_{kl}} + D_i N_j + D_j N_i. \tag{10.59}$$

Therefore, the embedding into the four-dimensional space-time is consistent with the equations on the hypersurface if we identify

$$M = N, \tag{10.60}$$
$$M_i = N_i, \tag{10.61}$$
$$2K_{ij} = G_{ijkl}\frac{\delta S}{\delta h_{kl}}, \tag{10.62}$$

or, equivalently [8],

$$\frac{\delta S}{\delta h_{ij}} = \sqrt{h}(h^{ij}K - K^{ij}), \tag{10.63}$$

$$h_{ij}\frac{\delta S}{\delta h_{ij}} = 2\sqrt{h}K. \tag{10.64}$$

In terms of the extrinsic derivative, the constraint equations and the rate equations take the form

$$H = \frac{1}{2}G_{ijkl}\frac{\delta S}{\delta h_{ij}}\frac{\delta S}{\delta h_{kl}} - \sqrt{h}R = K_{ij}\sqrt{h}\left(K^{ij} - Kh^{ij}\right) - \sqrt{h}R \tag{10.65}$$

$$= -\sqrt{h}\left[K^2 - K^{ij}K_{ij} + R\right] = 0, \tag{10.66}$$

$$H_i = -2D_j\left(h_{ik}\frac{\delta S}{\delta h_{kj}}\right) = -4\sqrt{h}D_j K = 0, \tag{10.67}$$

$$K_{ij} = \frac{1}{2N}\left(-\dot{h}_{ij} + D_i N_j + D_j N_i\right). \tag{10.68}$$

The next step is to show that this leads to equations that are invariant not only under spatial coordinate transformations, but also under more general space-time coordinate transformations. To do this, we introduce the following four-vectors and four-tensors [8]: Denote the unit normal n_μ to the hypersurface by

$$n_\mu = (0, 0, 0, -N), \tag{10.69}$$
$$n^\mu = (-N^k/N, 1/N), \tag{10.70}$$

and the projection tensor $h_{\mu\nu}$ by

$$h_{\mu\nu} = g_{\mu\nu} + n_\mu n_\nu. \tag{10.71}$$

The extrinsic curvature K_β^α can be written in terms of $h_{\mu\nu}$ and n_μ as

$$K_\beta^\alpha = h_\alpha^\mu \nabla_\mu n_\beta, \tag{10.72}$$

and it satisfies

$$K_\beta^\alpha n^\beta = 0. \tag{10.73}$$

The 3-space components of $h_{\mu\nu}$ and $K_{\alpha\beta}$ agree, of course, with h^{ij} and K_{ij}.
  With the help of these quantities, one can show that [8]

$$G_{\mu\nu} n^\mu n^\nu = \frac{1}{2} \left[K^2 - K^{\alpha\beta} K_{\alpha\beta} + R \right], \tag{10.74}$$

$$h_\alpha^\mu G_{\mu\nu} n^\nu = D_\beta K_\alpha^\beta - D_\alpha K, \tag{10.75}$$

where $G_{\mu\nu} \equiv R_{\mu\nu} - \frac{1}{2} R g_{\mu\nu}$ is the Einstein tensor. Therefore

$$H = 0 \quad \Leftrightarrow \quad G_{\mu\nu} n^\mu n^\nu = 0, \tag{10.76}$$

$$H_i = 0 \quad \Leftrightarrow \quad h_\alpha^\mu G_{\mu\nu} n^\nu = 0, \tag{10.77}$$

which means that the constraint equations have been rewritten in a way that shows explicitly that they are invariant under space-time coordinate transformations. Furthermore, the rate equation is nothing else but the definition of the extrinsic curvature, which can also be expressed in an invariant way, $K_\beta^\alpha = h_\alpha^\mu \nabla_\mu n_\beta$.
  This shows that we can take the equations originally defined on the hypersurface and rewrite them as equations that are invariant under space-time coordinate transformations. The reverse is also true: given the equations in space-time, we can do a splitting of space and time and recover the equations defined on the hypersurface. In other words, we have shown general covariance even though the formalism we started with was only required to be invariant under spatial coordinate transformations.

Appendix 2: Lemaître Coordinates From Gaussian Coordinate Conditions

We introduce Lemaître coordinates for the Schwarzschild black hole, discuss their physical interpretation, and show that they follow from assuming Gaussian coordinate conditions.

Physical Interpretation of the Lemaître coordinates

We follow the presentation of Stephani [19]. We start with the metric of a black hole in Schwarzschild coordinates,

$$ds^2_{Sch} = -\left(1 - \frac{2m}{R}\right) dT^2 + \frac{dR^2}{1 - \frac{2m}{R}} + R^2 d\Omega^2, \tag{10.78}$$

where $d\Omega^2 = d\theta^2 + \sin^2\theta d\phi^2$. Consider now the motion of test particles which fall freely and radially (i.e., $d\theta = d\phi = 0$). The equations of motion can be derived from the Lagrangian

$$L = \frac{1}{2} g_{\mu\nu} \frac{dx^\mu}{d\tau} \frac{dx^\nu}{d\tau}, \tag{10.79}$$

with $d\theta = d\phi = 0$, where τ is the proper time. The Lagrangian takes the form

$$L = -\left(1 - \frac{2m}{R}\right)\left(\frac{dT}{d\tau}\right)^2 + \frac{1}{1 - \frac{2m}{R}}\left(\frac{dR}{d\tau}\right)^2. \tag{10.80}$$

Since T is a cyclic coordinate, we get the conservation equation

$$\left(1 - \frac{2m}{R}\right)\frac{dT}{d\tau} = A = \text{const.} \tag{10.81}$$

To get a second conservation equation, we use $g_{\mu\nu}\frac{dx^\mu}{d\tau}\frac{dx^\nu}{d\tau} = -1$ and $d\theta = d\phi = 0$, which leads to

$$-\left(1 - \frac{2m}{R}\right)\left(\frac{dT}{d\tau}\right)^2 + \frac{1}{1 - \frac{2m}{R}}\left(\frac{dR}{d\tau}\right)^2 = -1. \tag{10.82}$$

Combining these two last two equations, we get

$$\frac{dR}{d\tau} = -\sqrt{A^2 - \left(1 - \frac{2m}{R}\right)}, \tag{10.83}$$

where we have chosen the negative sign of the square root because we are considering particles that fall radially into the black hole. Therefore we have the two relations

$$dT = \frac{A}{\left(1 - \frac{2m}{R}\right)} d\tau, \tag{10.84}$$

$$dR = -\sqrt{A^2 - \left(1 - \frac{2m}{R}\right)} d\tau. \tag{10.85}$$

We set $A = 1$, which corresponds to the case of particles that are at rest at infinity since

$$dT (R \rightarrow \infty; A = 1) = d\tau, \tag{10.86}$$

$$dR (R \rightarrow \infty; A = 1) = 0. \tag{10.87}$$

The Lemaître coordinates are defined in terms of new time and radial coordinates t and r which satisfy $dt(T, R) = d\tau$ and $dr = 0$ with $A = 1$. To achieve this, we set

$$dt = dT + \sqrt{\frac{2m}{R}} \frac{dR}{\left(1 - \frac{2m}{R}\right)}, \tag{10.88}$$

$$dr = dt + \sqrt{\frac{R}{2m}} dR = dT + \sqrt{\frac{R}{2m}} \frac{dR}{\left(1 - \frac{2m}{R}\right)}. \tag{10.89}$$

One can check that the two conditions $dt = d\tau$ and $dR = 0$ are satisfied. We can integrate the relations for dt and dR and find explicit forms for the coordinate transformation,

$$t(T, R) = T + 2\sqrt{2mR} + 2m \log \left| \frac{\sqrt{R} - \sqrt{2m}}{\sqrt{R} + \sqrt{2m}} \right|, \tag{10.90}$$

$$R(r, t) = \left[(r - t) \frac{3\sqrt{2m}}{2} \right]^{2/3}. \tag{10.91}$$

The line element of the Schwarzschild black hole metric in Lemaître coordinates takes the form

$$ds_L^2 = -dt^2 + \frac{2m}{R} dr^2 + R^2 d\Omega^2$$

$$= -dt^2 + \frac{4}{9} \left(\frac{9M}{2}\right)^{2/3} (r - t)^{-2/3} dr^2 + \left(\frac{9M}{2}\right)^{2/3} (r - t)^{4/3} d\Omega^2. \tag{10.92}$$

The Lemaître coordinates therefore have the following physical interpretation: the coordinate t corresponds to the proper time for particles which are at rest in the Lemaître coordinate system. Furthermore, $dR = d\theta = d\phi = 0$ holds for those particles which are initially at rest at infinity and then fall freely and radially.

Derivation of the Lemaître Coordinates From Gaussian Coordinate Conditions

We now derive the Lemaître coordinates from Gaussian coordinate condition, $N = 1$ and $N_r = 0$.

Using the rate equations with $N = 1$ and $N_r = 0$ and the momentum constraint, we can write $\dot{R}'$ as

$$
\begin{aligned}
\dot{R}' &= \left[-\frac{1}{R}\frac{\delta S}{\delta \Lambda} \right]' \\
&= \frac{R'}{R^2}\frac{\delta S}{\delta \Lambda} - \frac{1}{R}\left(\frac{\delta S}{\delta \Lambda} \right)' \\
&= \frac{R'}{R^2}\frac{\delta S}{\delta \Lambda} - \frac{1}{R}\frac{R'}{\Lambda}\frac{\delta S}{\delta R} \\
&= \frac{R'}{\Lambda}\left[-\frac{1}{R^2}\left(R\frac{\delta S}{\delta R} - \Lambda\frac{\delta S}{\delta \Lambda} \right) \right] \\
&= \frac{R'}{\Lambda}\dot{\Lambda}.
\end{aligned}
\tag{10.93}
$$

This is a differential equation involving R and Λ which has solution

$$
\Lambda = f(r)R',
\tag{10.94}
$$

where f is an arbitrary function of r.

We can determine f from the equation for $\dot{R}$. We first calculate $\dot{R}^2$ using Eq. (10.94), which leads to

$$
\dot{R}^2 = \left(\frac{1}{R}Q \right)^2 = f^2 + \frac{2m}{R} - 1.
\tag{10.95}
$$

Taking the time derivative leads to a differential equation for R,

$$
2\dot{R}\ddot{R} = -\frac{2m}{R^2}\dot{R},
\tag{10.96}
$$

which has solution

$$
R = \left(\frac{9m}{2} \right)^{1/3}[\rho(r) - t]^{2/3}
\tag{10.97}
$$

for some arbitrary function ρ of r. Then, to find f, we evaluate the expression for $\dot{R}^2$ again, this time using Eqs. (10.95) and (10.97),

$$\dot{R}^2 = \left(\frac{4m}{3}\right)^{2/3} (\rho - t)^{-2/3}$$

$$= f^2 + \frac{2m}{R} - 1$$

$$= f^2 + 2m \left[\left(\frac{9m}{2}\right)^{-1/3} (\rho - t)^{-2/3}\right] - 1$$

$$= \left(\frac{4m}{3}\right)^{2/3} (\rho - t)^{-2/3} + \left(f^2 - 1\right), \tag{10.98}$$

which forces $f = \pm 1$ so that $\Lambda = \pm R'$ or

$$\Lambda^2 = R'^2. \tag{10.99}$$

We now show that we end up with the Lemaître coordinates regardless of the choice of ρ, so we can simply set $\rho = r$. To see this, use

$$R = \left(\frac{9m}{2}\right)^{1/3} (\rho - t)^{2/3}, \qquad \Lambda = \pm \frac{2}{3} \left(\frac{9m}{2}\right)^{1/3} (\rho - t)^{-1/3} \rho', \tag{10.100}$$

and write down the line element with $N = 1$ and $N_r = 0$,

$$ds^2 = -dt^2 + \Lambda^2 dr^2 + R^2 d\Omega^2$$

$$= -dt^2 + \frac{4}{9} \left(\frac{9m}{2}\right)^{2/3} (\rho - t)^{-2/3} \left(\rho' dr\right)^2 + \left(\frac{9m}{2}\right)^{2/3} (\rho - t)^{4/3} d\Omega^2$$

$$= -dt^2 + \frac{4}{9} \left(\frac{9m}{2}\right)^{2/3} (\rho - t)^{-2/3} d\rho^2 + \left(\frac{9m}{2}\right)^{2/3} (\rho - t)^{4/3} d\Omega^2. \tag{10.101}$$

This is the line element expressed using Lemaître coordinates. Notice that setting $\rho = r$ in Eq.(10.100) leads to the relation

$$R\Lambda^2 = 2m, \tag{10.102}$$

therefore the mass of the black hole can be expressed in terms of R and Λ in a very simple way, $m = R\Lambda^2/2$.

References

1. Misner, C.W., Thorne, K.S., Wheeler, J.A.: Gravitation. Freeman, San Francisco (1973)
2. Kiefer, C.: Quantum Gravity. Oxford University Press, Oxford (2012)
3. Peres, A.: On Cauchy's problem in general relativity – II. Nuovo Cim. **XXVI**, pp 53–62 (1962)
4. Gerlach, U.H.: Derivation of the the Einstein equations from the semiclassical approximation to quantum geometrodynamics. Phys. Rev. **177**, 1929–1941 (1969)

5. Bergmann, P.G.: Hamilton-Jacobi and Schrödinger theory in theories with first-class Hamiltonian constraints. Phys. Rev. **144**, 1078–1080 (1966)
6. Giulini, D.: What is the geometry of superspace? Phys. Rev. D. **51**, 5630–5635 (1995)
7. Kiefer, C.: The semiclassical approximation to quantum gravity. In: Ehlers, J., Friedrich, H. (eds.) Canonical gravity: from classical to quantum, pp. 170–212. Springer, Berlin (1994)
8. Wald, R.: General Relativity. University of Chicago Press, Chicago (1984)
9. Reginatto, M.: Cosmology with quantum matter and a classical gravitational field: the approach of configuration-space ensembles. J. Phys.: Conf. Ser. **442**, 012009 (2013)
10. DeWitt, B.S.: Quantum gravity: the new synthesis. In: Hawking, S.W., Israel, W. (eds.) General relativity: an Einstein centenary survey, pp. 680–745. Cambridge University Press, Cambridge (1979)
11. Hamber, H.W., Williams, R.M.: On the measure in simplicial gravity. Phys. Rev. D **59**, 064014 (1999)
12. Fadeev, L.D., Popov, V.N.: Covariant quantization of the gravitational field. Sov. Phys.-Usp. **16**, 777–788 (1974)
13. Hall, M.J.W., Reginatto, M.: Interacting classical and quantum ensembles. Phys. Rev. A **72**, 062109 (2005)
14. Romano, J.D.: Spherically symmetric scalar field collapse: an example of the spacetime problem of time. arXiv:gr-qc/9501015v1 (1995)
15. Kuchař, K.V.: Geometrodynamics of Schwarzschild black holes. Phys. Rev. D **50**, 3961–3981 (1994)
16. Lau, S.R.: On the canonical reduction of spherically symmetric gravity. Class. Quantum Grav. **13**, 1541–1570 (1996)
17. Fischler, W., Morgan, D., Polchinski, J.: Quantization of false-vacuum bubbles: a Hamiltonian treatment of gravitational tunneling. Phys. Rev. D **42**, 4042–4055 (1990)
18. Brotz, T., Kiefer, C.: Semiclassical black hole states and entropy. Phys. Rev. D **55**, 2186–2191 (1997)
19. Stephani, H.: General Relativity. Cambridge University Press, Cambridge (1990)

Chapter 11
Coupling of Quantum Fields to Classical Gravity

Abstract We consider ensembles on configuration space that consist of quantum fields which interact with and are the source of a classical gravitational field. These are hybrid systems where gravity remains classical while matter is described by quantized fields. There are some well known arguments in the literature which claim that such models are not possible. However, an examination of the most prominent ones, that are detailed enough to allow scrutiny, indicates that the hybrid models considered here are not excluded by any of the consistency arguments. We illustrate the approach with two examples. Our first example is a cosmological model. We consider the case of a closed Robertson–Walker universe with a massive quantum scalar field and solve the equations using a particular ansatz which selects a highly non-classical solution, one in which the scale factor of the Robertson–Walker universe is restricted to discrete values as a consequence of the interaction of the classical gravitational field with the quantized scalar field. We discuss this cosmological model in two approximations, that of a minisuperspace model and that of a midisuperspace model. Our second example concerns black holes. We consider CGHS black holes and show that we recover Hawking radiation from the equations that describe a hybrid system consisting of a classical CGHS black hole in a collapsing geometry interacting with a quantized scalar field. We also show that the hybrid model provides a natural resolution to the well known problem of time in quantum gravity.

11.1 Introduction

There are a number of reasons to consider the coupling of quantum fields to classical gravity.

A full theory of quantum gravity is not yet available, and an approximation in which spacetime remains classical while matter is described in terms of quantum fields is often physically and computationally appropriate. Furthermore, since the quantization of gravity does not appear to follow from consistency arguments alone [1], it is of interest to investigate whether hybrid systems can provide a satisfactory description of matter and gravitation. The study of such systems may provide valuable clues that can be of help in the search for a full quantum theory of gravity.

© Springer International Publishing Switzerland 2016   243
M.J.W. Hall and M. Reginatto, *Ensembles on Configuration Space*,
Fundamental Theories of Physics 184, DOI 10.1007/978-3-319-34166-8_11

Finally, one must also consider the possibility that the gravitational field may not be quantum in nature [2–4]. For example, Butterfield and Isham, while putting forward the point of view that some type of theory of quantum gravity should be sought, have concluded that there is arguably no definitive proof that general relativity has to be quantized [5]. Dyson has argued that it might be impossible in principle to observe the existence of individual gravitons [6] and this has lead him to the conjecture that "the gravitational field described by Einstein's theory of general relativity is a purely classical field without any quantum behaviour" [7]. His observations regarding the impossibility of detecting gravitons have been confirmed by detailed calculations [8, 9]. If Dyson's conjecture is true, hybrid models become unavoidable.

In the standard approach used for coupling quantum fields to a classical gravitational field (i.e., semiclassical gravity [10]), the energy momentum tensor that serves as the source in the Einstein equations is set equal to the expectation value of the energy momentum operator $\widehat{T}_{\mu\nu}$ with respect to some quantum state Ψ:

$$^4R_{\mu\nu} - \frac{1}{2}g_{\mu\nu}\,^4R + \lambda g_{\mu\nu} = \frac{\kappa}{2}\,\langle\Psi|\,\widehat{T}_{\mu\nu}\,|\Psi\rangle, \tag{11.1}$$

where $^4R_{\mu\nu}$ is the curvature tensor, ^4R the curvature scalar, $g_{\mu\nu}$ the metric tensor of spacetime, λ the cosmological constant and $\kappa = 16\pi G$ (in units where $c = G = 1$). There is also a non-relativistic, Newtonian analogue of Eq. (11.1). It is known as the Schrödinger-Newton equation, and in this case a non-linear term is added to the standard Schrödinger equation with the density of matter being represented by the square of the wave-function [11]. Since we restrict to relativistic systems, we will not make any further remarks about the Schrödinger-Newton equation except to point out that it shares some of the problems of semiclassical gravity, as one would expect.

Semiclassical gravity presents a number of well known difficulties which are not encountered when the formalism of ensembles on configuration space is used to couple quantized fields to a classical gravitational field. Explicit examples of such hybrid models are discussed in this chapter. These models are counterexamples to some well known arguments in the literature which claim that such models are not possible. Since these arguments are invariably based on gedanken experiments which involve measurements of one sort or another, it is appropriate to address the question of whether such consistency arguments based on measurement theory alone can show that the gravitational field must be quantized. An examination of the most prominent arguments in the literature which are detailed enough to allow scrutiny indicates there is no logical necessity to quantize gravity [1]. To illustrate this, we give a brief critique of the arguments of DeWitt [12], Eppley and Hannah [13], Page and Geilker [14], and Feynman [15].

DeWitt has argued that that the quantum theory must be extended to all physical systems for consistency reasons, concluding that [12]

> …the quantization of a given system implies also the quantization of any other system to which it can be coupled (…) therefore, the quantum theory must immediately be extended to all physical systems, including the gravitational field.

DeWitt's argument is rather involved and we will only give its broad outline here; for a more detailed review and critique see Ref. [1]. DeWitt considers the measurement of a system by an apparatus and assumes that it proceeds via a particular type of coupling. He then assumes limits on the accuracy of two of the apparatus variables, expressed in the form of an inequality. He goes on to show that this results in limits on the accuracy of a pair of corresponding system variables which are coupled to those apparatus variables; that is, the corresponding system variables have to satisfy an inequality which is similar to the one assumed for the apparatus variables. Up to this point, there is nothing specific to quantum mechanics in his argument: the derivation makes use of classical Poisson brackets only, which indicates that quantum considerations do not play a fundamental role in his calculation. As a matter of fact, his argument is very general and it applies equally well to two classical systems that interact, or to a classical system that interacts with a quantum system, as long as his assumptions on the type of coupling are satisfied. The only place where quantum mechanical considerations enter into the argument is in a further assumption: that the limits in the accuracy of the two apparatus variables are quantum mechanical in nature, the result of an uncertainty relation, which leads in turn to a particular type of inequality for the corresponding system variables which are coupled to the apparatus variables. However, one cannot conclude from this particular inequality that the system that is being measured *has* to be quantized, as DeWitt does [1]. His argument therefore does not provide a proof that the quantum theory must be extended to all physical systems; in particular, it does not prove the logical necessity of quantizing the gravitational field.

Perhaps the most influential paper arguing for the necessity of quantizing the gravitational field is the article by Eppley and Hannah [13]. Their gedanken experiment involves the interaction of a classical gravitational wave with a quantum system. They do not propose any particular model for this interaction but argue nevertheless that it would lead to a violation of momentum conservation, to a violation of the uncertainty principle, or to the transmission of signals faster than light. It should be emphasized that their arguments do not seem to depend on any feature that is unique to gravity or gravitational waves (in particular, the calculations are carried out with a linearized wave), thus their arguments should hold for any classical wave. Their argument, if correct, would suggest that any system that interacts with a quantum system must also be a quantum system. Their gedanken experiment is, however, fatally flawed. Mattingly has shown that the experiment cannot be carried out even in principle [16]. The device that they propose, even if it could be built, would not be able to establish their claims, nor is it plausible that it could be built with any materials compatible with the values of c, $\hbar$, and G. Even more damaging, their detector would have to be so massive as to be within its own Schwarzschild radius [16]. Furthermore, Huggett and Callender have argued that the violations of physical principles are only present in the Copenhagen interpretation of quantum mechanics and are thus, at least partially, resolvable within alternative interpretations [17]. Finally, a careful analysis shows that even without the question of realizability or interpretation, one can not conclude from the Eppley and Hannah gedanken experiment that the gravitational must be quantized [1].

The paper of Page and Geilker [14] has also been widely cited in the literature in discussions on the necessity of quantizing gravity. They argue that semiclassical gravity is not a viable theory but, as they do not consider any other alternatives to semiclassical gravity, their paper has nothing to say regarding the more general question of whether the coupling of a classical gravitational field to quantized matter fields is possible in other hybrid theories, such as the one that we discuss in this chapter. Thus we will not discuss this paper further, except to point out that various difficulties with Page and Gilker's interpretation of their experiment [18] call into question the significance of their result.

Finally, Feynman, in his lectures on gravitation [15], has sketched a gedanken experiment which, he argues, show the necessity of quantizing the gravitational field. He considers a double slit experiment and places a "gravity detector" between the slits and the screen that is assumed to be far away from the slits. The interaction takes place via the gravitational field of the quantum system, which emits gravitational waves that are observed when they reach the detector. The bulk of his argument is given in the following statement:

> The position of the electron is described by an amplitude (...). If the gravity interacts through a field, it follows that the gravity field must have an amplitude also (...). But this is precisely the characteristic of a quantum field, that it should be described by an amplitude rather than a probability.

Feynman's argument, like those of Eppley and Hanna and DeWitt, makes no use of any feature that is particular to gravity. If the "gravity detector" were to be replaced by any other detector which interacts with the quantum system via a field, the discussion would not have to be modified in any essential way. Furthermore, his argument applies to a very restricted class of hybrid theories. The argument points to difficulties with semiclassical gravity, Eq. (11.1), where the gravitational field couples to $\langle \Psi | \widehat{T}_{\mu\nu} | \Psi \rangle$, the expectation value of the energy momentum tensor. With such a coupling, much of the information that is contained in the complex amplitude (i.e., the wave functional of the field) is discarded. But it is clear that Feynman's argument does not exclude hybrid systems of the type that we discuss in this chapter, where the total system is described in terms of a pair of functionals P and S or, equivalently, the complex functional $\Psi = \sqrt{P}e^{iS/\hbar}$.

We must conclude that the existing gedanken experiments do not show that gravity must be quantized. Although there are physical arguments which speak in favor of a quantum theory of gravity [19, 20], this is an issue that can only be decided by experiment.

There is no proof that general relativity has to be quantized and there are a number of reasons to consider the possibility that quantum matter fields interact with a classical gravitational field. A consistent description of these types of hybrid systems is possible using ensembles on configuration space.

In the remaining sections of this chapter, we consider ensembles on configuration space that consist of quantum fields which interact with and are the source of a classical gravitational field. In Sect. 11.2 we formulate the theory of such ensembles for both the general case and for the case of spherically symmetric gravity. For simplicity we limit ourselves to the case in which matter is in the form of a quantized scalar field, but the approach can be extended to other types of quantized matter fields. In the two sections that follow, we illustrate the approach with two examples. The first example is a cosmological model: in Sect. 11.3 we consider the case of a closed Robertson–Walker universe with a massive quantum scalar field and solve the equations using a particular ansatz, one which selects a highly non-classical solution. Our second example concerns black holes: in Sect. 11.4 we consider CGHS black holes and show that we recover Hawking radiation from the equations that describe a hybrid system consisting of a classical CGHS black hole in a collapsing geometry interacting with a quantized scalar field. We also show that the hybrid model provides a natural resolution to the well known problem of time in quantum gravity.

11.2 The Coupling of Classical Gravitational Fields to Quantum Matter Fields

In Chap. 10, we developed the formalism that is needed for the description of configuration space ensembles of classical gravitational fields. In this chapter, we show that one can extend this formalism to include in addition *quantum* matter fields which interact with and are the source of the classical gravitational field. This is done following the approach for hybrid systems developed in Chaps. 8 and 9. We consider the particular example in which matter is in the form of a quantized scalar field. We chose this case because quantized scalar fields are simpler than other quantized fields and therefore it is possible to develop the theory with a minimum of technical complications. The approach can also be applied to matter described by other types of quantized fields.

11.2.1 General Case

In Sect. 10.2.3, we saw that an appropriate ensemble Hamiltonian for classical gravitational fields is given by

$$\mathcal{H}_h^C = \int d^3x \, N \int Dh \, P H_h^C, \tag{11.2}$$

where N is the lapse function introduced in Eq. (10.2) and H_h^C is given by

$$H_h^C = \kappa G_{ijkl} \frac{\delta S}{\delta h_{ij}} \frac{\delta S}{\delta h_{kl}} - \frac{1}{\kappa} \sqrt{h} \, (R - 2\lambda), \tag{11.3}$$

where $\kappa = 16\pi$ (in units where $c = G = 1$), $G_{ijkl} = (2h)^{-1/2} \left(h_{ik}h_{jl} + h_{il}h_{jk} - h_{ij}h_{kl} \right)$ is the DeWitt supermetric, h_{ij} is the metric on a space-like hypersurface, h is the determinant of h_{ij}, R is the curvature scalar, λ is the cosmological constant, and D_j is the spatial covariant derivative.

The ensemble Hamiltonian of Eq. (11.2) leads to two functional equations: the Einstein–Hamilton–Jacobi equation for pure gravity and an equation that may be interpreted as a continuity equation under the assumptions that $S[h_{ij}]$ and $P[h_{ij}]$ satisfy the conditions $D_k \left(\delta S/\delta h_{kl} \right) = D_k \left(\delta P/\delta h_{kl} \right) = 0$, which correspond to invariance under spatial coordinate transformations, and $\frac{\partial S}{\partial t} = \frac{\partial P}{\partial t} = 0$. These and other details concerning the formulation of classical ensembles of gravitational fields are discussed in Sect. 10.2.

A *hybrid system* where a quantum scalar field ϕ couples to the classical metric h_{kl} requires a generalization of Eq. (11.2), of the form [21, 22]

$$\mathscr{H}_{\phi h} = \int d^3x \int Dh D\phi \, P N \left[H^C_{\phi h} + F_\phi \right], \tag{11.4}$$

where

$$H^C_{\phi h} = H^C_h + \frac{1}{2\sqrt{h}} \left(\frac{\delta S}{\delta \phi} \right)^2 + \sqrt{h} \left[\frac{1}{2} h^{ij} \frac{\partial \phi}{\partial x^i} \frac{\partial \phi}{\partial x^j} + V(\phi) \right] \tag{11.5}$$

is a purely classical term which now includes the coupling to a scalar field ϕ, and

$$F_\phi = \frac{\hbar^2}{4} \frac{1}{2\sqrt{h}} \left(\frac{\delta \log P}{\delta \phi} \right)^2 \tag{11.6}$$

is an additional, non-classical term that must be included in the ensemble Hamiltonian when the scalar field is quantized.

The Hamiltonian equations of motion for P and S that follow from $\mathscr{H}_{\phi h}$ are

$$\frac{\partial P}{\partial t} = \frac{\Delta \mathscr{H}_{\phi h}}{\Delta S}, \qquad \frac{\partial S}{\partial t} = -\frac{\Delta \mathscr{H}_{\phi h}}{\Delta P}, \tag{11.7}$$

where $\Delta/\Delta F$ denotes the variational derivative with respect to the functional F (see Appendix A of this book). With $\frac{\partial S}{\partial t} = \frac{\partial P}{\partial t} = 0$ (see Sect. 10.2), the equations of motion can be written as

$$\int d^3x N \left[H^C_{\phi h} - \frac{\hbar^2}{2\sqrt{h}} \left(\frac{1}{A} \frac{\delta^2 A}{\delta \phi^2} \right) \right] = 0, \tag{11.8}$$

where $A \equiv \sqrt{P}$, and

$$\int d^3x N \left[\frac{\delta}{\delta h_{ij}} \left(P G_{ijkl} \frac{\delta S}{\delta h_{kl}} \right) + \frac{1}{\sqrt{h}} \frac{\delta}{\delta \phi} \left(P \frac{\delta S}{\delta \phi} \right) \right] = 0. \tag{11.9}$$

The interpretation of Eqs. (11.8–11.9) is similar to the interpretation given in the analogous case of classical gravity that we discussed in Sect. 10.2 (compare to Eqs. (10.20–10.21) for a classical ensemble of gravitations fields), except that now we are dealing with a hybrid system. Equation (11.8) is a generalization of the Einstein–Hamilton–Jacobi equation for the case of gravity with a classical scalar field (and reduces to it when $\hbar \rightarrow 0$) and Eq. (11.9) may be interpreted as a continuity equation.

11.2.2 Midisuperspace Example: Spherically Symmetric Gravity

As we did in Chap. 10, we consider the midisuperspace model that corresponds to assuming a spherically symmetric spacetime and derive the corresponding equations.

We saw in Sect. 10.3 that the Einstein–Hamilton–Jacobi equation for the case of spherically symmetric vacuum gravity can be written in the form $H_{AR}^C = 0$ with

$$H_{AR}^C = -\frac{1}{R}\frac{\delta S}{\delta R}\frac{\delta S}{\delta A} + \frac{A}{2R^2}\left(\frac{\delta S}{\delta A}\right)^2 + V \tag{11.10}$$

(in units where $c = G = 1$), where V is given by

$$V = \frac{RR''}{A} - \frac{RR'A'}{A^2} + \frac{R'^2}{2A} - \frac{A}{2}. \tag{11.11}$$

The ensemble Hamiltonian of a hybrid system where matter is in the form of a minimally coupled quantized radially symmetric scalar field of mass m is given by

$$\mathscr{H}_{\phi AR} = \int dr \int D\phi DADR \, PN\left[H_{\phi AR}^C + F_\phi\right], \tag{11.12}$$

where

$$H_{\phi AR}^C = H_{AR}^C + \frac{1}{2AR^2}\left(\frac{\delta S}{\delta\phi}\right)^2 + \frac{R^2}{2A}\phi'^2 + \frac{AR^2m^2}{2}\phi^2, \tag{11.13}$$

is a purely classical term which now includes the coupling to a scalar field ϕ and

$$F_\phi = \frac{1}{8AR^2}\left(\frac{\delta\log P}{\delta\phi}\right)^2 \tag{11.14}$$

is an additional, non-classical term that must be included in the ensemble Hamiltonian when the scalar field is quantized. Equation (11.12) is the analogous of Eq. (11.4) for the case of spherically symmetric gravity.

Assuming again the constraints $\frac{\partial S}{\partial t} = \frac{\partial P}{\partial t} = 0$, the corresponding equations are

$$\int dr\, N \left[H^C_{\phi \Lambda R} - \frac{1}{2 \Lambda R^2} \left(\frac{1}{A} \frac{\delta^2 A}{\delta \phi^2} \right) \right] = 0, \tag{11.15}$$

where $A \equiv \sqrt{P}$, and the continuity equation

$$\int dr\, N \left[\frac{\delta}{\delta R} \left(P \frac{1}{R} \frac{\delta S}{\delta \Lambda} \right) + \frac{\delta}{\delta \Lambda} \left(P \frac{1}{R} \frac{\delta S}{\delta R} - P \frac{\Lambda}{R^2} \frac{\delta S}{\delta \Lambda} \right) \right.$$
$$\left. - \frac{\delta}{\delta \phi} \left(P \frac{1}{\Lambda R^2} \frac{\delta S}{\delta \phi} \right) \right] = 0. \tag{11.16}$$

11.3   Hybrid Cosmological Model

Our first example of the coupling of a quantum field to classical gravity is a cosmo-logical model. We consider the case of a closed Robertson–Walker universe with a massive scalar field. We solve the equations using a particular ansatz which selects a highly non-classical solution, one in which the scale factor of the Robertson–Walker universe is restricted to *discrete values* as a consequence of the interaction of the classical gravitational field with the quantized scalar field. This example indicates that some of the features that one would expect from a fully quantized theory, like discrete values for the scale factor, can already be present in the corresponding hybrid system.

We discuss this cosmological model in two approximations: first, using a min-isuperspace model in which the space of fields is replaced by a finite dimensional configuration space, and then using the midisuperspace model of spherical symmet-ric gravity that we discussed in the previous section. In both cases, we substantially follow the exposition given in Ref. [22].

11.3.1   Minisuperspace Hybrid Cosmological Model

The line element of a closed Robertson–Walker universe can be written in the form

$$ds^2 = -N^2(t)\, dt^2 + a^2(t)\, d\Omega^2, \tag{11.17}$$

where N is the lapse function, a is the scale factor and $d\Omega^2$ is the standard line element on S^3. This form corresponds to a special choice of foliation that is adapted to the symmetry of the model, and for this reason the shift vector does not appear in the line element. Symmetry reduction allows for a minisuperspace model in which the space of fields is replaced by a finite dimensional configuration space [19]. This leads to equations that are much simpler than the ones of the full theory. One must keep in mind however that the predictions of minisuperspace models have to be treated with some care, because the restriction to a finite number of degrees of freedom is a drastic reduction of the infinitely many degrees of freedom of a field theory. For this reason it is desirable to check the conclusions derived from the minisuperspace model. We do this in the next section where we consider the corresponding midisuperspace solution and show that it is in agreement with the main predictions of the minisuperspace solution.

The case of a closed Robertson–Walker universe with a massive scalar field is treated in some detail in Ref. [19] and we refer the reader to this monograph for a more complete discussion. Here we will take as our starting point the Hamilton–Jacobi equation derived in this reference. The configuration space has two coordinates, which we call a and ϕ. We already introduced the coordinate a in Eq. (11.17), it is the scale factor, and the coordinate ϕ represents the scalar field. For simplicity, we restrict to a potential term that is quadratic in ϕ. Then, the classical Hamilton–Jacobi equation takes the form $H_{\phi a}^C = 0$ with

$$H_{\phi a}^C = -\zeta \frac{1}{a}\left(\frac{\partial S}{\partial a}\right)^2 + \frac{1}{a^3}\left(\frac{\partial S}{\partial \phi}\right)^2 - \frac{1}{\zeta}a + \zeta\frac{\lambda a^3}{3} + m^2 a^3 \phi^2 \qquad (11.18)$$

(in units where $c = 1$), where m is the mass of the scalar field, the constant $\zeta = 8\pi G/3V_0$ is proportional to the gravitational constant G, and $V_0 = 2\pi^2$ is the volume of S^3 [19]. In order to simplify the equations of the minisuperspace model and make them more readable, we will choose units in this section for which $\zeta = 1$ (i.e., we will not set $G = 1$ in this section).

The ensemble Hamiltonian for the minisuperspace model describing a quantized field interacting with the gravitational field can be written in the form [21]

$$\mathcal{H}_{\phi a} = \int \mathrm{d}a\mathrm{d}\phi P\left[H_{\phi a}^C + F_\phi\right] = \int \mathrm{d}a\mathrm{d}\phi P\left[H_{\phi a}^C + \frac{\hbar^2}{4}\frac{1}{a^3}\left(\frac{\partial \log P}{\partial \phi}\right)^2\right]. \quad (11.19)$$

Setting $\zeta = 1$ in Eq. (11.18) and imposing the constraints $\frac{\partial S}{\partial t} = \frac{\partial P}{\partial t} = 0$, we get the equations

$$-\frac{1}{a}\left(\frac{\partial S}{\partial a}\right)^2 + \frac{1}{a^3}\left(\frac{\partial S}{\partial \phi}\right)^2 - a + \frac{\lambda a^3}{3} + m^2 a^3 \phi^2 - \frac{\hbar^2}{a^3}\frac{1}{A}\frac{\partial^2 A}{\partial \phi^2} = 0 \qquad (11.20)$$

and

$$-\frac{\partial}{\partial a}\left(\frac{P}{a}\frac{\partial S}{\partial a}\right) + \frac{\partial}{\partial \phi}\left(\frac{P}{a^3}\frac{\partial S}{\partial \phi}\right) = 0, \qquad (11.21)$$

where $A \equiv \sqrt{P}$.

An *exact solution* can be derived for the case $S = 0$. This is a highly non-classical solution, in that it leads to discrete values of the scale factor which exclude the singularity, as we show below.

When $S = 0$, Eq. (11.20) reduces to

$$-\frac{\hbar^2}{a^3}\frac{1}{A}\frac{\partial^2 A}{\partial \phi^2} - a + \frac{\lambda a^3}{3} + a^3 m^2 \phi^2 = 0 \tag{11.22}$$

while Eq. (11.21) is automatically satisfied. Equation (11.22) has the form of a time-independent Schrödinger equation with quadratic potential for the real function A (see the discussion on stationary ensembles in Sect. 8.3). The non-negative, normalizable solutions take the form

$$P_n(\phi, a) = \delta(a - a_n)\frac{\alpha_n}{\sqrt{\pi}2^n n!}\exp\left(-\alpha_n^2 \phi^2\right)[H_n(\alpha_n \phi)]^2 \tag{11.23}$$

where the H_n are Hermite polynomials, $\alpha_n^2 = a_n^3 m/\hbar$, and the a_n satisfy the condition

$$a_n - \frac{\lambda a_n^3}{3} = 2\hbar m\left(n + \frac{1}{2}\right) \tag{11.24}$$

for $n = \{0, 1, 2, \ldots\}$. If the term proportional to the cosmological constant λ can be neglected, the quantization condition takes the simple form $a_n = 2\hbar m\left(n + \frac{1}{2}\right)$.

Notice that while the transformation $A = \sqrt{P}$ leads, via Eq. (11.22), to a Schrödinger equation for A, it is not possible to introduce solutions that are linear superpositions of the A_n because the potential term in the equation ends up being a function of a_n.

This solution that has been derived in this section has some remarkable features. One can see that the coupling of the quantum field to a purely classical metric leads to a *quantization condition* for the scale factor a. Furthermore, the classical singularity at $a = 0$ is *excluded* from these solutions. This is also true of the corresponding midisuperspace solution that we consider in the next section.

11.3.2 Midisuperspace Hybrid Cosmological Model

We now discuss the same cosmological model as before but this time in terms of the midisuperspace formulation of spherically symmetric gravity. The equations that need to be solved are Eqs. (11.15–11.16). We want to consider a class of solutions that is analogous to the minisuperspace solutions described in the previous section. To do this, we look for a solution that satisfies the following two requirements: the condition $S = 0$ holds, and the solution is one that is adapted to a foliation of spaces of constant positive curvature with constant lapse function N.

With these two assumptions, Eq. (11.15) reduces to a single Schrödinger functional equation for $A = \sqrt{P}$,

$$-\frac{1}{2\Lambda R^2}\left(\frac{\delta^2 A}{\delta\phi^2}\right) + \left[\lambda\frac{\Lambda R^2}{2} + V + \frac{R^2}{2\Lambda}\phi'^2 + \frac{\Lambda R^2 m^2}{2}\phi^2\right]A = 0, \qquad (11.25)$$

while Eq. (11.16), is automatically satisfied. This is analogous to the situation that we encountered in the minisuperspace model.

To solve Eq. (11.25), one can apply standard techniques developed for the Schrödinger functional representation of quantum field theory [19, 23–25]. We derive here an explicit solution which corresponds to the lowest state of the minisuperspace model which was considered in the previous section. In this case, A will be a ground state Gaussian functional; i.e.,

$$A_{(0)} \sim \exp\left\{-\frac{1}{2}\int\int dy\,dz\,\Lambda_y\,\Lambda_z\,R_y^2\,R_z^2\,\phi_y\,K_{yz}\,\phi_z\right\}. \qquad (11.26)$$

Instead of $A_{(0)}$, one may also consider the excited states which solve the functional Schrödinger equation. We discuss the consequences of making this alternative choice at the end of this section.

The derivation of this solution is given in the Appendix to this chapter. One can show that the equation that determines the functional $A_{(0)}$ can be mapped to a functional Schrödinger equation in a space of constant curvature and the kernel K_{xy} can be expressed in the simple form

$$K_{xy} = \frac{1}{2a_0^4}\sum_n \sqrt{\gamma_n}\,\psi_x^{(n)}\psi_y^{(n)}, \qquad (11.27)$$

where the $\psi_r^{(n)}$ are solutions of a time-independent Schrödinger-type equation in a space of constant curvature,

$$-\frac{1}{\sin^2 r}\frac{\partial}{\partial r}\left(\sin^2 r\,\frac{\partial\psi_r^{(n)}}{\partial r}\right) + m^2 a_0^2\,\psi_r^{(n)} = \gamma_n\psi_r^{(n)}. \qquad (11.28)$$

The eigenvalues γ_n are given by

$$\gamma_n = n^2 - 1 + m^2 a_0^2, \quad n = 1, 2, 3\ldots. \qquad (11.29)$$

As shown in the Appendix to this chapter, in principle it is possible to derive an expression for a_0 which depends on the cosmological constant λ and on the energy $E_{(0)}$ of the quantized scalar field, where the latter is given by [25]

$$E_{(0)} \sim a_0^3\int dr\,\sin^2 r\,K_{rr}. \qquad (11.30)$$

However, $\int dr\,\sin^2 r\,K_{rr} \sim \sum_n \gamma_n$, which diverges. This is a consequence of the infinite zero-point energy of the quantum field. Thus it becomes necessary to use a renormalization procedure to extract a finite result for a_0. In analogy to the minisuperspace solution discussed in the previous section, one would expect $a_0 > 0$ although the precise value of a_0 will depend on the details of the renormalization procedure.

We will not discuss possible renormalization procedures for this solution here. However, we point out that for this particular case, the equation for $A_{(0)}$ is similar in form to a functional Schrödinger equation for a quantum scalar field in an Einstein universe, so it is possible to use previous results from the literature where such renormalization procedures have been carried out (e.g., Ref. [26]). Note that this is not the generic case: since the equations of hybrid cosmology are non-linear, in general they will not map to a Schrödinger-type functional equation. The simplification that is achieved here is a direct consequence of choosing $S = 0$. With a different ansatz, the equations can not be solved in this way.

A similar analysis may be carried out where the ground state functional $A_{(0)}$ is replaced by an excited state. Consider a first excited state $A_{(1)}$ specified by the eigenfunction $\psi_r^{(n)}$. This state will differ in energy from the (divergent) ground state energy by a finite amount $\Delta E(n)$ which depends on the value of γ_n, with $\Delta E(n) = \sqrt{\gamma_n}$ [25]. Notice that $\Delta E(n)$ can only take discrete values because γ_n is quantized, and this means that a_0 will also be restricted to *discrete values*. Therefore, the coupling of the quantum scalar field and classical gravitational field leads to the quantization of the radius of the universe, not only for the minisuperspace model but also for the midisuperspace model.

11.3.3 Discussion

It is of interest to consider solutions for the case of potentials that include other ϕ-dependent terms in addition to the term quadratic in ϕ that we have considered in the minisuperspace and midisuperspace models. In all cases, the ansatz $S = 0$ will lead to equations that reduce to the form of a time-independent Schrödinger equation with a modified potential term that is a function of a. Consider for simplicity the minisuperspace model. In this case, we will have an energy term given by $E = a - \lambda a^3/3$. If the solution of this Schrödinger equation only admits discrete energy levels E_n, we will be lead again to a quantization condition for the scale factor, of the form $E_n = a_n - \lambda a_n^3/3$. Thus, the quantization of the scale factor is a generic feature of such models. In particular, consider the case in which the modified potential remains non-negative and $\lambda = 0$. Then, the ground state energy E_n is strictly positive, and the minimum value of the scale factor is given by $a_0 = E_0$. Hence, the quantum fluctuations associated with the matter field in the ground state may be interpreted as being directly responsible for removing the classical singularity.

Referring again to the miniusperspace model with $S = 0$, we point out that the solutions $\{P_n\}$ that correspond to a given modified potential have an interesting property that might be of relevance to discussions of the problem of the arrow of time. It has previously been argued by Zeh and Kiefer that, with appropriate initial conditions, the solutions of the Wheeler–DeWitt equation for a quantized

Robertson–Walker model with small perturbations will have the property that the entropy, suitably defined, increases with increasing scale factor (this is connected to the asymmetry of the potential term with respect to the intrinsic time and, in particular, to the property that the potential vanishes as $a \to 0$) [19, 27, 28]. Thus, while there is no external time parameter in quantum cosmology, one may introduce an intrinsic time parameter defined in terms of the radius a (or any increasing function of a such as $\log a$). In a similar way, there is a natural ordering of the solutions $\{P_n\}$, in terms of a discrete time variable given by n, and this ordering leads to a "thermodynamic" arrow of time. This follows from the observation that the amount of structure associated with a solution P_n (as determined, for example, by counting the number of nodes in A_n for different values of n) increases with increasing n.

Hybrid cosmologies which allow for the coupling of quantum matter fields to classical gravitational fields can have solutions which display striking non-classical features, such as discrete values for the radius of the universe and the elimination of classical singularities.

11.4 CGHS Black Hole in the Presence of a Quantized Scalar Field

Our second example of the coupling of quantum fields to classical gravity concerns black holes. We consider the CGHS black hole [29], a particular black hole model which arises in dilaton gravity. This is a model of gravity in a two-dimensional spacetime which contains a scalar field ϕ, known as the dilaton, and a parameter λ which is similar to the cosmological constant of general relativity.

Despite its simplicity, the CGHS black hole captures many of the the essential features of general relativistic black holes [30, 31]. We show in particular that we recover Hawking radiation from the equations that describe a hybrid system consisting of a classical CGHS black hole in a collapsing geometry interacting with a quantized scalar field. We follow closely the methods in Demers and Kiefer [31].

11.4.1 CGHS Black Hole and Classical Massless Scalar Field

The action for the CGHS black coupled to a classical massless scalar field can be written in the form [31]

$$S = \int dx\, dt\, \sqrt{-g}\left[\frac{1}{G}\left(R\phi + 4\lambda^2\right) - \frac{1}{2}\left(\nabla f\right)^2\right], \qquad (11.31)$$

where G is the gravitational coupling constant, which is dimensionless in two dimensions, ϕ is the dilaton field, f the massless scalar field, and the parameter λ plays the role of the cosmological constant. The action of Eq. (11.31) can be derived from the one given in the original CGHS black hole paper [29],

$$S = \int dx\, dt\, \sqrt{-\tilde{g}}\left[\frac{1}{G}\, e^{-2\tilde{\phi}}\left(\tilde{R} + 4\left(\widetilde{\nabla}\phi\right)^2 + 4\lambda^2\right) - \frac{1}{2}\left(\widetilde{\nabla}f\right)^2\right], \qquad (11.32)$$

via the transformation $\phi = e^{-2\tilde{\phi}}$ and $g_{\alpha\beta} = e^{-2\tilde{\phi}}\tilde{g}_{\alpha\beta}$ [31]. This transformation has the advantage of eliminating the dilaton kinetic energy term which appears in Eq. (11.32).

We write the line element in the form [31]

$$ds^2 = e^{2\rho}\left[-\sigma^2 dt^2 + (dx + \xi\, dt)^2\right]. \qquad (11.33)$$

This form of the metric corresponds to a 1+1 decomposition of the two-dimensional space-time with lapse function σ and shift function ξ. Notice that Eq. (11.33) does not follow the standard convention for a 1+1 decomposition because there is an overall multiplicative factor $e^{2\rho}$.

One may introduce a Hamiltonian formulation using the action of Eq. (11.31) and the 1+1 decomposition of space-time defined by Eq. (11.33) [31, 32], and from this Hamiltonian formulation one can derive Hamilton–Jacobi equations for the system. We go directly to the Hamilton–Jacobi formulation of the theory, since the Hamiltonian formulation will not be needed. As expected, there are two constraint equations that need to be satisfied, a Hamiltonian constraint and a momentum constraint. The Hamiltonian constraint takes the form

$$H_{\phi\rho f}^{C} = -\frac{G}{2}\frac{\delta S}{\delta\phi}\frac{\delta S}{\delta\rho} + \frac{1}{2G}V_G + \frac{1}{2}\left(\frac{\delta S}{\delta f}\right)^2 + V_M = 0, \qquad (11.34)$$

where the potential terms are given by

$$V_G = 4\left(\phi'' - \phi'\rho' - 2\lambda^2 e^{2\rho}\right), \quad V_M = \frac{1}{2}f', \qquad (11.35)$$

and the primes are used to indicate derivatives with respect to x. The momentum constraint is given by

$$\rho'\frac{\delta S}{\delta\rho} - \left(\frac{\delta S}{\delta\rho}\right)' + \phi'\frac{\delta S}{\delta\phi} + f'\frac{\delta S}{\delta f} = 0. \qquad (11.36)$$

It is equivalent to the requirement that S must be invariant under coordinate transformations.

For future reference, we write down the rate equations for the fields [31],

$$\dot{\rho} = -\frac{\sigma G}{2}\frac{\delta S}{\delta \phi} + \xi \rho' + \xi',$$ (11.37)

$$\dot{\phi} = -\frac{\sigma G}{2}\frac{\delta S}{\delta \rho} + \xi \phi'.$$ (11.38)

$$\dot{f} = \frac{\sigma}{2}\frac{\delta S}{\delta f} + \xi \phi'.$$ (11.39)

In the next section we will consider a CGHS black hole coupled to a quantized massless scalar field, but before we do that it will be instructive to look at how one can define ensembles on configuration space for the purely classical system. It is straightforward to do this using the general procedure that we introduced in Chap. 10 for classical ensembles of gravitational fields.

An appropriate ensemble Hamiltonian is given by

$$\mathscr{H}^C_{\phi\rho f} = \int dx\,\sigma \int D\phi D\rho Df\; P H^C_{\phi\rho f},$$ (11.40)

with corresponding equations of motion

$$\frac{\partial P}{\partial t} = \frac{\Delta \mathscr{H}^C_{\phi\rho f}}{\Delta S}, \quad \frac{\partial S}{\partial t} = -\frac{\Delta \mathscr{H}^C_{\phi\rho f}}{\Delta P}.$$ (11.41)

Taking into consideration that σ is an arbitrary function of x and making use of the conditions $\frac{\partial S}{\partial t} = \frac{\partial P}{\partial t} = 0$, the equations of motion are the Hamiltonian constraint, Eq. (11.34), and a continuity equation which takes the form

$$\frac{G}{2}\left[\frac{\delta}{\delta\phi}\left(P\frac{\delta S}{\delta\rho}\right) + \frac{\delta}{\delta\rho}\left(P\frac{\delta S}{\delta\phi}\right)\right] - \frac{\delta}{\delta f}\left(P\frac{\delta S}{\delta f}\right) = 0.$$ (11.42)

One must also impose the restriction that S be invariant under coordinate transformations to ensure that the momentum constraint, Eq. (11.36), is automatically satisfied.

11.4.2 CGHS Black Hole and Quantized Massless Scalar Field

We now consider a hybrid system which consists of a classical CGHS black hole coupled to a quantized scalar field. To do this, we use the general procedure that we have developed in this chapter. An appropriate ensemble Hamiltonian is obtained by modifying the expression of Eq. (11.40) according to

$$\mathcal{H}_{\phi\rho f} = \int dx\, \sigma \int D\phi D\rho Df\, P\left[H^C_{\phi\rho f} + F_f\right] \tag{11.43}$$

where

$$F_f = \frac{1}{2}\frac{\hbar^2}{4}\left(\frac{\delta \log P}{\delta f}\right)^2. \tag{11.44}$$

As we have done before, we will require the conditions $\frac{\partial S}{\partial t} = \frac{\partial P}{\partial t} = 0$ and assume that S is invariant under coordinate transformations, so that the momentum constraint, Eq. (11.36), is satisfied.

Taking into consideration that σ is an arbitrary function of x, the equations that follow from the ensemble Hamiltonian of Eq. (11.43) are

$$H_{\phi\rho f} = -\frac{G}{2}\frac{\delta S}{\delta \phi}\frac{\delta S}{\delta \rho} + \frac{1}{2}\left(\frac{\delta S}{\delta f}\right)^2 + \frac{1}{2G}V_G + \frac{1}{2}f'^2$$
$$+ \frac{\hbar^2}{8}\left[\frac{1}{P^2}\left(\frac{\delta P}{\delta f}\right)^2 - \frac{2}{P}\frac{\delta^2 P}{\delta f^2}\right] = 0, \tag{11.45}$$

and

$$\frac{G}{2}\left[\frac{\delta}{\delta \phi}\left(P\frac{\delta S}{\delta \rho}\right) + \frac{\delta}{\delta \rho}\left(P\frac{\delta S}{\delta \phi}\right)\right] - \frac{\delta}{\delta f}\left(P\frac{\delta S}{\delta f}\right) = 0. \tag{11.46}$$

This last equation is identical to the continuity equation that we derived for the classical case in the previous section.

To find solutions to these equations we expand S in powers of G,

$$S = \frac{1}{G}S_0 + S_1 + \dots \tag{11.47}$$

This is a Born–Oppenheimer type of expansion with respect to the gravitational constant [31]. We first look for an approximate solution correct to order G^0. As shown below, it is convenient to use the product rule of probability theory to express P in the form

$$P = P[\rho, \phi]\, P[f|\rho, \phi] =: P_0[\rho, \phi]\, P_1[f, \rho, \phi] \tag{11.48}$$

where $P[f|\rho, \phi] = P_1[f, \rho, \phi]$ is the conditional probability of f given ρ, ϕ.

At order G^{-2} we get the equation

$$\frac{1}{2}\left(\frac{\delta S_0}{\delta f}\right)^2 = 0, \tag{11.49}$$

which implies that S_0 is not a functional of f, $S_0 = S_0[\rho, \phi]$.

At order G^{-1} we get the two equations

$$-\frac{G}{2}\frac{\delta S_0}{\delta\phi}\frac{\delta S_0}{\delta\rho} + \frac{1}{2G}V_G = 0, \tag{11.50}$$

$$-\frac{\delta}{\delta f}\left(P_0 P_1 \frac{\delta S_0}{\delta f}\right) = 0. \tag{11.51}$$

Equation (11.50) has the form of Eq. (11.34) but in the absence of the massless scalar field f. Thus S_0 solves the Hamilton–Jacobi functional equation for the CGHS black hole *without* matter fields. Equation (11.51) is automatically satisfied due to Eq. (11.49).

At order G^0 we get the two equations

$$-\frac{1}{2}\left(\frac{\delta S_0}{\delta\phi}\frac{\delta S_1}{\delta\rho} + \frac{\delta S_0}{\delta\rho}\frac{\delta S_1}{\delta\phi}\right) + \frac{1}{2}\left(\frac{\delta S_1}{\delta f}\right)^2 + \frac{1}{2}f'^2$$
$$+\frac{\hbar^2}{8}\left[\frac{1}{P_1^2}\left(\frac{\delta P_1}{\delta f}\right)^2 - \frac{2}{P_1}\frac{\delta^2 P_1}{\delta f^2}\right] = 0 \tag{11.52}$$

and

$$P_0\left[\frac{1}{2}\left(\frac{\delta P_1}{\delta\phi}\frac{\delta S_0}{\delta\rho} + \frac{\delta P_1}{\delta\rho}\frac{\delta S_0}{\delta\phi}\right) - \frac{\delta}{\delta f}\left(P_1\frac{\delta S_1}{\delta f}\right)\right]$$
$$+P_1\left[\frac{\delta}{\delta\phi}\left(P_0\frac{\delta S_0}{\delta\rho}\right) + \frac{\delta}{\delta\rho}\left(P_0\frac{\delta S_0}{\delta\phi}\right)\right] = 0. \tag{11.53}$$

We now show that we can recover the Schrödinger functional equation for a scalar field from the equations of up to order G^0. This requires fixing a gauge. We choose the conformal gauge, defined by $\sigma = 1$ and $\xi = 0$. Then, the rate equations for the fields, Eqs. (11.37) and (11.38), simplify to

$$\dot\rho = -\frac{1}{2}\frac{\delta S_0}{\delta\phi}, \dot\phi = -\frac{1}{2}\frac{\delta S_0}{\delta\rho}, \tag{11.54}$$

which leads immediately to

$$-\frac{1}{2}\left(\frac{\delta S_0}{\delta\phi}\frac{\delta S_1}{\delta\rho} + \frac{\delta S_0}{\delta\rho}\frac{\delta S_1}{\delta\phi}\right) = \left(\dot\rho\frac{\delta S_1}{\delta\rho} + \dot\phi\frac{\delta S_1}{\delta\phi}\right), \tag{11.55}$$

$$\frac{1}{2}\left(\frac{\delta P_1}{\delta\phi}\frac{\delta S_0}{\delta\rho} + \frac{\delta P_1}{\delta\rho}\frac{\delta S_0}{\delta\phi}\right) = -\left(\frac{\delta P_1}{\delta\phi}\dot\phi + \frac{\delta P_1}{\delta\rho}\dot\rho\right). \tag{11.56}$$

We have seen that S_0 solves the Hamilton–Jacobi functional equation for the CGHS black hole without matter fields, Eq. (11.50). Therefore, it is natural to choose P_0 so that it solves the continuity equation for an ensemble of CGHS black holes where there is no coupling to matter fields,

$$\frac{\delta}{\delta\phi}\left(P_0\frac{\delta S_0}{\delta\rho}\right) + \frac{\delta}{\delta\rho}\left(P_0\frac{\delta S_0}{\delta\phi}\right) = 0. \tag{11.57}$$

This is always possible because there are no other restrictions on P_0.

Using Eqs. (11.55–11.57), the equations of order G^0, Eqs. (11.52–11.53), take the simpler form

$$\left(\dot\rho\frac{\delta S_1}{\delta\rho} + \dot\phi\frac{\delta S_1}{\delta\phi}\right) + \frac{1}{2}\left(\frac{\delta S_1}{\delta f}\right)^2 + \frac{1}{2}f'^2$$
$$+ \frac{\hbar^2}{8}\left[\frac{1}{P_1^2}\left(\frac{\delta P_1}{\delta f}\right)^2 - \frac{2}{P_1}\frac{\delta^2 P_1}{\delta f^2}\right] = 0 \tag{11.58}$$

and

$$\left(\frac{\delta P_1}{\delta\phi}\dot\phi + \frac{\delta P_1}{\delta\rho}\dot\rho\right) + \frac{\delta}{\delta f}\left(P_1\frac{\delta S_1}{\delta f}\right) = 0. \tag{11.59}$$

If we now integrate Eqs. (11.58–11.59) with respect to the x coordinate and use the equalities

$$\dot S_1 = \int dx\left(\dot\rho\frac{\delta S_1}{\delta\rho} + \dot\phi\frac{\delta S_1}{\delta\phi}\right), \quad \dot P_1 = \int dx\left(\frac{\delta P_1}{\delta\phi}\dot\phi + \frac{\delta P_1}{\delta\rho}\dot\rho\right), \tag{11.60}$$

we are led to

$$\dot S_1 + \int dx\left\{\frac{1}{2}\left(\frac{\delta S_1}{\delta f}\right)^2 + \frac{1}{2}f'^2 + \frac{\hbar^2}{8}\left[\frac{1}{P_1^2}\left(\frac{\delta P_1}{\delta f}\right)^2 - \frac{2}{P_1}\frac{\delta^2 P_1}{\delta f^2}\right]\right\} = 0 \tag{11.61}$$

and

$$\dot P_1 + \int dx\left\{\frac{\delta}{\delta f}\left(P_1\frac{\delta S_1}{\delta f}\right)\right\} = 0. \tag{11.62}$$

The rate equations for P_1 and S_1, Eqs. (11.61–11.62), can be derived from the ensemble Hamiltonian

$$\mathcal{H}_f^Q = \int dx\int Df P_1\left[\frac{1}{2}\left(\frac{\delta S_1}{\delta f}\right)^2 + \frac{1}{2}f'^2 + \frac{\hbar^2}{8}\left(\frac{\delta\log P_1}{\delta f}\right)^2\right]. \tag{11.63}$$

We now go to the wavefunction representation via the complex canonical transformation $\Psi = \sqrt{P_1}e^{iS_1/\hbar}$, $\bar\Psi = \sqrt{P_1}e^{-iS_1/\hbar}$. The ensemble Hamiltonian $\mathcal{H}_f^Q$ is mapped to

$$\mathcal{H}_f^Q = \int dx\int Df\frac{1}{2}\left(\hbar^2\frac{\delta\bar\Psi}{\delta f}\frac{\delta\Psi}{\delta f} + f'^2\bar\Psi\Psi\right), \tag{11.64}$$

with a corresponding equation for Ψ given by

$$i\hbar\dot{\Psi} = \int dx\, \frac{1}{2}\left(-\hbar^2\frac{\delta^2}{\delta f^2} + f'^2\right)\Psi =: \hat{H}_f\Psi. \tag{11.65}$$

This is precisely the Schrödinger functional equation for a quantized free scalar field f on a flat background. In this way, the perturbative expansion to order G^0 leads to an approximate solution that describes a quantized scalar field on the background space-time of the CGHS black hole. Notice that the information about the gravitational field comes through the definition of the time via the variables ρ and ϕ [31], since the time that enters into the Schrödinger functional equation is a "gravitational time" defined via Eq. (11.60).

It is instructive to compare this derivation to the one in the paper by Demers and Kiefer which takes as its starting point the Wheeler–DeWitt equation of quantum gravity [31]. While the end result of the perturbative expansion to order G^0 is the same in both cases (i.e., the recovery of the Schrödinger functional equation), the derivation that we present here is perhaps more straightforward in that certain problematic terms which appear in the expansion of the Wheeler–DeWitt equation are absent from our analysis: namely, the terms that arise from the quantization of gravity, which do not appear in our model because in our case the CGHS black hole remains classical. Thus there are already differences between the hybrid model and quantum gravity after considering only the first three terms of the expansion.

11.4.3 CHGS Black Hole Formation Through Collapse of Matter and Hawking Radiation

The perturbative analysis of the previous section shows that the solution of the equations to order G^0 describes a quantized scalar field on the the background space-time of the CGHS black hole. Following Demers and Kiefer [31], we now consider a "collapsing" scenario and show that we recover Hawking radiation from the corresponding equations.

11.4.3.1 Formation of a CGHS Black Hole by Gravitational Collapse

We model the collapsing space-time in terms of a shockwave of classical matter which causes the formation of the black hole.

We first consider a black hole without matter fields, described in terms of the action

$$S = \int dx\, dt\, \sqrt{-\tilde{g}}\left[\frac{1}{G}\,e^{-2\tilde{\phi}}\left(\tilde{R} + 4\left(\widetilde{\nabla}\phi\right)^2 + 4\lambda^2\right)\right], \tag{11.66}$$

which corresponds to the action given by Eq. (11.32) for the case in which there is no scalar field f. It will be convenient to choose the conformal gauge, so that the line element can be put in the form

$$ds^2 = -e^{2\tilde{\rho}}dx^+dx^-, \tag{11.67}$$

where we introduced light-cone coordinates $x^{\pm} = t \pm x$.

The equations for $\tilde{\phi}$ and $\tilde{\rho}$ that follow from the action of Eq. (11.66) with the line element of Eq. (11.67) are given by [30]

$$-4\partial_+\partial_-\tilde{\phi} + 4\partial_+\tilde{\phi}\partial_-\tilde{\phi} + 2\partial_+\partial_-\tilde{\rho} + \lambda^2 e^{2\tilde{\rho}} = 0, \tag{11.68}$$

$$2\partial_+\partial_-\tilde{\phi} - 4\partial_+\tilde{\phi}\partial_-\tilde{\phi} - \lambda^2 e^{2\tilde{\rho}} = 0, \tag{11.69}$$

and

$$4\partial_+\tilde{\phi}\partial_+\tilde{\rho} - 2\partial_+^2\tilde{\phi} = 0, \tag{11.70}$$

$$4\partial_-\tilde{\phi}\partial_-\tilde{\rho} - 2\partial_-^2\tilde{\phi} = 0. \tag{11.71}$$

Equations (11.68–11.69) follow from the variation of the action, while Eqs. (11.70–11.71) are constraints that arise as a consequence of having set $g_{++} = g_{--} = 0$ in the metric. To find an explicit form for the CGHS black hole metric, we need to solve for $\tilde{\phi}$ and $\tilde{\rho}$.

The solutions to these equations take a simple form if we set $\tilde{\rho} = \tilde{\phi}$. This can be done without loss of generality, and one can check that it leads to the black hole (BH) solution [30, 31]

$$e^{-2\tilde{\rho}} = e^{-2\tilde{\phi}} = \frac{M}{\lambda} - \lambda^2 x^+ x^- \tag{11.72}$$

with line element

$$ds^2_{BH} = -\left(\frac{M}{\lambda} - \lambda^2 x^+ x^-\right)^{-1} dx^+ dx^-. \tag{11.73}$$

The parameter M is the ADM mass of the black hole. The linear dilaton vacuum (LDV) solution is obtained by setting $M = 0$, with line element

$$ds^2_{LDV} = \left(\lambda^2 x^+ x^-\right)^{-1} dx^+ dx^-. \tag{11.74}$$

To describe a collapsing CGHS geometry, it is necessary to introduce matter fields. It will be sufficient to add a term to the action of Eq. (11.66) which describes a massless classical field $\tilde{\theta}$. This leads to the action

$$S = \int dx \, dt \, \sqrt{-\tilde{g}} \left[\frac{1}{G} e^{-2\tilde{\phi}} \left(\tilde{R} + 4 \left(\widetilde{\nabla\phi} \right)^2 + 4\lambda^2 + \right) - \frac{1}{2} (\widetilde{\nabla\theta})^2 \right]. \quad (11.75)$$

We consider a scenario on which the collapse is produced by a shock wave of classical matter θ. The Penrose diagram for this collapsing scenario is shown in Fig. 11.1. The shock wave is imparted at $\lambda x^+ = 1$. The region with $\lambda x^+ < 1$ is the LDV region, the region with $\lambda x^+ > 1$ is the BH region. (The diagram also shows the overlapping slices at $t_y = t_v = 0$ which are used in the next section when comparing the vacuums states of the two regions.)

For this scenario, the solution takes the form [30, 31]

$$e^{-2\tilde{\rho}} = e^{-2\tilde{\phi}} = \frac{M}{\lambda} (1 - \lambda x^+) \Theta(\lambda x^+ - 1) - \lambda^2 x^+ x^-$$

$$= \frac{M}{\lambda} \Theta(\lambda x^+ - 1) - \lambda^2 x^+ \left[x^- + \frac{M}{\lambda^2} \Theta(\lambda x^+ - 1) \right] \quad (11.76)$$

where $\Theta(x)$ is the step function. Thus for $\lambda x^+ < 1$ we have the LDV solution while for $\lambda x^+ > 1$ we have the BH solution with black hole mass M (after shifting the coordinate x^- by M/λ^2), as required.

Having described the collapsing geometry, we now introduce coordinates in each region which can be used to define the notion of vacuum states for a quantized scalar field on this background geometry. This requires inertial coordinates; i.e., coordinates which exhibit explicitly the asymptotic flatness of the metric. In the LDV region we define light-cone coordinates $y^\pm = t_y \pm y$ according to

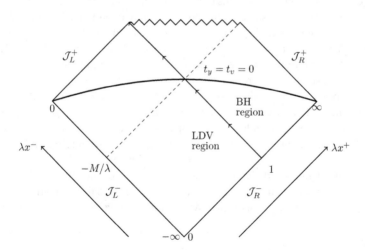

Fig. 11.1 Penrose diagram for collapsing CGHS black hole. The shock wave is indicated by the arrowed line, the horizon by the dashed line. The comparison between the vacuum states of the two regions is done at the overlapping slices labeled $t_y = t_v = 0$

$$\lambda x^+ = e^{\lambda y^+}, \, \lambda x^- = -\frac{M}{\lambda} e^{-\lambda y^-}, \tag{11.77}$$

and in the BH region we define light-cone coordinates $v^\pm = t_v \pm v$ according to

$$\lambda x^+ = 1 + \frac{\lambda}{M} e^{\lambda v^+}, \, \lambda x^- = -\frac{M}{\lambda} - e^{-\lambda v^-}. \tag{11.78}$$

One can check that the metric in each region is asymptotically flat for these choice of coordinates; i.e., $ds^2 \to -dy^+ dy^-$ in the LDV region and $ds^2 \to -dv^+ dv^-$ in the BH region.

11.4.3.2   Hawking Radiation

The main aim of this section is to show explicitly the emergence of Hawking radiation in the hybrid model.

The presence of Hawking radiation in the collapsing geometry can be established by considering solutions of Eq. (11.65). The main idea is to start from a vacuum state for the scalar field in the absence of a black hole, let it evolve according to the functional Schrödinger equation, and compare it with the vacuum solution in the geometry that corresponds to the case of a black hole which has formed by a collapse process. The notion of vacuum state is defined here with respect to inertial coordinates; i.e., coordinates which exhibit explicitly the asymptotic flatness of the metric. The collapsing geometry can be realized by assuming the presence of a shock wave of classical matter which forms the black hole, as we did in the previous section.

We use essentially the same approach that Demers and Kiefer applied to the Wheeler–DeWitt equation [31]. For this reason, we limit ourselves to a summary of results rather than a detailed derivation and refer the reader to the paper of Demers and Kiefer for a more thorough account which includes a number of intermediate steps that we have omitted here.

The key point of the analysis is the formulation of the appropriate *boundary conditions*. In the LDV region (where timelike and spacelike directions are labeled t_y and y), we impose the boundary condition $f(v) \to 0$ as $v \to \pm\infty$. This means that f has to vanish at the origin $y = 0$. In the BH region (where timelike and spacelike directions are labeled t_v and v), $f(v)$ has no restrictions at $v = 0$ [31].

Following Demers and Kiefer, we introduce Fourier transforms of the field f (and set $\hbar = 1$ in this section to facilitate the comparison to their equations). Due to the difference in boundary conditions, we get different expressions for f in the two regions,

$$\text{LDV region}: f(y) = \sqrt{\frac{2}{\pi}} \int_0^\infty dk \, F(k) \, \sin(ky), \tag{11.79}$$

$$\text{BH region}: f(v) = \sqrt{\frac{1}{2\pi}} \int_{-\infty}^\infty dk \, G(k) \, e^{ikv}, \tag{11.80}$$

where $F(k)$ and $G(k)$ are the Fourier transforms of f in the LDV and BH regions, respectively. Note that $F(k)$ must be real, $F(k) = \bar{F}(k)$, while $G(k)$ is complex.

The Schrödinger functional equation can be written in terms of F and G as

$$\text{LDV region : } i\frac{\partial \Psi}{\partial t_y} = \frac{1}{2}\int_0^\infty dk \left(-\frac{\delta^2}{\delta F^2} + k^2 F^2\right)\Psi, \tag{11.81}$$

$$\text{BH region : } i\frac{\partial \Psi}{\partial t_v} = \frac{1}{2}\int_{-\infty}^\infty dk \left(-\frac{\delta^2}{\delta G \delta \bar{G}} + k^2 |G|^2\right)\Psi. \tag{11.82}$$

At $t_y = t_v = 0$ (see Fig. 11.1), the space slices overlap and we can compare the solutions. For the ground state solutions (which take the form of Gaussian functionals) we get [31]

$$\text{LDV region : } \Psi_0[f, t_y = 0] = N \exp\left(-\frac{1}{2\hbar}\int_0^\infty dk\, kF^2(k)\right), \tag{11.83}$$

$$\text{BH region : } \Psi[f, t_v = 0] = N \exp\left(-\frac{1}{\hbar}\int_{-\infty}^\infty dk\, |k||G|^2(k)\right). \tag{11.84}$$

The state described by Eq. (11.83) is clearly different from the state described by Eq. (11.84). To establish the presence of Hawking radiation, it is necessary to make a quantitative comparison. To express one in terms of the other, we introduce a Bogolubov-type relation of the form [31]

$$F(k) = \int_{-\infty}^\infty dl\alpha(l)G(l), \, (k > 0). \tag{11.85}$$

One can write $\alpha(l)$ as an integral and also, at least approximately, in closed form (see Eq. 11.56 of Ref. [31]). Then, using the solution for $\alpha(l)$, the initial state described by Eq. (11.83) can be expressed in terms of $G(k)$ as

$$\Psi_0[f, t_y = 0] = N \exp\left(-\int_{-\infty}^\infty dk\, k \coth\left(\frac{\pi k}{2\lambda}\right)|G|^2(k)\right). \tag{11.86}$$

We are ultimately interested in the solution of Eq. (11.82) with the initial condition given by Eq. (11.86). It takes the form [31]

$$\Psi = N(t_v) \exp\left(-\int_{-\infty}^\infty dk\, k \coth\left(\frac{\pi k}{2\lambda} + ipt_v\right)|G|^2(k)\right). \tag{11.87}$$

The state described by Eq. (11.87) is known as a squeezed state [33].

With the wavefunctional of Eq. (11.87) it is now possible to evaluate the number operator $< n(k) >$ of the mode with wave number k for the vacuum state given by Eq. (11.84); i.e., for the case in which the black hole is present. At $t_v = 0$, it is given by [31]

$$< n(k) > = \frac{1}{\exp\left(2\pi |k|/\lambda\right) - 1},$$ (11.88)

which is a Planck distribution, as expected, with temperature $T_{BH} = \lambda/2\pi$.

We can therefore conclude that Hawking radiation emerges in the hybrid model that we have considered here. Due to the dynamical background (i.e., the collapsing geometry), an initial vacuum state does not remain a vacuum state but becomes instead a thermal state with respect to observers at a later time. In the Heisenberg picture the non-invariance of the vacuum is called 'particle creation' (or 'field excitation'), while in the Schrödinger picture that we have used for our analysis it corresponds to the process of squeezing quantum states [19].

The two-dimensional dilatonic CGHS black hole model is clearly a toy model which cannot capture the full complexity of general relativistic black holes, and one must be careful when extrapolating to the four-dimensional case. However, the results derived in this section do suggest that a hybrid system consisting of a classical black hole coupled to a quantized scalar field may be able to account for Hawking radiation in a consistent way, leading to an understanding of its emergence without having to quantize gravity first. A more definite statement will clearly require further investigation to see if a similar analysis can be successfully carried out for a more realistic black hole model.

11.4.4 Non-perturbative Approach and the Emergence of Time

One of the goals of this section is to look at an example which shows how time emerges in the hybrid formalism. As is well known, quantum gravity has a serious conceptual difficulty, known in the literature as the problem of time. This is a consequence of the drastically different concepts of time in general relativity and quantum theory: time is absolute in standard quantum theory and dynamical in general relativity, and any attempt to combine both theories to formulate a quantum theory of gravity leads to a difficult conceptual problem [34]. The problem of time derives from the timeless nature of the Wheeler–DeWitt equation and concerns among other questions the problem of whether a physical concept of time may be introduced at a fundamental level in quantum gravity. There are three main solutions to this problem: the choice of a concept of time before quantization, the identification of a concept of time after quantization, and the option of a timeless theory [19].

As emphasized by Kiefer, the application of the semiclassical approximation to the Wheeler–DeWitt equation leads to the emergence of time in the semiclassical regime, as a semi-classical concept [34, 35]. In the case of the hybrid formulation that we discuss here, there is no problem of time: since the gravitational field remains classical, it is possible to introduce a "gravitational time" that is generally valid. Thus the gravitational field acts as a clock.

We saw in the previous sections how time emerges when we consider an approximate solution based on an expansion in powers of G. We now consider instead a non-perturbative approach. We write S and P in the form

$$S = S_0[\rho, \phi] + S_1[\rho, \phi, f], \quad P = P_0[\rho, \phi]P_1[\rho, \phi, f] \tag{11.89}$$

and look for a solution of Eqs. (11.45) and (11.46) that is based on these expressions. We want to be able to interpret the system so that we can distinguish between a classical gravitational sector and a quantum scalar field sector. To achieve this, we need to choose an appropriate gauge. Once more, we consider the conformal gauge, which as we discussed before leads to a simplification of the rate equations for the variables ϕ and ρ of the gravitational field. Taking into consideration that $S = S_0 + S_1$, we get

$$\dot{\rho} = -\frac{G}{2}\left(\frac{\delta S_0}{\delta \phi} + \frac{\delta S_1}{\delta \phi}\right), \quad \dot{\phi} = -\frac{G}{2}\left(\frac{\delta S_0}{\delta \rho} + \frac{\delta S_1}{\delta \rho}\right). \tag{11.90}$$

We will use the time associated with this particular choice of gauge to define a time-dependent functional Schrödinger equation for the quantized scalar field, using the same approach that we used before when we considered the expansion in powers of G. Notice that Eq. (11.90) differs from Eq. (11.54) because we now consider the exact solution instead of attempting to find a perturbative solution.

We now consider Eqs. (11.45) and (11.46), use the expressions given in Eq. (11.89), and group terms to allow for an interpretation that distinguishes between the classical gravitational sector and the quantum scalar field sector. Taking Eq. (11.90) into consideration, Eq. (11.45) can be written as

$$\left\{-\frac{G}{2}\frac{\delta S_0}{\delta \phi}\frac{\delta S_0}{\delta \rho} + \frac{1}{2G}V_G\right\} + \left\{\left(\frac{\delta S_1}{\delta \phi}\dot{\phi} + \frac{\delta S_1}{\delta \rho}\dot{\rho}\right) + \frac{1}{2}\left(\frac{\delta S_1}{\delta f}\right)^2 + \frac{1}{2}f'^2\right.$$
$$\left. +\frac{G}{2}\frac{\delta S_1}{\delta \phi}\frac{\delta S_1}{\delta \rho} + \frac{\hbar^2}{8}\left[\frac{1}{P_1^2}\left(\frac{\delta P_1}{\delta f}\right)^2 - \frac{2}{P_1}\frac{\delta^2 P_1}{\delta f^2}\right]\right\} = 0, \tag{11.91}$$

and Eq. (11.46) as

$$P_1\frac{G}{2}\left\{\left[\frac{\delta}{\delta \phi}\left(P_0\frac{\delta S_0}{\delta \rho}\right) + \frac{\delta}{\delta \rho}\left(P_0\frac{\delta S_0}{\delta \phi}\right)\right] + \left[\frac{\delta}{\delta \phi}\left(P_0\frac{\delta S_1}{\delta \rho}\right) + \frac{\delta}{\delta \rho}\left(P_0\frac{\delta S_1}{\delta \phi}\right)\right]\right\}$$
$$+ P_0\left\{-\left(\frac{\delta P_1}{\delta \phi}\dot{\phi} + \frac{\delta P_1}{\delta \rho}\dot{\rho}\right) - \frac{\delta}{\delta f}\left(P_1\frac{\delta S_1}{\delta f}\right)\right\} = 0. \tag{11.92}$$

We would like to set each of the expressions in Eqs. (11.91) and (11.92) which are in curly brackets separately equal to zero, but this is clearly not possible in the general case. The problem is caused by the first expression in curly brackets that appears in Eq. (11.92): S_1 is a functional of f, while P_0 and S_0 are not. However, we *can* solve the equations step by step (see below), and at the same time introduce well defined classical and quantum sectors, whenever

$$\frac{\delta S_0}{\delta \phi} >> \frac{\delta S_1}{\delta \phi}, \frac{\delta S_0}{\delta \rho} >> \frac{\delta S_1}{\delta \rho}. \tag{11.93}$$

This condition amounts to assuming that the dependence on the gravitational variables is much weaker for S_1 than it is for S_0. Such a condition does not seem unreasonable: one can expect S_0 to account for most of the gravitational field degrees of freedom and for S_1 to act as a correction term required by the presence of the quantized scalar field. If the inequalities in Eqs. (11.93) hold, we can neglect the terms that contain S_1 in the first expression in curly brackets which appears in Eq. (11.92). We will assume this from now on.

Thus we are left with two pairs of functional equations that can be solved separately. The first pair of functional equations is

$$-\frac{G}{2}\frac{\delta S_0}{\delta \phi}\frac{\delta S_0}{\delta \rho} + \frac{1}{2G}V_G = 0, \quad \frac{\delta}{\delta \phi}\left(P_0\frac{\delta S_0}{\delta \rho}\right) + \frac{\delta}{\delta \rho}\left(P_0\frac{\delta S_0}{\delta \phi}\right) = 0, \tag{11.94}$$

identical to the functional equations for a classical ensembles of CGHS black holes described in terms of $P_0[\phi, \rho]$ and $S_0[\phi, \rho]$. The second pair of functional equations are

$$\left(\frac{\delta S_1}{\delta \phi}\dot{\phi} + \frac{\delta S_1}{\delta \rho}\dot{\rho}\right) + \frac{1}{2}\left(\frac{\delta S_1}{\delta f}\right)^2 + \frac{1}{2}f'^2$$

$$+\frac{G}{2}\frac{\delta S_1}{\delta \phi}\frac{\delta S_1}{\delta \rho} + \frac{\hbar^2}{8}\left[\frac{1}{P_1^2}\left(\frac{\delta P_1}{\delta f}\right)^2 - \frac{2}{P_1}\frac{\delta^2 P_1}{\delta f^2}\right] = 0, \tag{11.95}$$

$$-\left(\frac{\delta P_1}{\delta \phi}\dot{\phi} + \frac{\delta P_1}{\delta \rho}\dot{\rho}\right) - \frac{\delta}{\delta f}\left(P_1\frac{\delta S_1}{\delta f}\right) = 0. \tag{11.96}$$

If we now integrate Eqs. (11.95–11.96) with respect to the x coordinate and use the equalities

$$\dot{S}_1 = \int dx\left(\dot{\rho}\frac{\delta S_1}{\delta \rho} + \dot{\phi}\frac{\delta S_1}{\delta \phi}\right), \dot{P}_1 = \int dx\left(\frac{\delta P_1}{\delta \phi}\dot{\phi} + \frac{\delta P_1}{\delta \rho}\dot{\rho}\right), \tag{11.97}$$

we are led to

$$\dot{S}_1 + \int dx\left\{\frac{1}{2}\left(\frac{\delta S_1}{\delta f}\right)^2 + \frac{1}{2}f'^2 + \frac{G}{2}\frac{\delta S_1}{\delta \phi}\frac{\delta S_1}{\delta \rho}\right.$$

$$\left. +\frac{\hbar^2}{8}\left[\frac{1}{P_1^2}\left(\frac{\delta P_1}{\delta f}\right)^2 - \frac{2}{P_1}\frac{\delta^2 P_1}{\delta f^2}\right]\right\} = 0, \tag{11.98}$$

$$\dot{P}_1 + \int dx\left\{\frac{\delta}{\delta f}\left(P_1\frac{\delta S_1}{\delta f}\right)\right\} = 0. \tag{11.99}$$

We can now go to the wavefunction representation via the complex canonical transformation $\Psi = \sqrt{P_1}e^{iS_1/\hbar}$, $\bar{\Psi} = \sqrt{P_1}e^{-iS_1/\hbar}$. Then, Eqs. (11.98–11.99) can be combined and put in the form of a *non-linear functional Schrödinger equation*,

$$i\hbar\dot{\Psi} = \hat{H}_f \Psi = \frac{1}{2}\left(-\hbar^2\frac{\delta^2}{\delta f^2} + f'^2 + \frac{G}{2}\Delta\right)\Psi. \tag{11.100}$$

where the non-linear correction term Δ is given by

$$\Delta = \frac{\delta S_1}{\delta\phi}\frac{\delta S_1}{\delta\rho} = -\frac{1}{4}\left(\frac{\delta\ln\bar{\Psi}}{\delta\phi} - \frac{\delta\ln\Psi}{\delta\phi}\right)\left(\frac{\delta\ln\bar{\Psi}}{\delta\rho} - \frac{\delta\ln\Psi}{\delta\rho}\right) \tag{11.101}$$

and the time is the gravitational time associated with the conformal gauge.

Thus, to solve the equations, we first derive expressions for P_0 and S_0 by solving Eq. (11.94), and then solve for P_1 and S_1 by solving Eqs. (11.98–11.99) or, equivalently, the non-linear functional Schrödinger equation, Eq. (11.100). The non-trivial interaction between gravitational and matter fields comes about via the dependence of P_1 and S_1 on ϕ and ρ, and it can be seen explicitly in Eqs. (11.90) and (11.97), which are used to define the notion of time, and in the non-linear correction term Δ given by Eq. (11.101) which appears in Eq. (11.100).

Thus, under the conditions of Eq. (11.93), it is possible to introduce well defined classical and quantum sectors. Furthermore, the time that is used to define the non-linear functional Schrödinger equation arises in a natural way from the classical gravitational sector of the hybrid system. In the case of the hybrid formulation that we discuss here, as opposed to the case of quantum gravity, there is no problem of time: the gravitational field acts as a clock and it is possible to introduce a "gravitational time" that is generally valid.

The formalism of ensembles on configuration space can be applied to classical black holes that interact with quantum matter fields. The case of a classical CGHS black hole and a quantized scalar field can be solved, with Hawking radiation being predicted for a collapsing space-time geometry. The well known problem of time in quantum gravity has a natural resolution in the hybrid approach.

Appendix: Ground State Gaussian Functional Solution

We follow the presentation given in App. C of Ref. [22]. To solve Eq. (11.15) for the case $S = 0$,

$$\int drN\left[-\frac{1}{2AR^2}\left(\frac{1}{A}\frac{\delta^2 A}{\delta\phi^2}\right) + \lambda\frac{\Delta R^2}{2} + V\right.$$

$$+ \frac{R^2}{2\Lambda}\phi'^2 + \frac{\Lambda R^2 m^2}{2}\phi^2 \Big] = 0, \tag{11.102}$$

consider the ansatz

$$A \sim \exp\left\{-\frac{1}{2}\int\int dy\, dz\, \Lambda_y\, \Lambda_z\, R_y^2\, R_z^2\, \phi_y\, K_{yz}\, \phi_z\right\}. \tag{11.103}$$

Evaluate $\frac{\delta^2 A}{\delta\phi^2}$ and collect terms that have the same power of ϕ. This leads to the following pair of equations for the kernel K_{xy},

$$\int dr\, N\left[\lambda\frac{\Lambda R^2}{2} + V(R,\,\Lambda) + \frac{\Lambda_r R_r^2}{2}K_{rr}\right] = 0 \tag{11.104}$$

and

$$\int dr\, \Lambda_r R_r^2\, N\left[\frac{\phi_r'^2}{2\Lambda_r^2} + \frac{m^2}{2}\phi_r^2 - \frac{1}{2}\int\int dydz\, \Lambda_y R_y^2 \Lambda_z R_z^2\, \phi_y K_{yr}K_{rz}\phi_z\right] = 0. \tag{11.105}$$

After an integration by parts in Eq. (11.105), the equations for K_{xy} take a form which is standard in the context of the Schrödinger functional representation of quantum field theory in curved spacetimes [19, 23–25].

Assume a foliation of spaces of constant positive curvature and a constant lapse function N and look for a solution valid under these conditions. Note that the Λ and R that appear in the line element of Eq. (10.24) satisfy $\sqrt{h} = \Lambda R^2$ and $h^{rr} = \Lambda^{-2}$, where h^{kl} is the inverse metric tensor on the three-dimensional spatial hypersurface of constant curvature. Then, Eq. (11.105) can be written in the form

$$\int dr\sqrt{h_r}\left[\frac{1}{2}h^{rr}\frac{\partial\phi_r}{\partial r}\frac{\partial\phi_r}{\partial r} + \frac{m^2}{2}\phi_r^2\right.$$
$$\left. - \frac{1}{2}\int\int dydz\sqrt{h_y}\sqrt{h_z}\,\phi_y K_{yr}K_{rz}\phi_z\right] = 0, \tag{11.106}$$

and the kernel K_{xy} satisfies

$$\int dr\sqrt{h_r}K_{yr}K_{rz} = \left[-\frac{1}{\sqrt{h_y}}\frac{\partial}{\partial y}\left(h^{yy}\sqrt{h_y}\frac{\partial}{\partial y}\right) + m^2\right]\delta(y,z), \tag{11.107}$$

where $\delta(y,z) = \frac{1}{\sqrt{h_y}}\delta(y-z)$ is the delta function on the hypersurface.

To get an explicit expression for K_{xy} that solves Eq. (11.107), introduce a fixed, particular set of coordinates for the line element of Eq. (10.24). Let

$$N = 1, \quad N_r = 0, \quad \Lambda = a_0, \quad R = a_0\sin r, \tag{11.108}$$

where $r \in [0, 2\pi)$ and a_0 can be interpreted as the scale factor of a closed Robertson–Walker universe. Then the solution of Eq. (11.107) is given by

$$K_{xy} = \frac{1}{2a_0^4} \sum_n \sqrt{\gamma_n}\, \psi_x^{(n)} \psi_y^{(n)}, \qquad (11.109)$$

where the basis functions $\psi_r^{(n)}$ are solutions of a Schrödinger-type equation in a space of constant curvature,

$$-\frac{1}{\sin^2 r} \frac{\partial}{\partial r} \left(\sin^2 r\, \frac{\partial \psi_r^{(n)}}{\partial r} \right) + m^2 a_0^2\, \psi_r^{(n)} = \gamma_n \psi_r^{(n)}. \qquad (11.110)$$

The $\psi_n(r)$ satisfy orthonormality and completeness relations. The eigenvalues γ_n are given by

$$\gamma_n = n^2 - 1 + m^2 a_0^2, \quad n = 1, 2, 3 \ldots. \qquad (11.111)$$

Given the solution of Eq. (11.105), a_0 can be expressed in terms of the cosmological constant and the energy E of the quantized scalar field using Eq. (11.104), since [25]

$$E \sim a_0^3 \int dr\, \sin^2 r\, K_{rr}. \qquad (11.112)$$

However, $\int dr\, \sin^2 r\, K_{rr} \sim \sum_n \gamma_n$, which diverges. This is a consequence of the infinite zero-point energy of the quantum field. Therefore, to extract a finite result for a_0 it becomes necessary to introduce renormalization procedures.

References

1. Albers, M., Kiefer, C., Reginatto, M.: Measurement analysis and quantum gravity. Phys. Rev. D **78**, 064051 (2008)
2. Rosenfeld, L.: On quantization of fields. Nucl. Phys. **40**, 353–356 (1963)
3. Carlip, S.: Is quantum gravity necessary? Class. Q. Grav. **25**, 154010 (2008)
4. Boughn, S.: Nonquantum gravity. Found. Phys. **39**, 331–351 (2009)
5. Butterfield, J., Isham, C.: In: Callender, C., Huggett, N. (eds.) Spacetime and the Philosphical Challenge of Quantum Gravity. Physics meets philosophy at the Planck scale. Cambridge University Press, Cambridge (2001)
6. Dyson, F.: Is a graviton detectable? Int. J. Mod. Phys. A **28**, 1330041 (2013)
7. Dyson, F.: The world on a string. N. Y. Rev. Books **51**(8), 16–19 (2004)
8. Rothman, T., Boughn, S.: Can gravitons be detected? Found. Phys. **36**, 1801–1825 (2006)
9. Boughn, S., Rothman, T.: Aspects of graviton detection: graviton emission and absorption by atomic hydrogen. Class. Quantum Grav. **23**, 5839–5852 (2006)
10. Parker, L., Toms, D.: Quantum Field Theory in Curved Spacetime: Quantized Fields and Gravity. Cambridge University Press, Cambridge (2009)
11. Diósi, L.: Gravitation and quantum-mechanical localization of macro-objects. Phys. Lett. A **105**, 199–202 (1984)

12. DeWitt, B.S.: The Quantization of Geometry. In: Witten, L. (ed.) Gravitation: an introduction to current research, pp. 266–381. Wiley, New York (1962)

13. Eppley, K., Hannah, E.: The necessity of quantizing the gravitational field. Found. Phys. **7**, 51–68 (1977)

14. Page, D.N., Geilker, C.D.: Indirect evidence for quantum gravity. Phys. Rev. Lett. **47**, 979–982 (1981)

15. Feynman, R.P.: The Feynman Lectures on Gravitation. Addison-Wesley, Boston (1995). Reading

16. Mattingly, J.: Why Eppley and Hannah's thought experiment fails. Phys. Rev. D **73**, 064025 (2006)

17. Huggett, N., Callender, C.: Why quantize gravity (or any other field for that matter)? Philos. Sci. **68**, S382–S394 (2001)

18. Mattingly, J.: Is quantum gravity necessary? In: Kox, A.J., Eisenstaedt, J. (eds.) The universe of general relativity (Einstein Studies Volume 11), pp. 327–338. Birkhäuser, Boston (2005)

19. Kiefer, C.: Quantum Gravity. Oxford University Press, Oxford (2012)

20. Kiefer, C.: Conceptual problems in quantum gravity and quantum cosmology. ISRN Math. Phys. **2013**, 509316 (2013)

21. Hall, M.J.W., Reginatto, M.: Interacting classical and quantum ensembles. Phys. Rev. A **72**, 062109 (2005)

22. Reginatto, M.: Cosmology with quantum matter and a classical gravitational field: the approach of configuration-space ensembles. J. Phys.: Conf. Ser. **442**, 012009 (2013)

23. Jackiw, R.: Diverse Topics in Theoretical and Mathematical Physics. World Scientific, Singapore (1995)

24. Hatfield, B.: Quantum Field Theory of Point Particles and Strings. Perseus Books, Cambridge (1992)

25. Long, D.V., Shore, G.M.: The Schrödinger wave functional and vacuum states in curved space-time. Nucl. Phys. B **530**, 247–278 (1998)

26. Herdeiro, C.A.R., Ribeiro, R.H.: Sampaio. M.: Scalar Casimir effect on a D-dimensional Einstein static universe. Class. Quantum Grav. **25**, 165010 (2008)

27. Kiefer, C., Zeh, H.D.: Arrow of time in a recollapsing quantum universe. Phys. Rev. D **51**, 4145–4153 (1995)

28. Zeh, H.D.: The Physical Basis of the Direction of Time. Springer, Berlin (1992)

29. Callan, C.G., Giddings, S.B., Harvey, J.A., Strominger, A.: Evanescent black holes. Phys. Rev. D **45**, R1005–R1009 (1992)

30. Strominger, A.: Les Houches lectures on black holes. arXiv:hep-th/9501071v1 (1995)

31. Demers, J.-G., Kiefer, C.: Decoherence of black holes by Hawking radiation. Phys. Rev. D **53**, 7050–7061 (1996)

32. Louis-Martinez, D., Gegenberg, J., Kunstatter, G.: Exact Dirac quantization of all 2D dilaton gravity theories. Phys. Lett. B **321**, 193–198 (1994)

33. Kiefer, C.: Hawking radiation from decoherence. Class. Quantum Grav. **18**, L151–L154 (2001)

34. Kiefer, C.: Does time exist in quantum gravity? arXiv:0909.3767v1 [gr-qc] (2009)

35. Kiefer, C.: The Semiclassical Approximation to Quantum Gravity. In: Ehlers, J., Friedrich, H. (eds.) Canonical gravity: from classical to quantum, pp. 170–212. Springer, Berlin (1994)

Appendix A
Variational Derivatives and Integrals

Definitions and useful properties of functionals, variational derivatives and functional integrals are collected here, and illustrated for the well known example of a classical scalar field.

A.1 Functional Derivatives

A functional, $F[f]$, is a mapping from a set of functions on configuration space to the real or complex numbers. We will denote the value of f at x by f_x. The functional derivative of $F[f]$ is defined via the variation of F with respect to f, i.e.,

$$\delta F := F[f + \delta f] - F[f] = \int dx \, \frac{\delta F}{\delta f_x} \delta f_x \tag{A.1}$$

for arbitrary infinitesimal variations $f \to f + \delta f$. Thus the functional derivative is a field density, $\delta F / \delta f$, having the value $\delta F / \delta f_x$ at position x. Note that this definition is analogous to the definition of the partial derivative of a function $g(x)$ via

$$g(x + \varepsilon) - g(x) = \varepsilon \cdot \nabla g(x)$$

for arbitrary infinitesimal variations $x \to x + \varepsilon$. It follows directly from Eq. (A.1) that the functional derivative satisfies product and chain rules analogous to ordinary differentiation.

The choice $F[f] = f_{x'}$ in Eq. (A.1) yields

$$\delta f_{x'} / \delta f_x = \delta(x - x'). \tag{A.2}$$

Moreover, if the field depends on some parameter, t say, then choosing $\delta f_x = f_x(t + \delta t) - f_x(t)$ in Eq. (A.1) yields

© Springer International Publishing Switzerland 2016
M.J.W. Hall and M. Reginatto, *Ensembles on Configuration Space*,
Fundamental Theories of Physics 184, DOI 10.1007/978-3-319-34166-8

$$\frac{dF}{dt} = \frac{\partial F}{\partial t} + \int dx \, \frac{\delta F}{\delta f_x} \frac{\partial f_x}{\partial t} \tag{A.3}$$

for the rate of change of F with respect to t. As another useful example, for a functional of the form $F = \int dx \, g(x, f, \nabla f)$ one has

$$
\begin{aligned}
\delta F &= \int dx \, [g(x, f + \delta f, \nabla f + \nabla(\delta f)) - g(x, f, \nabla f)] \\
&= \int dx \left[\frac{\partial g}{\partial f} \delta f + \frac{\partial g}{\partial \nabla f} \nabla(\delta f) \right] \\
&= \int dx \left[\frac{\partial g}{\partial f} - \nabla \cdot \frac{\partial g}{\partial \nabla f} \right] \delta f,
\end{aligned}
$$

where integration by parts has been used in the last line, assuming that the variations and/or the derivatives of g vanish at infinity. Hence from Eq. (A.1) one has

$$\delta F / \delta f = \partial g / \partial f - \nabla \cdot \partial g / \partial (\nabla f) \tag{A.4}$$

for this case. This formula is easily extended when g also depends on higher derivatives of f.

A.2 Functional Integrals

Functional integrals correspond to integration of functionals over the space of functions (or equivalence classes thereof). We will consider here a measure Df on this vector space which is *translation invariant*, i.e., $\int Df \equiv \int Df'$ for any translation $f' = f + h$ (which follows immediately, for example, from approaches to functional integration based on discretising the space of functions). In particular, this property implies the useful result

$$\int Df \, \frac{\delta F}{\delta f} = 0 \quad \text{for} \quad \int Df \, F[f] < \infty. \tag{A.5}$$

Equation (A.5) follows by noting that the finiteness condition and translation invariance imply

$$0 = \int Df \, (F[f + \delta f] - F[f]) = \int dx \, \delta f_x \left(\int Df \, \frac{\delta F}{\delta f_x} \right) \tag{A.6}$$

for arbitrary infinitesimal translations, where we use f_x to denote the value of f at x.

Thus, for example, if $F[f]$ has a finite expectation value with respect to some probability density functional $P[f]$, then Eq. (A.5) yields the "integration by parts" formula

$$\int Df \, P \frac{\delta F}{\delta f} = - \int Df \frac{\delta P}{\delta f} F. \tag{A.7}$$

Moreover, from Eq. (A.5) the total probability, $\int Df \, P$, is conserved for any probability flow satisfying a continuity equation of the form

$$\frac{\partial P}{\partial t} + \int dx \, \frac{\delta}{\delta f_x} [P V_x] = 0, \tag{A.8}$$

providing that the average flow rate, $\langle V_x \rangle$, is finite.

Finally, consider a functional integral of the form

$$I[F] = \int Df \, \xi(F, \delta F / \delta f), \tag{A.9}$$

where ξ denotes any function of some functional F and its functional derivative. Variation of $I[F]$ with respect to F then gives, to first order,

$$
\begin{aligned}
\Delta I &= I[F + \Delta F] - I[F] \\
&= \int Df \left\{ \frac{\partial \xi}{\partial F} \Delta F + \int dx \, \frac{\partial \xi}{\partial (\delta F / \delta f_x)} \frac{\delta(\Delta F)}{\delta f_x} \right\} \\
&= \int Df \left\{ \frac{\partial \xi}{\partial F} - \int dx \, \frac{\delta}{\delta f_x} \left[\frac{\partial \xi}{\partial (\delta F / \delta f_x)} \right] \right\} \Delta F \\
&\quad + \int dx \int Df \, \frac{\delta}{\delta f_x} \left\{ \left[\frac{\partial \xi}{\partial (\delta F / \delta f_x)} \right] \Delta F \right\}.
\end{aligned} \tag{A.10}
$$

Assuming that the functional integral of the expression in curly brackets in the last term is finite, this term vanishes from Eq. (A.5), yielding the result

$$\Delta I = \int Df \, \frac{\Delta I}{\Delta F} \Delta F \tag{A.11}$$

analogous to Eq. (A.1), where the variational derivative $\Delta I / \Delta F$ is defined by

$$\frac{\Delta I}{\Delta F} := \frac{\partial \xi}{\partial F} - \int dx \, \frac{\delta}{\delta f_x} \left[\frac{\partial \xi}{\partial (\delta F / \delta f_x)} \right]. \tag{A.12}$$

A.3   Example: Ensemble Hamiltonian for a Classical
   Scalar Field

To illustrate the application of the above concepts, we consider a classical scalar field, its corresponding ensemble Hamiltonian, and the equations of motion that follow from this ensemble Hamiltonian.

Field theories present well known mathematical and conceptual difficulties. As a consequence, the equations are formal in nature and it is necessary to examine each individual field theory to establish the validity of the ensemble formalism when applied to particular cases. This is done whenever particular field theories are discussed in the monograph. In the case of a field theory, the configuration is described by some field $\phi(x)$ (which may comprise multi-component fields that that carry a set of indices, which we omit here to simplify the notation). The probability density $P[\phi]$ is a functional, defined over the space of fields. We can consider a Hamiltonian description of the time evolution of the field, and introduce an auxiliary quantity $S[\phi]$ that is canonically conjugate to $P[\phi]$, as discussed in Chap. 5.

As an example, consider a classical scalar field ϕ. In the classical Hamilton–Jacobi formulation reviewed in Chap. 5, the equation for ϕ is given by

$$\frac{\partial S}{\partial t} + \int dx \left[\frac{1}{2} \left(\frac{\delta S}{\delta \phi} \right)^2 + |\nabla \phi|^2 + V(\phi) \right] = 0. \tag{A.13}$$

The ensemble Hamiltonian $\mathscr{H}$ is the functional integral corresponding to the mean energy of the field,

$$\mathscr{H}[P, S] := \int Df\, dx\, P \left[\frac{1}{2} \left(\frac{\delta S}{\delta \phi} \right)^2 + |\nabla \phi|^2 + V(\phi) \right]. \tag{A.14}$$

The Hamiltonian equations of motion for the dynamical variables P and S,

$$\frac{\partial P}{\partial t} = \frac{\Delta \mathscr{H}}{\Delta S}, \qquad \frac{\partial S}{\partial t} = -\frac{\Delta \mathscr{H}}{\Delta P}, \tag{A.15}$$

reduce via Eq. (A.12) to the Hamilton–Jacobi equation, Eq. (A.13), as required, and to the continuity equation

$$\frac{\partial P}{\partial t} + \int dx\, \frac{\delta}{\delta \phi} \left(P \frac{\delta S}{\delta \phi} \right) = 0. \tag{A.16}$$

Comparison with Eq. (A.8) shows that the local rate of change of the field is $V = \delta S/\delta \phi$.

Index

A

Action principle, 4
Aharonov, Y., 32
Aleksandrov, I.V., 193

B

Bell, J.S., 11, 141
Bergmann, P.G., 107
Bhattacharyya, A., 117
Bialynicki-Birula, I., 96
Black hole
    ensembles on configuration space, 232
    Hamilton-Jacobi formulation, 232
    Lemaître metric, 234, 237, 239, 240
Bohm, D., 141
Bohr, N., 174
Butterfield, J., 244

C

Callender, C., 245
Campbell, L.L., 119, 120
Caro, J., 193
Casimir functions, 144, 154
Casimir operator, 144, 155
Cencov, N.N., 119, 120
CGHS black hole, 255
    action, 255
    collapsing space-time, 261
    ensemble Hamiltonian, 257
    with a quantized scalar field, 257
        ensemble Hamiltonian, 257
        gravitational time, 266, 269
        Hawking radiation, 264, 269
        perturbative expansion, 258, 261
Classicality, 165, 207–209, 216

Complex structure, 122
Condon, E.U., 214
Configuration space, 4
    discrete, 7
Constraints, 40
Copenhagen interpretation, 177

D

Decoherence
    effective, 211, 214, 216
    ineffective, 211
Demers, J.-G-, 255, 261, 264
DeWitt measure, 227
DeWitt supermetric, 225, 227, 248
DeWitt, B.S., 244–246
Diosi, L., 192
Dirac, P.A.M., 58
Doebner, H.-D., 128
Dressel, J., 32
Dyson, F., 162, 244

E

Einstein-Hamilton-Jacobi equation, *see* Hamilton-Jacobi formulation, gravitation
Electromagnetic field, 102
    Lorentz gauge, 102
        exact uncertainty quantization, 103
    radiation gauge, 103
        exact uncertainty quantization, 104
Elze, H.-T., 192
Ensemble Hamiltonian
    as generator of time translations, 23
    classical particle, 5
    conservation of probability, 10

© Springer International Publishing Switzerland 2016
M.J.W. Hall and M. Reginatto, *Ensembles on Configuration Space*,
Fundamental Theories of Physics 184, DOI 10.1007/978-3-319-34166-8

Printed in the United States
By Bookmasters